METHODS OF
CLINICAL EXAMINATION:
A PHYSIOLOGIC APPROACH

METHODS OF CLINICAL EXAMINATION: A PHYSIOLOGIC APPROACH

Third Edition

By 19 Contributing Authors

Edited by
Richard D. Judge, M.D.
Clinical Professor of Postgraduate Medicine,
The University of Michigan Medical School, Ann Arbor

George D. Zuidema, M.D.
Professor and Director, Department of Surgery,
The Johns Hopkins University School of Medicine, Baltimore

Little, Brown and Company
Boston

CONTRIBUTING AUTHORS

ROBERT D. CURRIER, M.D.
Professor of Medicine (Neurology) and Chief, Division of Neurology,
University of Mississippi School of Medicine, Jackson

JOHN C. FLOYD, Jr., M.D.
Professor of Internal Medicine, Division of Endocrinology and Metabolism,
The University of Michigan Medical School, Ann Arbor

ROBERT A. GREEN, M.D.
Professor of Internal Medicine and Associate Dean for Student Affairs,
The University of Michigan Medical School, Ann Arbor

JAMES R. HAYWARD, D.D.S.
Chief, Department of Oral Surgery, The University of Michigan School of
Dentistry, Ann Arbor

JOHN W. HENDERSON, M.D., Ph.D.
Professor and Chairman, Department of Ophthalmology, The University
of Michigan Medical School, Ann Arbor

J. WILLIS HURST, M.D.
Professor and Chairman, Department of Medicine, Emory University
School of Medicine, Atlanta

RICHARD D. JUDGE, M.D.
Clinical Professor of Postgraduate Medicine, The University of Michigan
Medical School, Ann Arbor

THEODORE M. KING, M.D., Ph.D.
Director of Gynecology and Obstetrics, The Johns Hopkins University
School of Medicine, Baltimore

GEORGE H. LOWREY, M.D.
Professor of Pediatrics and Associate Dean, Student Affairs, School of
Medicine, University of California, Davis

PETER J. LYNCH, M.D.
Associate Professor of Medicine (Dermatology) and Chief, Division of
Dermatology, University of Arizona College of Medicine, Tucson

MURIEL C. MEYERS, M.D.
Professor of Internal Medicine, The University of Michigan Medical School;
Associate Director, Thomas Henry Simpson Memorial Institute, Ann Arbor

WILLIAM W. MONTGOMERY, M.D.
Professor, Department of Otolaryngology, Harvard Medical School;
Surgeon in Otolaryngology, Massachusetts Eye and Ear Infirmary, Boston

LEE H. RILEY, Jr., M.D.
Associate Professor, Orthopedic Surgery, The Johns Hopkins University
School of Medicine, Baltimore

GERHARD SCHMEISSER, Jr., M.D.
Professor, Orthopedic Surgery, The Johns Hopkins University School of
Medicine, Baltimore

RALPH A. STRAFFON, M.D.
Head, Department of Urology, The Cleveland Clinic Foundation, Cleveland

RON J. VANDEN BELT, M.D.
Clinical Instructor of Internal Medicine, Division of Cardiology,
University of Michigan Medical School, Ann Arbor

H. KENNETH WALKER, M.D.
Assistant Professor of Medicine (Neurology), Department of Medicine,
Emory University School of Medicine, Atlanta

JOHN G. WEG, M.D.
Associate Professor of Internal Medicine and Physician-in-Charge,
Pulmonary Division, University of Michigan Medical School, Ann Arbor

GEORGE D. ZUIDEMA, M.D.
Professor and Director, Department of Surgery, The Johns Hopkins
University School of Medicine, Baltimore

PREFACE

Although the term *physical diagnosis* has gradually become obsolete, the subject itself remains as important today as it was fifty years ago, and there is undeniably a continuing need for a well-structured introduction to the methods of clinical examination. Yet, when dealing with such fundamentals as history-taking and physical examination, one might justifiably question the need for innovation. Shouldn't the same rules apply to today's physician as applied to his grandfather?

We believe the answer is clearly no. In the first edition of our textbook, the approach to clinical examination was significantly modified, while the traditional emphasis on bedside examination was preserved. In subsequent editions, including this third edition, we have augmented these modifications in order to keep the content as up to date as possible. A number of factors guided the changes in approach. Modern instrumentation has improved our understanding of basic human physiology and pathophysiology. We therefore departed from a regional or anatomic orientation of material in favor of grouping by physiologic systems. Clinical medicine enters today's medical curricula at an earlier stage than heretofore, which prompted a further segmentation of the material into units that correlate with various basic sciences. Each chapter now contains sections on *technique of examination* and *normal findings* that can be interlocked with anatomy and physiology. Glossaries are provided to assist the preclinical student with clinical terminology. Separate sections on *cardinal symptoms* and *abnormal findings* can be correlated later with pathology and microbiology. The increasing importance of resuscitation and the growing incidence of trauma led to a greater emphasis on methods of examining the acutely ill or acutely injured patient.

Automation and data-processing modified the approach to history-taking and record-keeping. In our opinion, the problem-oriented medical system is a logical outgrowth of these innovations, and in the third edition we expanded the traditional source-oriented system of medical recording with a problem-oriented system. (An example of a complete problem-oriented medical record appears in the Appendix.) Also, the third edition includes an expanded chapter on circulatory evaluation, more correlation between pulmonary pathophysiology and physical findings, and a completely revised chapter on examining the skin. Six new contributors have been added to our roster of authors.

In the past there has been a tendency to see the introduction to methods of clinical examination as the sacred domain of the internist. Percussion and auscultation were emphasized to such an extent that the student might have questioned the legitimacy of an x-ray as a diagnostic tool. In this text we have included brief sections on special techniques of examination in order to place bedside methods in proper perspective. We have also continued to enlist contributing authors from a variety of surgical and medical subspecialties. It is our hope to promote balance at the same time as introducing the most modern diagnostic methods now available.

In the last analysis, however, no text on this subject can be a substitute for bedside experience. Clinical examination, by its very nature, must be a matter of self-education on the wards and in the clinics. With this in mind, illustrations have been selected for the sole purpose of demonstrating techniques and maneuvers. Representations of patients who display the classic signs of advanced disease have been intentionally omitted. Once pointed out at the bedside, these signs are easy enough to remember; but if anything, they may tend to blunt one's diagnostic ability, since the true test of the physician lies in the early identification of more subtle manifestations of disease.

We hope the beginning student will find this text of use in the development of his clinical vocabulary and in reviewing methods of examination prior to starting practical work on the wards with a tutor. Later, as an upperclassman, he may wish to review certain sections while serving his clinical clerkships. Finally, the practicing physician may find a few pearls buried within these pages; for the art of medicine is seldom if ever mastered, even by the expert.

We are grateful to the many contributors to this text and to Professor Gerald Hodge, Professor Denis Lee, and Professor Margaret Brudon of the Medical Illustration Unit of the University of Michigan, as well as to Mr. Alfred Teoli and Mr. Grant Laskbrook, for their excellent contributions to the illustrations. We thank Dr. W. Proctor Harvey, who generously allowed us to use illustrations obtained from the phonocardiographic files at Georgetown University Hospital. In addition, we thank Mrs. Sandra Bergman and Ms. Patricia Ingram for their assistance in preparing the manuscript and the staff of Little, Brown and Company — Mr. Fred Belliveau, Mrs. Lin Richter, Mrs. Rhoda McIntyre, and Mr. David Rollow — for their cooperation, encouragement, and support.

R. D. J.

G. D. Z.

CONTENTS

METHODS OF
CLINICAL EXAMINATION:
A PHYSIOLOGIC APPROACH

INTRODUCTION 1

Richard D. Judge

This text has but one purpose — to start you on your way to becoming a clinician. It is an introduction to the technique of diagnosis. Until now, as a student, you have been a "receiver." School and college have focused on providing *you* with knowledge and experience; your achievements were duly rewarded. Now you take your first step as a "giver"; to practice medicine is to serve. There are no grades behind the closed door of the examining room — only you and your patient. From here on you must evaluate yourself. At this moment, you are beginning a lifetime of self-education. Your success or failure in making the adjustment of the next year or two will determine to a large extent the kind of physician you will be twenty years after graduation.

There is a very wise German aphorism which says, "Only as a physician does one become a physician"*[10]. You cannot learn the art or science of medical practice from this or any other book. You cannot learn it in the library. You must do it at the bedside. To function there where the action is going on requires some basic skills — how to talk to sick people, what to ask, how to touch them, what to look for. These techniques are not mastered in a year nor in a lifetime of practice; but the method can and must be mastered — as rapidly as possible — for, with the preclinical sciences, it forms the foundation on which you will build your clinical superstructure. You may wade into deeper water only as you learn to swim. The major objective of this text is to teach you to swim in deep water as quickly and as efficiently as possible.

To think of "physical diagnosis" as a separate discipline is not only obsolete but dangerous. One cannot compartmentalize the history and physical nor separate physical diagnosis from clinical medicine or surgery without disrupting the basic logic on which medical practice is based. The diagnostic process is a continuum which begins invariably with a set of accurate observations. These may be verbal, physical, or laboratory — usually a blend of all three. To separate them is entirely artificial; they interweave themselves into a seamless fabric. The clinician does not cease to communicate with the patient on completing the history nor does he stop observing after the physical examination. His bedside impressions continue to guide both the diagnostic and the therapeutic program indefinitely. In fact, it is old-fashioned "physical diagnosis" which usually guides treatment in this day of the autoanalyzer. Penicillin is discontinued when the fever has subsided, the sputum is clear, and cough has disappeared. The diuretic program is modified as the weight drops, the orthopnea regresses, the respiratory rate normalizes, and the edema disappears. Sound medical practice requires

---

*Unterlagen für die Neuregelung der ärztlichen Prüfungsordnung. (Berlin, 1922.) From Flexner, A., *Medical Education,* New York: Macmillan, 1925. P. 10.

bedside observation as never before. It is here that we apply the scientific facts learned in the classroom and translate them into action. It is here also that compassion and understanding in medicine find major expression.

THE CLINICIAN'S JOB

The word *clinician* comes from the Greek κλινικος, meaning *bed*. As Feinstein has so well summarized it, "The clinician is the doctor at the sufferer's bedside, the doctor who accepts responsibility for the life entrusted to him by the patient, the doctor who plans the strategy and executes the tactics of therapeutic care" [8]. (Note that the clinician is not defined as the person making the diagnosis. He may or may not be, but he invariably assumes responsibility for the patient's care.) A significant part of the clinician's job is therefore by definition deeply interrelated with the individual patient and his personal life. He does not treat a disease; he treats patients who are ill. The term *disease* is an impersonal one which describes a cluster of interrelated phenomena as found on any page of any textbook of pathology or surgery. *Illness,* on the other hand, results from the interaction of a disease and a specific patient. It is anything but impersonal. The clinical phenomena caused by an illness include subjective sensations called "symptoms" and objective changes called "signs." In practice, symptoms must be dredged from a morass of information and misinformation, and signs must be disentangled from a welter of confusing anatomic and physiologic variables. Here, then, is another important aspect of the clinician's job.

But the essence of the clinician's job is even more fundamental. As often pointed out by Biörck [3], the primary responsibility of the clinician is one of *making decisions.* Every telephone call, every patient visit, every ward round, every laboratory report demands a rapid succession of decisions, many of which entail shattering consequences. Often the clinician must decide on the basis of limited or incomplete data; he is usually pressed for time and under considerable pressure. All kinds of decisions confront him, both diagnostic and therapeutic: what shall we do, what shall we say, when shall we proceed?

Many irrational factors inevitably become involved in the decision-making process since emotional interaction between patient and physician is always a contributing element. The clinician must identify and try to understand these factors. He must learn to know himself and why he acts as he does; he must learn to know his patients and why they act as they do. Finally, and very importantly, he must learn as much as he can about the process of his own work, the "anatomy and physiology" of the decision-making process.

THE PHYSICIAN

The physician's self-image is usually that of a man of science. *Medicine,* however, used to appear with arts, crafts, and hobbies under the Dewey Decimal System of library classification. There is a story that Laplace, when asked why, since medicine was not a science, he proposed admitting physicians to the Academy of Sciences, answered, "This is why: to get them among the men of science!" [2] We physicians are relative newcomers in the realm of science. Western medicine for many centuries was a combination of mysticism and empiricism. The kind of accurate observation which characterizes true science has always been a part of the process, but this part was relatively minor until the eighteenth century.

The art of medicine without science is quackery. On the other hand, the scientific method is not inconsistent with a humanitarian attitude. Humanity and science are quite compatible. Even the medical investigator may be motivated primarily by a desire to help his fellowman.

From the patient's point of view, there is no substitute for interest, acceptance, and especially empathy on the part of the physician. These may, indeed, be more important factors to him in selecting a physician than pure scientific ability, for these are the qualities that communicate to him that his doctor is *trying,* that he *wants to help.* The patient will reveal his serious problems only to someone who he feels is accepting him unconditionally. In most instances, therefore, it is wise to conceal any moral judgments you might have about the attitudes or behavior of your patient. Try to avoid betraying feelings of surprise, disdain, or hopelessness by a facial expression or a passing remark.

Empathy is a very desirable characteristic for the physician. It has been defined as appreciative perception or the ability to feel as another is feeling. In expressing concern for the patient, however, the doctor must maintain a significant degree of detachment, for sympathetic attachment destroys objectivity. If objectivity is surrendered, judgment becomes biased. This is one reason for the universal rule that a physician should avoid caring for members of his own family under ordinary circumstances.

Empathy requires sensitivity. Each of us has a variable measure of sensitivity, which is perhaps partially hereditary. The question is often asked, "How does one develop sensitivity?" One answer is, of course, to look beyond medicine to philosophy, art, or religion. Time spent in cultivating an appreciation for music or poetry may lead in a very direct way to the type of sensitivity and perception which increases diagnostic acumen in the examining room. There is a solid basis for this. Great writers, composers, and artists are by nature geniuses at observing. For this reason alone, they are worthy of study. In addition, a physician who develops an appreciation for subtle differences in form, color, harmony, or cadence is learning something indispensable. Art, for him, is also a means of educating the senses. A man who is familiar with a grace note will have little difficulty with a split-second sound.

As a student you will approach the bedside with a certain amount of anxiety and dread. Do not be concerned about your initial sense of awkwardness. This is natural. Try to "break the barrier" as quickly as possible. Fix your attention on your ultimate role, and start to swim. Remember that as a practicing physician you will be regarded as an expert with considerable authority and influence over the lives of your fellowmen. What you display in the way of confidence will greatly affect your patient's confidence in you. As Galen put it, "He heals the best in whom the most people have the greatest confidence." Your age, sex, personal appearance, voice, attitude, and personality all influence the patient's behavior toward you. It is important, therefore, that you begin by learning all that you can about yourself. Only by developing your individual talents along lines that are natural will you present a genuine image. Of course, there are innumerable approaches. One physician may use a touch of humor, another a friendly sincere attitude, another dignified reserve, and so on. Whatever seems natural is usually proper. Nevertheless, the knack of really "reaching" the patient requires thoughtful practice regardless of the route selected. Study those of your teachers who seem to have this ability and never let a pearl slip by, be it from a teacher, a colleague, a house officer, or nurse. Every usable phrase or idea should go into a "pearl book" carried in your pocket for this purpose. Later you can work the best of these into your own personal approach to patients.

The list of desirable characteristics for a physician would be too long to include here. Honesty and integrity are always mentioned. Of course they are indispensable, but foolish honesty can undermine confidence. It can also be exceedingly cruel. Honesty must be tempered by mercy and practiced tactfully. The hardest three words for the physician are, "I don't know." It is one of the stiffest tests of his honesty. If you come upon a fellow physician who is so proficient that he never needs to say, "I don't know," beware! Honesty also applies to the nonprofessional side of your life. A physician who leaves no time for his wife and children is not being honest.

One last characteristic of the physician will be mentioned. The physician must be critical. Doubt is the foundation of scientific medicine. Therefore, you must endeavor to develop the same critical attitude at the bedside that an investigator uses in his laboratory. By preserving critical judgment in your reasoning, in your reading, in everything you do, you are less likely to become complacent. Complacency more than anything else will stymie your growth, and the lessons of medicine must be learned and relearned throughout a lifetime of practice.

In summary then, if you wish to be classified as a "good clinician" twenty years after graduation you should be:

1. A good observer
2. A good critic
3. A good communicator
4. A good decision-maker
5. A good student — now and later

THE PATIENT

The term *patient* is derived from the Latin word *pati* — to suffer. A person becomes a patient by seeking medical aid. Biörck [4] has suggested an interesting idea. He has pointed out that there are really three patients: (1) the patient as he is, (2) the image of the patient in the physician's mind, and (3) the patient on the record. Each may differ in significant respects, but the clinician is involved with all three. The physician may never know the *real patient* — that is, as he really is in his home and at his work. In addition the *real patient* may withhold information, at times critical and important information. Some of this information may become apparent only through the chart — for example, a past history of a high uric acid level or borderline fasting blood sugar level. On the other hand, the chart may hold important facts about previous illness which have been forgotten by the real patient. The image of the patient in the physician's mind is often a mixture of the real patient and the patient on the record modified by the effect of his own personality. The interplay is complex and everchanging from day to day as the chart thickens and the physician and patient (hopefully) get to know each other better.

Not all patients are suffering, but all are anxious. This we know from the personal experience of being patients ourselves. The patient may not recognize his anxiety — indeed, he may deny it — but unless he is anxious, there is probably something fundamentally wrong with him. This is true for the simple periodic health checkup as well as for the potentially serious illness. Part of this anxiety stems from a natural fear of disease, with its many implications. Most patients are also anxious about the physician as well. Will he be competent? Will he be friendly? Will he cause discomfort? Will he find something serious? When the patient and physician are of different sexes, there is bound to be some tension involved in discussing bowel function or in examining the genital area or the rectum.

A confident, gentle approach will do much to minimize these anxieties. The experienced clinician will frequently discuss them openly with the patient during the course of the interview, particularly when the patient seems excessively worried. Some early reassurance may be necessary during history taking. In fact, therapy may begin during history taking in the form of calming the excited patient. However, in attempting to calm the patient you should avoid implying that "there is nothing to be worried about" until you are convinced that this in fact is the case. Obviously, if serious disease is eventually discovered, you will have created a psychological impasse.

The patient's anxiety may find expression in many ways other than simple agitation. Anger is a frequently used defense. The angry patient is a common and difficult problem. You must avoid the temptation of meeting such anger with hostility of your own. On the other hand, you cannot ignore it. One helpful approach is to get the angry patient to discuss his hostility. The physician usually finds that the real source of anger is not the result of something that he himself has done. Even when it is, once out in the open the hostility tends to be lessened, particularly if the physician is wise enough to be reasonable and take action to rectify the situation.

Another expression of the patient's anxiety may be extreme dependence or even overt affection. The former can be very trying and requires great patience. You should make every effort to maintain a matter-of-fact relationship with overly dependent patients, and you should guard against displaying anger or exasperation. The affectionate patient is an even more difficult problem. Fortunately it is very rare. Changing the subject at such times is usually all that is necessary.

If a patient begins to cry during the course of your interview, and this will happen not infrequently, it is a mistake to feel that it is always necessary to rush in and try to stop the tears. More often it is preferable to remain as objective as possible and try to communicate the fact that she (or he) should not feel embarrassed by the outbreak, that you understand why she (or he) is upset. Wait for the storm to pass. Afterward, you may want to explore the reasons for the patient's reaction, particularly in the event that the patient is of an ethnic background which would suggest stoicism rather than emotionalism.

Each patient will behave differently under the pressures of the examining room. Regardless of how the patient reacts, it is the obligation of the physician to try to understand why he feels the way he does. Only those of us who have been seriously ill ourselves can really understand the patient's point of view. Each one of us has only one life, and an illness is a very serious event within the course of that life. It is, therefore, reasonable that experiences connected with an illness be considered extremely significant and many times become highly tinged with emotion. Keep in mind the fact that the patient's behavior may provide a unique opportunity to obtain a degree of insight into his personality, since it may in fact typify his behavior with others. By analyzing this pattern as you go along, it may be possible to obtain very important information. The patient occasionally reveals more about himself through his behavior than through anything he tells you.

THE MEDICAL INTERVIEW

Automation

It is becoming increasingly clear that some of the information included in the routine medical history can be provided by the patients themselves through questionnaires. This information may be obtained through self-administered questionnaires, nonphysician assistants, or some sort of computerized system. The fact that automated processing of medical data is acquiring wide acceptance by no means minimizes the importance of the physician in his capacity as history-taker. What it does suggest is that questionnaires can improve the efficiency of interviewing by supplying certain noncritical information. Baseline data concerning past medical history, family history, and parts of the review of systems, tabulated and processed in advance, can serve as points of reference from which the physician may delve more comprehensively if such delving is indicated. On the other hand, the present illness and patient profile are critical areas which probably will always require direct professional evaluation. To the extent that time saved from routine, repetitive questioning can be applied to the more personal aspects of the patient's physical, psychologic, and environmental problems, automation is having a salutary influence on medical practice. Thus, there is good reason for you, with all your working life before you, to make yourself well acquainted with the type of logic and procedure that is common to both computer technique and conventional technique. The methods you will be using twenty years after graduation may well be a blend of the two.

As patient loads increase and more sophisticated apparatus becomes available, medical data recording may develop along several lines as suggested by Biörck [4].

Physician:
Information → physician memory
Information → physician memory → handwritten record
Information › physician memory › secretary › typed record
Information → physician memory → dictaphone → secretary → typed record
Information → physician memory → dictaphone → secretary → punch tape → magnetic memory storage → automatically typed record

Self-Administered Questionnaire:
Information → secretary → punch tape → magnetic memory storage → automatically typed record → *physician* → physician memory storage → further information → physician memory → dictaphone → and so forth

Value of the Medical History

In order to understand the logical role of the medical history in the diagnostic process, try to imagine for a moment what it would be like if the sequence of history and physical examination were reversed. It is, of course, unthinkable. The physician, a relative stranger, would be doing a "blind" examination on a bewildered, hostile subject. Yet, through the consideration of this turnabout, several deceptively simple yet vitally important objectives of the medical history tend to reveal themselves. These include:

1. Establishing contact — getting to know and understand the patient and his environment. This is frequently referred to as "developing rapport." It is the first step in history taking and the first step toward making the diagnosis.

2. Eliciting valuable diagnostic information. It has been demonstrated that an experienced physician can closely predict the final diagnosis in about half of his patients based on a careful history alone. Most of us are not that proficient, yet none will deny the immense diagnostic value of verbal observations.

3. Giving focus to the physical examination. The history is the first signpost on the road to diagnosis. It puts the physician on guard and makes him particularly watchful for certain physical findings. Every intelligent physician modifies his examination according to historical findings. This fact should not invite omissions or sloppy observation.

4. Gaining insight into the functional status. Severity of the patient's symptoms is especially useful as an early index of the seriousness of the illness. It points the way to subsequent disposition — emergent, urgent, routine.

5. Indicating appropriate laboratory studies. Isolated symptoms or historical facts often require special laboratory follow-up. For example, a family history of diabetes requires at least a two-hour postprandial blood sugar test and probably a glucose tolerance test; food-relieved epigastric discomfort might prompt an upper gastrointestinal x-ray even in the absence of any associated physical findings.

6. Initiating therapy. The seeds of future therapy are sown during the history. Make no mistake, the patient is sizing you up. Whether he accepts your recommendations or throws away your prescription is, in part, being decided.

Technique of Interviewing

Interviewing, like medicine itself, is partly an art and partly a science. It is a skill that is mastered not from a book but from experience in the examining room. The routine outline presented in this chapter lists the various topics that must be covered in a complete medical history. By following such an outline, you assure yourself that your history is complete. Yet no instructor can teach you how to frame your questions effectively. This you must discover for yourself. The psychologic relationships involved in the interaction between physician and patient during this process of history taking is far more complex than a simple outline might imply. Before proceeding to the mechanics of medical history taking, let us consider a few of the general principles that are common to all types of information-gathering interviews and apply them to the problem at hand.

As a physician you will have several fundamental objectives during the medical interview. These will include motivating the patient to communicate, controlling the interaction, and measuring the significance of the patient's responses.

Motivation and Memory. The patient is usually motivated by a desire for relief from his symptoms. This is a very sound positive force which favors a successful outcome. The major intrinsic deterrent is his fear of the consequences of his illness. This factor can produce a conflict that tends to make him an unreliable reporter. The physician himself introduces an extrinsic factor, namely, the effect of his own appearance, attitude, and pattern of questioning. By being calm and sympathetic and by showing genuine interest, you as a physician tend to minimize the forces that prompt the patient to distort or withhold information. A critical factor in this regard is the way in which you formulate your questions. For example, the patient will almost universally respond in a biased manner to a question which conveys the possibility of serious consequences. This is not necessarily due to conscious falsification of information. The response is simply affected by his subconscious fears. After you have created an atmosphere in which the patient feels safe to communicate fully without fear of being judged, you can further motivate him by involving him more deeply in the topic at hand. As the patient forgets himself and focuses on the problem, more delicate areas can be explored without fear of his withdrawal.

Even if the physician and the patient were able to develop a relationship in which all sources of bias and all barriers to communication were eliminated, the problem of memory would still be an important limiting factor. The patient cannot tell the physician about something he has forgotten. The subject of memory is a complex one. It is estimated that about fifteen trillion bits of information are handled by our ten billion or so brain cells during an average seventy-year lifespan. How this is coded and transmitted is still the subject of intense investigation. Many experiments suggest that a ribonucleic acid (RNA) molecule or set of molecules represents the molecular engram which is the permanent memory trace. In fact, it is impossible to prove that anything once learned is ever completely forgotten. For practical purposes, however, remembering and forgetting are subject to a number of well-known influences: intelligence, emotion, organic disturbances such as trauma, and falsification. The ability to remember may be modified by the pertinence of the information, by interference from other factors, and by repression or distortion. In everyday life, forgetting to perform a task can at times be traced back to some conflict, while remembering events in vivid detail may correlate with positive emotional experiences or at least ones which the individual considers significant. At Guy's Hospital, I once watched a consultant elicit incredibly detailed information from a patient. When she left the room he turned and said, "Whatever it was that happened to her on that day I'm not sure, but she considers it terribly important. She remembers everything!"

Direct questions may help to improve the patient's recall by reminding him of information he has forgotten to volunteer. The review of systems is designed to bring out accessory symptoms overlooked during the presenting illness. When you run up against an important memory block, you may try two tricks to help the patient remember. One is to lead the discussion around the general area, in hopes that some correlation will help bring back the forgotten material. A second method is to put the subject aside (into the "subconscious oven") for a while, and then return and repeat the question in slightly different terms. You may be surprised to have the patient say, "Now I remember. . . ."

Tension more often develops with patients of the opposite sex and with certain obvious subjects, such as sex, alcoholism, mental illness, epilepsy, and venereal disease. When you sense that your patient is disturbed, you might just as well desist since the possibilities of distorted or falsified responses are very great indeed. This is time for an explanation or total change of subject. If the area is critical, you must of course try again, at which time you can explain the importance of the information. If the patient remains abnormally defensive, do not fret. You have uncovered a very important fact. Probably it is somehow related to the rest of the problem.

Control. Because of the limitations of time, the interview should deal almost exclusively with relevant information. The physician must, therefore, control the process. You begin by taking the initiative. You introduce yourself. You place the patient in a comfortable position. Since privacy is indispensable, relatives and friends are courteously excused except when the patient is a child or when an interpreter is needed. To "break the ice," a moment or two spent on some irrelevant subject helps the patient to compose himself. When the business of history taking begins in earnest with some introductory question, the physician makes it clear at this point that the patient has his undivided attention. This may be done by some gesture such as leaning forward, setting aside his pen, or, in particular, by establishing eye contact.

Eye contact is a valuable method of controlling the interview situation. It establishes a sense of communication with the patient. It also helps the physician to concentrate on what he is being told and minimizes the patient's tendency to ramble. By carefully watching the patient's facial expression, the physician can immediately sense whether or not his questions are being understood. Needless to say, staring hypnotically at the patient will make him unbearably self-conscious. This should be avoided. With very shy patients, eye contact should be used with discretion since it can be detrimental. However, when properly used, it can be a major influence in guiding the course of the interview.

The physician's most effective tool for controlling the interview process is the way he frames his questions. The experienced physician uses all types of questions to gather his information. In evaluating the responses, he is conscious of the possible introduction of bias into the interview by the question itself. Certain types of general questions are neutral and virtually free from biasing effects. Others may be strongly weighted in one direction or another. It is only when the physician is unaware of the potential distorting influence of this or that type of question that he is likely to get into trouble. Let us consider several basic types of questions.

The neutral question should be used whenever possible. It is structured so that it does not suggest any particular response as being more acceptable to the physician or more beneficial to the patient. A neutral question can be open or closed. The open neutral question simply establishes a topic: "Tell me more about your headaches." The closed neutral question incorporates several alternative answers in the question: "Are your headaches more likely to occur in the morning, afternoon, or evening?"

The simple direct question is always closed because it requires simply a yes or no answer: "Do your headaches upset your stomach?" Direct questions may or may not be neutral, depending on such factors as voice inflection, context, and previous questions. Although direct questions will speed the interview, too many of them tend to overwhelm the patient and put words into his mouth. They are indispensable but require moderation.

The leading question is one which tempts the patient to give one answer rather than another. Though it automatically introduces bias, it may yield special information unobtainable by any other means. This technique is particularly useful in testing the reliability of a series of questions by loading the final query: "Would you say that your headaches come on only when you are feeling very tired?" Most physicians use occasional leading questions.

The loaded question is usually interjected to study the reaction of the patient, since it is so heavily biased that the answer itself is unimportant: "Do you ever think you might be better off dead?" Such a question is rarely if ever needed under ordinary circumstances. It would be directed to a depressed patient only after laying considerable groundwork. This shock technique would be primarily to assess his response to the suggestion of suicide.

Supplementary remarks are brief comments leveled at stimulating the patient to proceed. They tell the patient that he is doing well and should continue. They may consist of a simple assertion such as, "I see," or "Umm." A simple pause is sometimes an effective way of encouraging the patient to go ahead. Certain neutral phrases such as "Anything else," "How do you mean," or, "Tell me more about that" have the same positive effect.

By utilizing these different techniques selectively, you will gradually set a pattern that becomes intelligible to the patient. A head nod or an encouraging murmur is a reward that tells him that the topic is relevant and that he should continue. When the response is inadequate, you probe with a direct question. If the patient wanders too far afield, you may have to interrupt and change the subject. Interruption should be used only as a last resort, for if the patient's feelings are hurt he will surely retreat. This must be avoided if possible. By carefully observing the effect of your remarks on the patient, your questions should improve as you proceed.

Measuring and Communicating. As you elicit your information you simultaneously estimate its significance. You probe for precise temporal relationships and try to determine the relative severity of the various complaints. Certain symptoms considered extremely important by the patient may be discarded as irrelevant in the light of your insight and experience, while others which might be considered trivial by the patient are retained by you as significant. The interrelationships between symptoms must be determined as the interview proceeds, and you must decide whether any two or three complaints have a common cause or whether any single complaint has more than one cause. This process of probe and measure, probe and measure, continues throughout the interview.

You and your patient share a varying degree of mutual vocabulary. The patient's intelligence, background, experience, and formal education are important factors that must be assessed during the early phase of the interview. Then you will proceed accordingly. By framing your questions in understandable language, you obviously increase the likelihood of a coherent answer. In addition, however, you communicate something of great importance to the patient, namely, that you are capable of understanding his background and experience.

It may be necessary to probe to be sure the patient has a clear understanding of what is being asked of him; he will rarely admit he does not understand a specific question. More often, he will agree with the physician and thus introduce misinformation into the interview. For example:

Physician: Does the pain cause you to feel nauseated?
Patient: Yes (blank stare).
Physician: Do you understand what I mean by nauseated?
Patient: Not exactly.

Similes that draw comparisons to well-known, mutually appreciated terms can be of great value. Bile in the urine gives the appearance of tea. Blood in the vomitus looks like coffee grounds. Digested blood in the stool may appear like sticky tar. The absence of bile pigment in the stool gives it a pale appearance not unlike clay. Does the pain tingle? Does it come in surges? Did it drill into the back? Did it seem to come like a flush? By consciously developing such a vocabulary of commonly understood adjectives, verbs, and nouns, the physician insures a minimum of misunderstanding during the course of his history taking.

A second and entirely different vocabulary which requires careful development is your medical vocabulary. As a physician, you share a large and very specialized vocabulary with your colleagues. This vocabulary has been developed because of the preciseness of its terminology. In the course of professional discussions and case presentations, therefore, it is a great time-saver to use specific medical terms. "Intermittent claudication," for example, not only denotes pain in the extremities, but also brings to mind all the special characteristics of ischemic pain. The same is true of such terms as "tenesmus" or "angina pectoris." The use of these terms precludes the necessity for laborious descriptive phraseology, thus shortening case presentations and medical case notes.

Pitfalls in History Taking

As you develop your personal technique of history taking, try to watch that of others as often as possible. Jot down pearls such as especially effective questions for future use. You will be surprised to find that your senior teachers are not necessarily the most adept interviewers.

The major errors of interviewing can be categorized as suggested by Barbee and co-workers [1], and you will find it worthwhile to cross-examine yourself from time to time. The following outline can be used:

1. Data Collection
   Omissions of important questions
   Omissions of pertinent negatives
   Failure to elicit temporal relationships precisely
   Failure to follow up important leads
2. Structure
   Beginning too fast; not putting patient at ease
   Allowing the patient to ramble
   Needless repetition of questions (forgetting)
   Poor transitions (hopping between unrelated areas)
   Covering delicate areas too early
3. Validation of Data
   Using technical terms not understood by patient
   Not allowing patient to finish his answers
   Swamping patient with too many direct questions
   Failing to find out patient's interpretation of his symptoms
4. Physician Attitude
   Acting too friendly or not friendly enough
   Acting preoccupied, not listening
   Not enough eye contact
   Not enough interest in emotional factors
   Too much interest in emotional factors

Recording the History

During the interview, no attempt should be made to record the complete history in final form. There are several good reasons for this. In the first place, it is literally impossible; it distracts the patient and disrupts the procedure. Jotting down reminders, dates, ages, and numbers is not only acceptable but indispensable, however. Second, the medical record is not meant to be a repository for raw data. The information must be suitably condensed, logically sequenced, and converted as far as possible into crisp, pertinent, medical terminology, before recording it. This requires time and thought.

It should be pointed out that your case notes *as a student* have for their primary aim self-education, and for their secondary aim information storage. To achieve the former, they must be comprehensive and complete. They are, in fact, a "five-finger exercise" in memorization and self-instruction [6]. The checklists which you thus imprint on your memory are the ones you will use for the rest of your professional life. There must be no omissions. For this reason, try to follow the same symptom sequence within any system.

The secondary aim of your case notes must also be kept in mind. Remember that information storage is the purpose of the medical record. To be valuable, your notes must be usable. A cumbersome five-page presentation of an illness will rarely be reread, even by yourself. It defeats its own purpose. Therefore, systematize and concentrate your thoughts, even as a student.

THE DIAGNOSTIC PROCESS

The technique of doctoring, that is, the "anatomy and physiology" of the decision-making process, centers around the identification of a specific medical problem. This in turn provides a frame of reference from which to determine (1) what to do and (2) what to expect. Most patients have several problems, each of which requires diagnostic or management plans. The Problem-Oriented Medical Information System will be considered in detail in the next chapter.

Although giving an illness a diagnostic label confers a certain degree of reassurance on both the physician and the patient, it is not necessarily a prerequisite for treatment and it does not always accurately predict the prognosis. Other important decisions precede and follow a diagnosis. Since we are looking at the clinician primarily as a decision-maker, let us review the diagnostic process with this in mind.

Elements of diagnostic logic can be identified in various ways. One useful sequence is as follows: (1) observe, (2) describe, (3) interpret, (4) verify, (5) diagnose, and (6) act.

Observation. This includes getting to know the patient and his background, taking the initial history, doing the initial physical examination, and obtaining the routine laboratory studies. THE DECISION: what is physiologic and what is pathologic; grading the severity and significance of the abnormal findings.

Description. This essentially involves tabulating the observations. To construct a logical record it is necessary to eliminate irrelevant material, condense and concentrate the relevant findings into usable form, systematize these findings into logical clusters or patterns. THE DECISION: what to keep and what to throw away. All abnormalities may not be related to the patient's illness.

Interpretation. This involves comparing the tabulated data with a known body of knowledge. Disease patterns are described in the textbooks and in the literature. They are to a varying extent carried in the mind of the clinician. THE DECISION: what is the degree of correspondence between the patient's findings and the known patterns of disease; what are the logical possibilities; is there one, are there several, or possibly is there no significant correlation with a known disease.

Verification. This involves a plan or course of action which includes suitable laboratory studies and diagnostic procedures. The first step is to discuss the plan with the patient and obtain his permission to proceed. As each test is completed, the results are tabulated and further bedside observations are made. THE DECISION: selecting the proper diagnostic tests and determining their priority; interpreting the results.

Diagnosis. This involves applying the final label. THE DECISION: which disease or diseases account for the illness.

Action. This involves determining a course based on the diagnosis. THE DECISION: selecting the proper treatment, whether it be surgical or nonsurgical.

The clinician's responsibility does not end with the initiation of treatment. He must observe and record the results of therapy, modify it, frequently discontinue it after a period of time, and summarize the record for the sake of future doctors who will be taking care of the patient. He must also advise the patient and inform him concerning his present status and his future prospects. The problem-oriented medical system provides a useful method of tabulating and retrieving medical information. This system will be considered in detail in Chapter 2.

THE CASE PRESENTATION

An important function of the students and house staff is the presentation of the history, physical findings, and laboratory data to other members of the staff, either on ward rounds, in the outpatient clinic, or at special conferences. It will be necessary to further summarize your thoughts in making such a presentation, for, if it becomes too long, its effectiveness may be lost completely. Proper organization of the material coupled with a clear, concise exposition is essential. At these times, limit yourself to medical terminology as much as possible. Ultimately, five or six minutes should suffice for any presentation.

If the patient is presented at the bedside, he should be introduced to the attending physician in a suitable manner, after which the history is begun. An introductory statement is in order, such as: "Mr. Jones is a carpenter, 56 years of age, who developed exertional dyspnea two days ago." It is unnecessary to point out that the patient is male or female, white or Negro, when he is present. A few additional sentences which give insight into the patient's social and economic background are frequently worthwhile at this point. This will provide an accurate setting into which you will insert the illness.

Historical data which may embarrass the patient must be omitted, to be added after the group leaves the bedside; in some instances, this will include the entire history. A special effort must always be exercised to avoid using medical terms that convey a poor prognosis to the patient or that may in other ways prove upsetting. Such terms as leukemia, cancer, Hodgkin's disease, and tuberculosis may be replaced with suitable euphemisms as "mitotic figures," "Dorothy Reed's disease," or "Koch's bacillus infection."

It should not be necessary to point out that a neat appearance and a dignified manner are all-important considerations for the student as well as the physician. In the presence of patients, smoking is generally considered to be in bad taste, despite the fact that you will see occasional exceptions to this rule during your medical experience.

REFERENCES

1. Barbee, R. A., Feldman, S., and Chosy, L. W. The quantitative evaluation of student performance in the medical interview. *J. Med. Educ.* 42:238, 1967.
2. Bernard, C. *An Introduction to the Study of Experimental Medicine.* New York: Macmillan, 1925. P. 205.
3. Biörck, G. The doctor's job. *Brit. J. Med. Educ.* 1:47, 1966.
4. Biörck, G. The three patients. *Acta Med. Scand.* Suppl. 445:463, 1966.
5. Bird, B. *Talking with Patients.* Philadelphia: Lippincott, 1955.
6. Brooke, B. F. The logic of hospital case notes. *Lancet* 1:738, 1962.
7. Collen, M. F. Automated multiphasic screening and occupational data. *Arch. Environ. Health* (Chicago) 15:280, 1967.
8. Feinstein, A. R. *Clinical Judgement.* Baltimore: Williams & Wilkins, 1967. P. 21.
9. Feinstein, A. R. Compassion, computers, and the regulation of clinical technology. *Ann. Intern. Med.* 66:789, 1967.
10. Flexner, A. *Medical Education.* New York: Macmillan, 1925.

THE PROBLEM-ORIENTED MEDICAL INFORMATION SYSTEM

2

H. Kenneth Walker
J. Willis Hurst

Those of you reading this book stand at the threshold of your clinical career. A valuable attribute to begin to develop now is a *system* for continuing education which you will use throughout your professional life. A serious misconception is to feel your future education will be derived largely from the literature, lectures, meetings, and the like. On the contrary, the overwhelming majority of your learning experiences will occur when you see your patients and go through the process of identifying and solving their problems. Patients seek medical care because of problems. As previously suggested, your role as a clinician will be to characterize these problems through your history-taking and physical and laboratory examinations. You will then use appropriate diagnostic and therapeutic measures either to solve the problems or to make them more bearable. The remainder of this chapter is devoted to a system built around the problems of patients, including the display of information about these problems in the medical record.

The medical record contains the positive and negative information you have collected, displays your formulation of the problems and plans for approaching them, and finally shows your progress in solving them. The format for the standard, source-oriented record has not changed since the first written description in 1847 [2]; case reports in journals today are identical with case reports from Johns Hopkins at the turn of the century. These traditional records present two difficulties:

1. The *data base* is not *defined* — that is, specified in advance. For example, one can pick two records dealing with a given problem from the same clinician's practice and find 25 pieces of data present in one record and 50 in the other. Neither the clinician nor anyone else can say why he collected 25 pieces of data on one patient and 50 pieces on another patient with the same problem.

2. Information is placed in the record according to the *source* from which it originated — laboratory, x-ray, physicians, nurses, etc. This convention makes information difficult to retrieve from the record.

In the mid-fifties Lawrence L. Weed began the development of the Problem-Oriented Medical Record (POMR) in an effort to overcome the difficulties of the traditional source-oriented medical record [6]. The Problem-Oriented Medical Information System (Fig. 2-1) has three components:

1. The Problem-Oriented Medical Record
2. Audit of the record to identify discrepancies
3. Educational program to correct discrepancies

THE PROBLEM-ORIENTED MEDICAL RECORD

A Problem-Oriented Medical Record has four components:

1. Defined Data Base
2. Complete Problem List
3. Initial Plans titled and numbered by problem
4. Follow-up: Progress Notes titled and numbered by problem

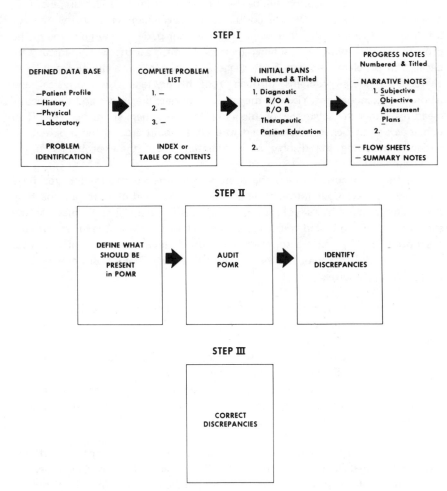

FIGURE 2-1.   Problem-Oriented Medical Information System. Step I, The Problem-Oriented Medical Record; Step II, Audit; Step III, Educational Programs.

The Defined Data Base

The Defined Data Base has four parts:

Patient profile
History
Physical examination
Laboratory examination

The Data Base may be *problem-specific* or *comprehensive.* Problem-specific Data Bases are constructed for specific problems, such as diabetes, hypertension, cholecystitis, and depressive neurosis. A comprehensive Data Base is used when you wish to identify all problems that a given patient has.

All Data Bases must be *defined*: before a Data Base is collected you must state exactly each piece of information you will seek. The record should display clearly one of three possibilities for each bit of data:

Searched for and found
Searched for and not found
Not searched for

You will define your own comprehensive and problem-specific Data Bases. The content will vary markedly depending upon the composition of your patient population and your goals for patient care.

Patient Profile. The aim of the profile portion of the history is to summarize the patient's life, to get some idea of his present position in his environment, and to discover what stimuli in his environment may be contributing to his disease. It is neither necessary nor desirable to ask in every case about each item mentioned below. However, in general the patient profile should provide information about how the patient spends his usual day, his life style, travel and occupational history, and any other information that will help you and the patient make plans together for his life after discharge.

Life history. Birthplace, position in family, socioeconomic status of family, changes in residence, education, jobs held.

Marital status. Compatibility and adjustment.

Occupation. Nature of present and previous work, adjustment to working situation, interpersonal friction, presence of occupational hazards such as exposure to metals, especially lead, arsenic, and chromium, and to poisonous gases and benzol.

Finances. It is important to know whether the patient's financial status may be contributing to his present illness and whether his illness will impose a financial hardship on him.

Personality. Personality type and reaction to environment. Patients can usually give their own appraisal of themselves in terms such as "easygoing," "tense," "high-strung," "chronic worrier," and so forth. Such remarks may be of help to the physician and serve as a starting point to allow the patient to speak more freely about himself.

Habits. Diet, regularity of eating and sleeping, exercise, use of tobacco and alcohol, drugs taken regularly such as laxatives, barbiturates, and tranquilizers.

Religion. Religious affiliation, attitude toward religion.

History. You must become highly skilled at discovering problems by taking a history from the patient. The first meeting between you and your patient is crucial. As previously pointed out, you should begin by introducing yourself and stating your goals. The second step is to inquire about the comfort of the patient, thereby letting him know your concern for him. This is followed by a short period of casual conversation which will help him relax.

Chief complaint (C.C.). The next step is to get from the patient a spontaneous statement of his problems with minimal guidance from you. This is referred to as the Chief Complaint (C.C.). One technique is to ask the patient why he came to seek medical help. The patient generally responds with a few sentences stating one or more problems. He then stops and waits for you to indicate how he should proceed. He stops because he is unsure whether you want him to continue talking or whether you want him to stop and let you ask the questions throughout the remainder of the interview. A common error at this point is for the clinician to seize immediately upon the initial problem stated by the patient and asked closed (yes—no) questions. The difficulty with this method is that the patient can have other problems he may not be given the opportunity to express; you may never discover them because you may not ask the right questions. A useful technique you might use at the point when the patient hesitates is to ask, "And what else is bothering you?" Or lean forward with an anticipatory look on your face, waiting for the pressure of silence to arouse further statements. The goal is to ask the type of questions or use the kind of nonverbal stimuli that encourage the patient to speak in *paragraphs* rather than to answer yes or no to specific questions. The difficulty with yes—no questions at this stage of the interview is that an infinite number of possible questions is needed in order to assure that all pertinent information is obtained. The "paragraph technique" is faster and more reliable.

Present illness (P.I.). At the end of this phase of the interview you have a statement of all the patient's problems as seen by him. You now take each problem and characterize it to your satisfaction by asking the patient relevant questions. This part of the history-taking is referred to as the Present Illness (P.I.). Begin by asking nonspecific questions about the problem and end with very specific questions. For example, the problems as stated by the patient might be headaches, double vision, and weakness of the right arm. You could begin by saying, "Tell me more about your headaches," and end the characterization of this problem by asking very specific questions, e.g., "How long do your headaches last?" Then proceed to the next problem: "Tell me more about your double vision."

Great care must be exercised in deciding how to ask specific questions. Often you can obtain more reliable information by provoking the patient into contradicting you, such as asking an emaciated patient, "Have you gained any weight?" You should note very carefully and consciously how patients respond to varying ways of phrasing a question and modify future questions on the basis of your cumulative experience in the art of eliciting information. Individual patients vary greatly in their ability to convey information about their problems. An example is the number of words required by different people to give one fact — or the "density of facts" per unit time. At one end of the spectrum is the highly verbal patient who delivers one fact every ten minutes. At the other end is the concrete patient who, when asked what brought him to the office, simply took off his shirt and pointed to his dermatitis. You must learn how to get the most and best information from each patient by conscious and frequent experiments with formulating questions in different ways and evaluating which way yields the best qualitative and quantitative information.

The data on the present illness must be written in chronologic sequence and in readable form. The onset of the illness should be carefully dated. This date should be that of the appearance of the first symptom that can be included logically as a part of the patient's present illness. In referring to a chronologic date, you should not use such expressions as "last Wednesday" but either a specific calendar date or a specific number of days between onset and history recording. If the onset was acute, the state of health immediately prior to it should be determined and a careful description given of the earliest symptoms. Any relevant circumstances that may have led to the onset of illness should also be noted. In the case of acute infections, inquiry should be made concerning possible exposure and incubation period. Symptoms at the onset, such as chills, fever, headache, gastrointestinal disturbances, cough, and regional pain, and any other general or local symptoms should be listed.

When the present illness has progressed in attacks separated by intervals of good health, it is necessary to obtain the history of a typical attack — onset, duration, and associated symptoms such as pain, chills, fever, jaundice, and hematuria. Inquiry should also be made as to whether these attacks were precipitated by any particular activity or by other factors such as diet, exercise, or excitement. When the symptomatology indicates a specific organ or system or suggests a particular disease, you should obtain all the data, both positive and negative, that would be relevant for this anatomic area of the physiologic system. It is also advisable to include such constitutional symptoms as anorexia, weight loss, fatigue, and fever that may accompany the present illness.

If the patient has previously seen a physician who obtained laboratory studies or prescribed medications for the present illness, this information should be recorded. If the patient cannot give the details of the drugs and dosages taken, it may be desirable to contact the physician directly.

Review of systems (R.O.S.). The review of systems is intended to detect disease in areas other than those covered primarily in the present illness and also to be certain that one has not missed some other manifestation of disease that was not covered in the present illness report. This systemic review consists of a logical sequence of questions concerning the major anatomic areas of the body, beginning at the head and working downward. For specific questions by system, see the sample POMR in the appendix.

Family history (F.H.)

Present Status of Parents and Siblings. Age and health status of each living member of the immediate family; if any have died, age at the time of death and the supposed cause of death.

Similar Illness or Symptom in the Family. If anyone in the patient's family has experienced an illness similar to that under investigation in the patient, this history should be carefully explored.

Commonly Inherited Diseases. Diabetes, hypertension, heart disease, renal disease, cancer, allergy, mental illness.

Past history (P.H.)

Childhood Diseases. Scarlet fever, measles, mumps, whooping cough, chickenpox, polio, and diphtheria. Special attention should be given to the possibility of rheumatic fever, with questions about attacks of joint pains or previously detected heart murmurs or chorea. A history of "strep throat" should be investigated and any renal complication recorded.

Major Illnesses. Pneumonia, hepatitis, malaria, or any other serious illness, any complications or sequelae from illness.

Allergies. Asthma, hay fever, hives, contact dermatitis, food idiosyncrasies, serum reactions, drug reactions.

Accidents. Serious injuries associated with unconsciousness, fractures, or penetrating wounds, any sequelae from such accidents.

Operations. All surgical procedures with the date of performance and any complications of surgery; in previous surgery of the pelvis, list the indications, e.g., pain, bleeding, tumor.

Pregnancies and Deliveries. Number of pregnancies, any abnormalities during pregnancy or at the time of delivery.

Immunizations. Vaccination against smallpox, immunization for diphtheria, tetanus, polio, measles, pertussis.

In summary, by the end of the history the physician should have constructed in his mind or on a sheet of scratch paper an incomplete Problem List for the patient. He should have characterized each problem as completely as possible from the history. Further clarification of each problem and additional problems may be discovered during the physical and laboratory examinations as given in succeeding chapters. A Complete Problem List can be constructed only after the historical data is integrated with the physical examination and the laboratory examination.

Additional detailed information on medical history-taking is available in the text by Morgan and Engel [5].

Physical Examination. Record all important findings, both normal and abnormal. The patient should be undressed and properly draped. The examination should be carried out as gently as possible, all unnecessary exposure, exhaustion, or chilling of the patient being carefully avoided. If the patient is acutely ill, it may be advisable to postpone the general physical examination and limit yourself only to the important findings necessary for guidance of treatment. Findings should be recorded as accurately as possible and in quantitative terms such as millimeters or centimeters. Precision in description stimulates precision in observation. Immediate recording (in shorthand, on scratch paper) reduces mistakes and distortions.

This general outline can be followed: Inspection of the head, neck, back, and chest should be done, whenever possible, with the patient in the sitting posture. The patient should then be asked to lie down. His blood pressure may be rechecked at this time, and the heart, lungs, breasts, axillae, abdomen, genitalia, rectum, and extremities are examined. A major portion of the neurologic examination can be done while the patient is still recumbent, but certain features (such as analysis of gait) necessitate getting the patient out of bed. At this time it is convenient to complete the orthopedic examination. The female patient may then be prepared for pelvic and rectal examination; this sequence permits the examiner to leave the room and begin to write or dictate the results of the examination while the nurse is preparing the patient.

For a detailed outline of the physical examination see the appendix.

Laboratory Examination. A few of the more important laboratory determinations will be considered in subsequent chapters. A comprehensive discussion of laboratory evaluation is beyond the scope of this text.

The Complete Problem List

After collecting the Defined Data Base you are ready to construct a Complete Problem List. This sheet is always placed at the beginning of the medical record and serves as an index or table of contents. A *problem* is defined as anything that requires diagnostic or management plans or that interferes with the quality of life as perceived by the patient. Problems can come from any of several categories:

Examples

Anatomic	Enlarged heart, hernia
Physiologic	Hypertension, jaundice
Symptomatic	Dyspnea, abdominal pain
Etiologic	Diabetes, viral hepatitis
Demographic	Tobacco abuse
Social	Unheated house, inadequate money
Psychiatric	Depression, schizophrenia

Each problem should be assigned a number. The terminology or way in which a problem is stated will vary according to the quantity and quality of the information collected and the experience and ability of the clinician. The Problem List must be complete, that is, each problem must be entered.

The date the problem is entered tells anyone seeking information from the record that the initial statement about that problem can be found within the record on that exact date. Each problem requiring further diagnostic steps has an arrow placed after the problem. When the problem is solved, the date of solution is placed over the arrow and indicates the date of the Progress Note where a detailed discussion can be found.

7-1-72

6-28-72   1. Jaundice ──────▶ obstruction of common duct by gallstone

The initial statement can be found within the record on 6-28-72. An arrow was placed after the problem to indicate that further resolution of the problem was necessary. The date over the arrow indicates that on 7-1-72 there is a Progress Note summarizing the evidence that allowed the problem of jaundice to be resolved as being caused by common duct obstruction by a gallstone.

Each problem must be numbered. That number stays with the problem forever, even though the statement of the problem may change. All notes relating to the problem in the example just given will have the number *1* whenever the problem occurs in the record. From 6-28-72 until 7-1-72 number 1 was stated as "jaundice." From 7-1-72 on, number 1 will be followed by "common duct obstruction by gallstone." This ensures that all data concerning a specific problem are always associated with a particular number throughout the chart. Furthermore, the sequence of events leading to the solution of the problem is apparent. Thus information in the medical record is accessible.

Initial Plans

The next step is to take each active problem on the Complete Problem List and generate titled and numbered Initial Plans. The three components of Initial Plans are:

Diagnostic. Listed here are the *rule-outs.* The most likely or important possibility to be ruled out or confirmed is listed first, followed by the other rule-outs in order of likelihood or priority. After each rule-out come your plans for gathering more information pertinent to it, listed in the order you will execute them.

Therapeutic
1. Plans for specific drugs or therapy, including exact dosage
2. Parameters to be followed in order to determine response to therapy
3. Plans for discovering or monitoring side effects of therapy

Patient Education. What the patient and/or his family have been told about the problem. One aspect of this component is to obtain the agreement of the patient about the presence or formulation of the problem, if appropriate.

Progress Notes

The solving of problems is displayed in the record by three types of follow-up notes:

Narrative Progress Notes. Each numbered and titled note is followed by four headings:

SOAP
Subjective: Information described by the patient.
Objective: Information such as physical findings and laboratory data.
Assessment: The clinician's interpretations of the subjective and objective data.
Plans: Measures to be undertaken for diagnostic, therapeutic, and patient education reasons.

Flow Sheets. Each specific problem has a numbered and titled flow sheet when appropriate. The recording, following, and interpretation of certain types of problems are done better by flow sheets than by narrative description. Interrelationships among various bits of data are more easily perceived; efficiency of recording data is improved. Good examples of conditions for which flow sheets are very helpful include diabetic ketoacidosis, congestive heart failure, renal failure, and acute or chronic pulmonary disease.

Summary Statements. Such statements as interval notes and discharge and death notes are written in a problem-oriented fashion, with *subjective, objective, assessment,* and *plan* components. Prognosis is listed under *assessment.* Drug therapy and clinic appointments are listed under *plan.*

Brevity is important in all notes. An important advantage of the POMR is that all information is readily accessible to any user. Duplicate statements of data are unnecessary and a waste of time.

In summary, the POMR specifies how data is to be displayed in the medical record. All POMRs have the information placed in a similar fashion and linked to specific problems. This uniformity of the method of display of information makes the data easily retrievable. The linkage of data to specific problems also preserves the interrelationships among various data.

The steps required in preparing a POMR are as follows:

1. Collect a Defined Data Base.
2. Make a Complete Problem List with each problem numbered and titled. This list should be the first page of the patient's record.

3. Write a numbered and titled Initial Plan for each problem with diagnostic, therapeutic, and patient education components. Write problem-oriented orders.

4. Give follow-up in the form of numbered and titled Progress Notes with *subjective, objective, assessment,* and *plan* components. All material entered in the record should be problem-oriented.

At the end of the book there is an example of a POMR (Appendix).

AUDIT

Audit is the process of taking your Defined Data Base, defined Initial Plans, and Progress Notes for specific problems and comparing this *defined standard of care* with what you in fact delivered as reflected in the POMRs of your patients. *Audit is the comparison of what you did with what you promised.* There are several different levels of audit [3].

1. Was defined task carried out?   Thoroughness
2. Does record accurately reflect patient findings?   Reliability
3. Was reasoning correct, wise, and elegant?   Analytic sense
4. $\dfrac{\text{1, 2, or 3}}{\text{Time, money, space, or energy}}$ Efficiency

Thoroughness

Checking the POMR to see if the defined standards were adhered to. In order to arrive at this point you will go through the following sequence:

1. Data Bases must be defined for comprehensive care and specific problems. Record design must indicate clearly each bit of data searched for and found, or not found, or not searched for.

2. What is normal and abnormal for each piece of information that is to be collected must also be defined. That is, it must be decided beforehand what will and will not be a problem. This must be done in an all-or-nothing fashion in order for the clinician to operate effectively. You will need to consult the literature and the remainder of this book frequently in order to arrive at these *problem definitions.*

3. Management protocols must be set up, specifying Initial Plans and follow-up for individual problems. Again, you will need to consult the literature and keep these up to date as medical information changes.

4. Medical care is then delivered and displayed in a POMR. The process of audit can now begin.

5. Auditing or inspection of the Data Base of the POMR is performed to determine if all the information in the Defined Data Base was searched for.

6. Inspection of the POMR is performed to make sure all abnormalities are accounted for — that is, stated either as problems or as attributes of a problem.

7. Inspection of the Initial Plans and Progress Notes for each problem is then performed and it is determined whether they follow the agreed-upon management protocols.

8. A list is made of all discrepancies discovered in the above steps.

Reliability

Check the patient's history and physical examination and audit the POMR to see if the findings on the patient are accurately reflected in the record. This is an audit of history and physical examination techniques.

Analytic Sense

Inspect the POMR and ask the question: "Was the reasoning correct, wise, and elegant?"

Efficiency

Take thoroughness, reliability, or analytic sense as the numerator and time, money, space, or energy as the denominator. The commonest way of expressing efficiency is to take thoroughness as the numerator and time as the denominator. It is equally and often more important when dealing with large numbers of patients or patients who are not wealthy to divide thoroughness by the cost.

CORRECTION OF DISCREPANCIES
(Continuing Education of the Clinician)

To summarize, you start out by defining in a very precise fashion your standard of medical care. After delivering the care you analyze your records and note the discrepancies between what you promised and what you delivered. The next step is to develop educational programs aimed at the elimination of discrepancies. The educational programs may take many forms. The lecture system is used as a rule but has many deficiencies. Lectures have a place, but they can never substitute for the disciplined action of one who has the spirit of learning. Your own ability to analyze a discrepancy and determine the best method of correction is the key to your continuing education.

Remember that the everyday practice of medicine — identifying, characterizing, and solving problems — is one of the most challenging forms of scientific research you can engage in. Lawrence Weed [6] and Stuart Graves [3] have shown a useful way of looking at medicine as a science in relationship to the POMR.

Phenomena to be observed   Defined Data Base
Hypotheses   Complete Problem List
Testing the hypotheses   Initial Plans and Progress Notes

You must consciously and continuously improve yourself in the context of what you do each day. Constant *correlation* of history with findings on the physical and laboratory examinations, with basic science information, and finally with follow-up are prerequisites to your continued growth as a clinician. Initial Plans for specific problems need updating as more is learned about mechanisms and treatment of diseases. Follow-up will change in a similar fashion. The level of resolution of a problem as initially stated reflects the state of your medical knowledge and experience. The increasing speed with which you solve problems reflects increasing efficiency as you gain experience and knowledge.

You will grow in knowledge and ability if you look at continuing education as that which occurs as you encounter problems in patients each day. The medical literature and experience of others is important and useful as it is brought to bear in a very specific fashion on the problems you encounter daily. For example, there is little sense in attending a two-day symposium on porphyria if the chances are you will see only one case in your lifetime, while at the same time you understand poorly many of the problems of the patients you see every day.

Your experience must be accessible to you. The great usefulness of the Weed system is that this becomes possible. You must lay plans so you can plot your education according to the needs of your patients and their problems in your practice. *Your records should be planned so you can use your past experience with particular problems to help solve those same problems when they reappear in other patients.* If a 75-year-old patient with gallstones walks into your office eight years from now, you should be able to review rapidly your experience with gallstones and use that experience to help solve his particular problem. This is one of the best forms of clinical research — your experience with problems in your own patient population. Arrange your record system so you can keep up with the number and frequency of problems of your patients. Then plan your continuing education around these problems.

As you see patients and master the techniques of physical diagnosis with the guidance of the rest of this book, make up Problem Cards for each problem you see in a patient. Jot down the name of the patient and the hospital number. If the problem is unresolved, go back later and find out how it was resolved. Make up a new Problem Card for what the problem was resolved to. You will be pleased by the accessible and retrievable experience you will have accumulated over the short period of the next two years. Your future education starts today!

REFERENCES

1. Bjorn, J., and Crose, H. *Problem-Oriented Practice.* Chicago: Modern Hospital Press, McGraw-Hill, 1970.
2. Crompton, S. *Medical Reporting and Case Taking.* London: Isaac Pitman, 1847.
3. Graves, S. Clinical Investigation. In H. K. Walker, J. W. Hurst, and M. F. Woody (Eds.), *Applying the Problem-Oriented System.* New York: Medcom, 1973.
4. Hurst, J. W., and Walker, H. K. *The Problem-Oriented System.* New York: Medcom, 1972.
5. Morgan, W. L., and Engel, G. L. *The Clinical Approach to the Patient.* Philadelphia: Saunders, 1969.
6. Weed, L. L. *Medical Records, Medical Education, and Patient Care.* Cleveland: Case Western Reserve University Press, 1971.

BEGINNING THE EXAMINATION

3

Richard D. Judge

THE ART OF OBSERVING

The ability to observe accurately is a great asset in most fields of endeavor; it is indispensable in the practice of medicine. The physical examination is simply a series of observations. What factors play a role in observation? How are they developed? What are the pitfalls? Before considering what to look for during the examination, let us turn our attention briefly to the act of looking itself, the art of observing.

Goethe epitomized the process when he said, "We see only what we know." In this simple statement he managed to incorporate both of the important basic elements of observation, the sensory or perceptual act and the correlative or conceptual process. The first step is to perceive. The second is to relate the sensory stimuli to some relevant knowledge or past experience. Both abilities are indispensable, and both are irrevocably interdependent. From our reading, from our lectures and clinical experience, we accumulate the body of knowledge which gives significance to what we perceive. This background is concerned with more than the simple interpretation of findings; it is part of the act of observing itself. Let us consider an example. An inexperienced student at the foot of the bed of a patient with clubbed digits is asked to describe anything he sees which might be significant. Rarely, he may point out the peculiar rounded appearance of the nails without being able to explain its significance. Far more often, however, the finding will go unnoticed. It does not register because it has no meaning for him. The lesson here is a simple one. In your eagerness to begin doctoring with "real patients," do not lose sight of the importance of your preclinical scholarship. Without basic knowledge derived from careful study, you cannot expect to be an artful observer. You see only insofar as you understand.

Observations are mediated by the senses. The physician in particular uses sight, hearing, touch, occasionally smell, and very often his sixth sense — intuition. Unfortunately, the conveniences of twentieth-century living tend to cause atrophy of our sensory capabilities. We are, in general, far less perceptive of our environment than were our great-grandparents. For this reason, you may find it discouragingly difficult at first. You may rationalize away the importance of bedside observation and fall back on the laboratory. Why do a careful cardiac examination? Simply order a chest x-ray and ECG. But keep in mind that while laboratory tests help to make diagnostic and therapeutic decisions, they really only supply *additional observations.* These, admittedly, are often regulated and measured by machines. *Yet, no new senses have been provided.* Certain tendencies for error may have been reduced, but others have been increased. The physician still follows the same sequence based on scientific method — observe, describe, verify, decide, act; wherever there is scientific method, accurate observation is invariably the first step.

Abraham Flexner defended bedside observation fifty years ago as follows [3].

Science resides in the intellect, not in the instrument. To call a careful and correct bedside observation clinical and a laboratory examination scientific, as if there were some qualitative distinction between the two, is absurd .... The term "scientific" cannot be denied to an accurate observation at the bedside, if it is conceded to a similarly accurate observation made by means of a microscope; nor can it be denied to a correct description of a process observed in a patient, while conceded to the correct description of a process observed in a rabbit or guinea pig. The clinic ... is not saved to science by laboratory methods; it includes them as simply additional weapons with which to do better what scientific clinicians have always done, viz. observe, explore, unravel.

Medical observation is complex; it must therefore be deliberate and systematic. It requires concentration. An ill-timed distraction may cause an oversight; the oversight, a missed diagnosis. It is always helpful to focus your attention on one thing at a time. Looking at the hand is not enough. You must scrutinize in turn the nails, the skin texture, the color, the hair distribution, and so forth. Your outline helps in this respect by limiting your field of observation. It makes you less susceptible to errors of omission.

The mental phase of observation may be both conscious and subconscious. What attracts us more than anything else is change. Like the continuous sound which is "heard" only when it suddenly stops, the physical abnormality comes to capture our attention almost automatically. At this point we begin to think. We look more carefully. We may reposition the patient, have him breathe more deeply, or perform some other maneuver aimed at accentuating the finding or facilitating its analysis. We also begin to think ahead, for one finding may be a signpost that tells us to look carefully for other specific possibilities. One or two spider angiomas noticed during history taking tells the experienced observer to watch particularly for an enlarged liver, palmar erythema, clubbing, spleno-megaly, testicular atrophy, dilated abdominal veins, external hemorrhoids, hyperactive reflexes, or a flapping tremor. Of course, the first signpost on the way to diagnosis is the patient's history itself.

Another important point to remember is that not only may we miss a perfectly obvious physical finding, but we frequently invent quite false observations. The following story is a paraphrase of one by W. I. B. Beveridge [2]:

During a psychology congress in Göttingen, a man suddenly rushed into the meeting followed by another with a revolver. There was a scuffle; a shot was fired and both men rushed out. The whole episode lasted about twenty seconds. The chairman immediately asked everyone present to write down what they had seen. The observers were not told that the incident had been carefully rehearsed and photographed. Of the forty reports presented, only one had less than 20 per cent error about principal facts, and 25 had more than 40 per cent error. The most interesting feature was that in over half the accounts, *10 per cent or more of the details were pure invention.*

Why do we invent false observations? There are two basic reasons. First of all, our senses may play tricks on us. We are all familiar with optical illusions. Concentration and careful scrutiny help us to avoid this pitfall. Repetition is also invaluable. A colleague once told me that whenever an auscultatory finding was questionable he went back and listened to the heart two more times. If he was still certain of it after three examinations, he recorded it. Following this recommendation may save you the occasional embarrassment of having to correct an illusory observation.

A second source of invention has its origin in the mind itself. We all have a tendency to distort what we observe according to our preconceived expectations. In this respect, Goethe's aphorism, "We see only what we know," has a special and more subtle significance. Our past experience may prejudice our perception. Our minds may subconsciously fill in the gaps with erroneous material. The special danger period for interjecting such subjective information into the record is when you are writing out your work-up. By jotting down notes during the examination and writing out your findings while they are still fresh in your mind, you will minimize this tendency.

In summary, the three major sources of inaccurate observation are (1) oversight, (2) forgetting, and (3) bias. *Oversights* are minimized by habituation to method and by fractionating observations into logical, sequential small units. *Forgetting* is minimized by jotting down immediate notes and reminders, and by transferring the information to the permanent record as promptly as possible. *Bias* is a lifelong challenge. It may be lessened by an understanding of how it distorts. By repeated self-analysis, by being on guard, you may minimize the tendency, but "before the cock crows" you will fall prey to your prejudices. We all do, unfortunately.

PREPARING THE PATIENT

The physical examination should be made as easy as possible for the patient. He usually expects it to be a relatively distasteful experience. If the physician is considerate and gentle the patient should feel, when it is all over, that most of his or her fears on that score were unfounded. The ideal examining room is private, warm enough to avoid chilling, and free from distracting noise and sources of interruption. Natural light is preferable whenever possible. The examining table may be placed with the head against the wall, but both sides (particularly the right) and the foot should be accessible to the examiner.

The first crisis concerns the problem of undressing. In the outpatient department the patient disrobes with the physician out of the room. Ideally the female patient should be assisted by a nurse or attendant. If none is available the physician should take a moment to explain how the patient should drape herself. It is always stressed that the patient must disrobe completely. This is certainly true, and it is folly to try to examine the heart through a nightgown or the abdomen through a slip. But it is equally true that respect for the patient's modesty is an important factor which greatly affects the subsequent physician-patient relationship. Tell the patient to leave on his or her underpants. They can be moved aside later, when necessary. This simple concession will not interfere with 90 percent of the examination, and it will greatly reduce anxiety. In addition, a sheet should always be available to further drape the patient from the waist down. A towel or special gown may be used to cover the female chest. This combination allows complete examination without prolonged, embarrassing exposure.

Begin and end by washing your hands. When possible this should be done in the patient's presence. Your movements should be deliberate and methodical but always gentle. Your attitude should be basically objective but not serious. An occasional smile or distracting comment is a great help in relieving tension. Try to avoid surprise. Preface each maneuver with a simple direction or explanation. When the time comes to inspect the breasts prepare the patient with a simple statement such as, "Now we must remove this towel for a moment." The same is true for examining the abdomen. This gives a sense of purpose to your actions which practically eliminates embarrassment. The pelvic examination, of course, takes special preparation, and a nurse must *always* be present.

At times it will be necessary to cause the patient some discomfort. In these instances it is particularly important to explain the necessity of going ahead. For example, "I'm sorry if this hurts but it's important." The examiner should not hold back because of pain, but on the other hand he must balance the importance of the observation against the degree of discomfort and proceed as deftly as possible.

THE TOOLS OF EXAMINATION

The student's bag should include the following:

Pen light	Reflex hammer
Oto-ophthalmoscope with several ear specula of	Tuning fork (128 or 256 cps)
varying sizes	Pin
Sphygmomanometer (aneroid type)	Notebook
Stethoscope with bell and diaphragm	Goniometer
end-pieces	Tape measure

He should have easy access to:

Tongue blades	Finger cots
Cotton	Disposable tissues
Rubber or disposable plastic gloves	Paper cups
Lubricant	Vaginal speculum

VITAL SIGNS

The vital signs include height, weight, temperature, blood pressure, pulse rate, and respiratory rate. These data are recorded together at the top of each record. They are usually taken in advance by an attendant or nurse. Many physicians recheck the blood pressure and most of them retake the pulse as part of their own examination. The role of paramedical assistants is growing, however, and the attitude toward many of the simple aspects of the physical examinations is being modified in much the same way as automation has modified history taking. The time is at hand when the physician can no longer afford the luxury of misusing precious time.

The temperature may be taken orally or rectally, and in the United States the Fahrenheit scale is usually used. The method of thermometry needs no explanation. The oral thermometer is left in place for several minutes to insure accuracy, although when fever is present it will be observed that the mercury column climbs quite rapidly (15 to 30 seconds) to within a few tenths of the final reading. Falsely low levels may result from incomplete closure of the mouth, breathing through the mouth, leaving the thermometer in place for too short a time, or the recent ingestion of cold substances. Falsely elevated levels may result from inadequate shaking down of the thermometer, previous ingestion of warm substances, recent strenuous activity, or even a very warm bath. Rectal temperature averages one-half to one degree higher than oral. The axillary temperature is so inaccurate that it will not be considered.

In most persons there is a diurnal (occurring every day) variation in body temperature of one-half to two degrees. The lowest ebb is reached during sleep, at which time the temperature may fall as low as 96.5 to 97°F. As the patient begins to awaken, the temperature slowly rises. The "fast starter" who jumps from his bed and reaches his peak efficiency early in the day usually has reached 98°F. or more by the time he awakens and may peak to his maximum level of 98.6 to 99.2°F. before noon. The "slow starter" oftens shows a lower temperature on rising and may not peak until late afternoon or early evening. His efficiency may be greatest later in the day.

There is a well-known temperature pattern in the menstruating female which reflects the effects of ovulation. The morning temperature falls slightly just prior to menstruation and continues at this level until the mid-point between the two periods. Twenty-four to thirty-six hours prior to ovulation there may be a further drop, and coincident with ovulation the morning temperature rises and remains at a somewhat higher level until just before the next menses.

You will note that the upper limit of normal on the standard thermometer is 98.6°F. *It must be remembered that this is an arbitrary value which applies primarily to patients at bedrest.* Many normal people show higher levels when active, and on a hot summer day in the outpatient clinic readings of 99.4°F. are not at all uncommon in perfectly normal individuals. Even higher values can be recorded in children following hard play.

If the oral temperature exceeds 99°F. in a patient at bedrest, it is referred to as fever. Fever may be classified for descriptive purposes as intermittent, in which there is a return to normal or subnormal levels one or more times daily; remittent, in which diurnal fluctuations occur but normal levels are never attained; continuous, in which only minor variations are recorded; and relapsing, in which periods of fever for several days' duration are punctuated by spontaneous periods of normal temperature. The diagnostic value of classifying fever in this manner is relatively limited. Intermittent fever commonly accompanies acute inflammation; relapsing fever may result from neoplasm, particularly lymphoblastoma; but many exceptions to any rule regarding temperature patterns are encountered in clinical practice. When the temperature returns to normal after a period of fever it is said that the patient has become afebrile.

Subnormal temperature is common in many individuals. The student has only to note the frequency of values in the 97°F. range recorded on his basal metabolism reports. Subnormal readings are commonly found in patients in coma due to drug intoxication or metabolic acidosis.

GENERAL APPEARANCE

After the vital signs, you will make a series of observations which characterize the patient's general appearance. A number of separate elements are involved in "looking at the patient." Among these are:

Position. Restless or still, comfortable or uncomfortable, flat or upright, breathless, legs flexed, hands fidgety or tremulous, how ill.

Expression. Cheerful, anxious, morose, unintelligent, exhausted, angry, excited or indifferent, facial symmetry, swelling, chronologic age vs. apparent age, unusual mannerisms.

Conscious State. Alert or drowsy, asleep, stuporous or comatose, delirious, varying level of consciousness.

Intelligence. As expected in view of education, work, age and social class, good or bad historian, memory for recent and remote events, attentive or listless, cooperative or evasive.

Temperament. Introspective or extroverted, friendly or hostile, stolid or nervous, worrier, perfectionist.

Additional aspects of the general appearance which are considered in detail further on in the text include (1) habitus, (2) dress, (3) complexion, (4) gait, (5) nails, and (6) hair.

Develop the habit of casually observing the patient's hands. They are highly informative. It is said that when a physician was called to the harem to treat one of the Pashida's wives, he was allowed to examine only the hand of the patient extended between the folds of the drapery [1]. This procedure was not entirely without diagnostic value. One need only watch a concert pianist to recognize the close relationship between brain and hand. Speech is intimately associated with the hand and gestures may convey more meaning than words. The types of rings (and other jewelry) have a special message. Nicotine stains also convey a certain impression as does the general hygiene. We meet one another with a handshake. The hand reflects our activities: the rough, weathered farmer's or laborer's hand, often mutilated by exposure to injury; the soft, sensitive hand of the adolescent girl; the reddened, rough hands of the overbusy housewife. The hand is affected by hundreds of different clinical diseases. As Ask-Upmark has pointed out, "[the hand] reveals in the healthy man or woman, much of the character and of the profession. . . . The words handle, handicap and handicraft are significant. What would Helen Keller have done without her hands? . . . The hand is extended to bless, to pray, and to take the oath. It is also used for action and extended in friendship and help. It is the ultimate condition for creating art, but also for our existence as human beings" [1]. Watch the hands with care.

Note also the wrinkles in the faces of older patients. They may provide a clue about background. Abraham Lincoln once observed that a person over forty is responsible for his facial appearance. Certainly longstanding effects of happiness or unhappiness, productivity or frustration, self-indulgence or service, may write themselves indelibly into the facial lines. Remember also that what you observe within the first two minutes after the patient enters your office may be all-important in recognizing depression and other emotional disorders. Later the patient's complaints may obscure the issue. If you suspect that the patient's personality is unusual, if his smile lacks conviction or his expression lacks animation, seek out more information from family or friends. It is sometimes difficult to assess a change in appearance or personality without third-party assistance. This is equally true for psychologic and physiologic changes.

REFERENCES

1. Ask-Upmark, E. *Bedside Medicine.* Stockholm: Almquist and Wiksell, 1963. P. 27.
2. Beveridge, W. I. B. *The Art of Scientific Investigation.* New York: Random House, 1957. P. 133.
3. Flexner, A. *Medical Education.* New York: Macmillan, 1925. P. 7.

ENDOCRINE SYSTEM

4

John C. Floyd, Jr.

The physical manifestations of disordered endocrine gland function warrant special consideration. The recognition of such manifestations alone may establish the diagnosis of an endocrinopathy and at the least can lead the physician to suspect that such a disorder is present. For this reason the physical findings in several of the more important endocrine syndromes will be briefly summarized. Complete details of the physical manifestations and the complex laboratory indications of deficient or excessive function of the endocrine glands more properly are given later in the medical curriculum. Although in clinical parlance eponyms are used frequently to denote endocrine syndromes, only a few eponyms will be given in the glossary, and generally they will not be used in subsequent discussion. Illustrations depicting the striking appearance of patients affected by advanced or long-standing endocrine abnormalities are not included in this chapter. The identification of these disorders at much earlier stages in their evolution is to be emphasized. The author believes that a gallery of illustrations of advanced pathologic conditions may bias against searching for the subtleties of early disease the detection of which can lead to early diagnosis.

GLOSSARY

Acromegaly A disorder in the adult resulting from an excess production and secretion of growth hormone by the pituitary gland, characterized by overgrowth of bony, cartilaginous, and soft tissues, especially noticeable in acral parts.

Addison's disease A disorder resulting from chronic underproduction of cortisol and aldosterone by the adrenal cortex, characterized by hyperpigmentation, asthenia, and low blood pressure.

Cushing's syndrome A disorder resulting from chronic overproduction of cortisol by the adrenal cortex, characterized by thinned skin and wasted muscles, accumulation of fat on the trunk, easy bruising, plethora, and high blood pressure.

Exophthalmos Prominence or protuberance of the eyes, frequently associated with hyperthyroidism.

Goiter Enlargement of the thyroid gland.

Graves' disease A disorder characterized by exophthalmos, goiter, and overproduction of thyroid hormone, the latter producing thyrotoxicosis.

Hirsutism A state of increased amounts of body and facial hair, especially in the female.

Hyperparathyroidism, hypoparathyroidism Overproduction of parathyroid hormone and hypercalcemia in the former state, underproduction and hypocalcemia in the latter. Both are characterized by abnormalities in neuromuscular excitability as well as by abnormalities of the eye and of bone.

Hypogonadism Absent or reduced function of the testis or ovary characterized by diminished germ cell production and/or maturation and by decreased production of sex hormones. Underdevelopment of secondary sexual characteristics or regression of developed secondary sexual characteristics results.

Myxedema   A disorder resulting from underproduction of thyroid hormone, characterized by puffiness of soft tissues, slowing of body movements, and deepening of the voice.

Obesity   The accumulation of an excessive amount of fat.

Virilism   A state of masculinization in the female.

TECHNIQUE OF EXAMINATION AND NORMAL FINDINGS

The endocrine glands that normally may be palpable are the thyroid gland and the gonads. The latter will be considered in subsequent chapters.

In spite of the general inaccessibility of the endocrine glands, much concerning their function can be appreciated by noting the appearance of the patient. Age, sex, and racial descent are associated with variation in physical findings and must be taken into account. The hormones secreted by endocrine glands regulate, in part, biologic processes that play important roles in growth, maturation, and production of energy, in regulation of body temperature, pulse rate, and blood pressure, and in the maintenance of a constant internal environment (homeostasis). The skin and its appendages mirror endocrine function particularly well; skin color, pigmentation, texture, and thickness, the quantity and distribution of scalp, facial, and body hair, and the amount and distribution of subcutaneous fat are all significantly influenced by endocrine gland function. The state of development of other readily observable secondary sex characteristics, the breast and external genitalia, also reflect endocrine gland function. Hence, careful and systematic observation of the patient's body proportions, configuration, secondary sex characteristics, and skin and its appendages is essential in the physical examination, especially as it relates to the endocrine system.

Male and female possess readily observable differences in their genitalia and in breast development. Except during puberty glandular tissue is not palpable in the breast of the male. Female contours in general are more rounded than those of the male due to a generalized distribution of fat and relative underdevelopment of musculature. In the female, the hips are broader than the shoulders and the thighs tend to approximate in the midline. The calves tend to curve outward. The upper border of the pubic hair is horizontal (female escutcheon), without an extension of hair toward the umbilicus, although a few dark hairs may be present on the abdomen and about the areolae. The male presents a more rugged appearance due to relatively less fat and more sharply outlined musculature. His shoulders are broader than his hips. There is usually a space between the thighs. The calves are curved inward. Pubic hair extends upward from the pubic area in a triangular pattern toward the umbilicus (male escutcheon).

Habitus refers to the individual's build or body type. It is important to recognize that there are several normal types. The *asthenic* individual (ectomorph) is slender and may appear to be underweight. The shoulders are narrow and the anteroposterior diameter of the chest is reduced. The costal angle is acute. The bone structure is delicate, and the arms and legs, hands and feet are long in relation to those of normal subjects of other body types. The abdomen is flat, the buttocks are small, and the musculature of the extremities is light. The *sthenic* individual (mesomorph) is relatively square and athletic in appearance. The bone structure is heavy, the neck muscular, the shoulders broad, the pectoral muscles large, the extremities heavily muscled, and the buttocks large. The *pyknic* individual (endomorph) is heavy, soft, and rounded, due to an accumulation of body fat. Bone structure may or may not be heavy. The neck is thick but not necessarily because of well-developed musculature. The abdomen is protuberant, and the costal angle is wide. The breasts are fatty, and the arms and legs are relatively short, with short fingers and thick wrists. The buttocks, thighs, and calves are usually heavy and fat.

The determination of skeletal proportions aids in the evaluation of the level of growth and development and may point to the kind of process that has led to an aberration of body proportions. The more commonly determined skeletal measurements are:

1. Span — from fingertip to fingertip (upper extremities in abduction).
2. Lower Skeletal Segment — heel (floor) to top of symphysis pubis.
3. Upper Skeletal Segment — top of symphysis pubis to crown (also given by subtracting lower segment from height).
4. Skeletal or Body Ratio — obtained by dividing the Upper by the Lower Skeletal Segment. In the newborn this ratio is about 1.70. Normally the ratio steadily declines to about 1.0 at age 9—10 years and remains at this value throughout the remainder of life.

The estimation of desirable body weight is important, since desirable weight is associated with lower mortality rate. The tables of desirable body weights published by the Metropolitan Life Insurance Company are given in Table 4-1.

In the beginning of the examination the patient is asked to expose his head, neck, and upper chest and to sit erect, hands in his lap and with shoulders relaxed. By sitting facing the patient the examiner can make careful and systematic note of size and configurations of skull, facial bone, jawbone, soft tissues of the face, preauricular and supraclavicular areas, scalp, ears, nose, lips, tongue, and state of dentition. He can also make note of facial acne, facial and scalp hair distribution and types, degree of pigmentation of the skin, and presence of any pigmentation of buccal mucosa and gums. From the same position the examiner can ascertain the presence of arcus cornea, band keratopathy, cataract, and tarsal plate calcification and can begin the examination of the thyroid. As the examination progresses to other parts of the anatomy the same kind of careful assessment of skin, pigmentation, body and sex hair, nails, fat distribution, and bony configuration is continued. In addition, careful observations are made of the genitalia and the breasts, nipples, and areola with special regard to development and degree of pigmentation. In the assessment of endocrine gland function by physical examination the physician must ascertain

Table 4-1. Desirable Weights for Men and Women (According to Height and Frame, Age 25 and Over)

Height	Weight in Pounds (in indoor clothing)		
	Small Frame	Medium Frame	Large Frame
Men (in shoes — 1" heels)			
5' 2"	112—120	118—129	126—141
3"	115—123	121—133	129—144
4"	118—126	124—136	132—148
5"	121—129	127—139	135—152
6"	124—133	130—143	138—156
7"	128—137	134—147	142—161
8"	132—141	138—152	147—166
9"	136—145	142—156	151—170
10"	140—150	146—160	155—174
11"	144—154	150—165	159—179
6' 0"	148—158	154—170	164—184
1"	152—162	158—175	168—189
2"	156—167	162—180	173—194
3"	160—171	167—185	178—199
4"	164—175	172—190	182—204
Women (in shoes — 2" heels)			
4' 10"	92— 98	96—107	104—119
11"	94—101	98—110	106—122
5' 0"	96—104	101—113	109—125
1"	99—107	104—116	112—128
2"	102—110	107—119	115—131
3"	105—113	110—122	118—134
4"	108—116	113—126	121—138
5"	111—119	116—130	125—142
6"	114—123	120—135	129—146
7"	118—127	124—139	133—150
8"	122—131	128—143	137—154
9"	126—135	132—147	141—158
10"	130—140	136—151	145—163
11"	134—144	140—155	149—168
6' 0"	138—148	144—159	153—173

Source: Prepared by the Metropolitan Life Insurance Company. Derived primarily from data of the *Build and Blood Pressure Study,* Society of Actuaries, 1959.

whether lactation is present. This is easily done by manually applying pressure to the milk ducts and nipple. The examiner may detect thickening or thinning of the skin by picking up the skin of the dorsum of the hand or forearm and rolling it gently between his thumb and forefinger. Muscle size can be appreciated by palpating the forearm muscles or quadriceps femoris as the patient contracts the muscle. Some physical signs which ordinarily are the result of excessive or deficient hormone secretion sometimes may also be observed in the presence of normal hormone production and activity. A eunuch, for example, may have no beard because of a lack of testicular androgen production, while the male American Indian with normal testicular androgen production has no beard by virtue of his genetic background. Some women with normal rates of androgen production (for females) may demonstrate hirsutism as a familial or ethnic characteristic.

The thyroid gland may be examined by inspection and palpation. When it is located substernally it may be detected by percussion of the chest. Movement of the thyroid produced by the act of swallowing aids in its inspection and palpation. Sips of water allow repetitive swallowing. Face the patient and observe the base of his neck as he swallows. Repeat the observation with the patient slightly extending his neck. The normal thyroid gland usually is not visible. An enlargement may be evident as a subtle fullness which glides upward transiently upon deglutition. This is more easily appreciated when the neck is illuminated by obliquely directed light.

The thyroid gland is frequently not palpable in normal patients. However, in the average asthenic individual it is felt as a vague layer of tissue which glides briefly beneath the fingers, rising slightly with swallowing.

The examiner may palpate the thyroid gland from a position in front of or behind the subject. In either case flexion of the subject's neck toward the lobe being examined results in relaxation of the corresponding sternocleidomastoid muscle, and this facilitates palpation. When the examiner is behind the patient, he lightly places the tips of the first two or three fingers of both hands on either side of the patient's trachea slightly below the level of the thyroid cartilage (Figure 4-1A). Both lobes are surveyed simultaneously as the patient swallows. A light, rotary motion of the examiner's fingers will help to delineate nodules and irregularities. Next, palpate each side separately. Flex the neck to the side being examined. The first two fingers of the left hand are used to palpate the right lobe, while the right hand is placed behind the sternocleidomastoid muscle to evert the gland as much as possible (Figure 4-1D). The left lobe is similarly examined with the neck flexed slightly to that side (Figure 4-1C). With each maneuver the patient is asked to swallow. Palpation should be gentle, because vigorous pressure may cause soreness, choking, or cough, making further examination difficult.

With the examiner in front of the subject, the right lobe is examined as shown in Figure 4-2A. The right thumb is used to displace the larynx and the gland to the side being examined. With the left first and second fingers placed behind the sternocleidomastoid muscle, the examiner attempts to palpate the underlying thyroid tissue between these fingers and the thumb of that hand. The left side is examined by exchanging the relative positions of the examiner's hands (Figure 4-2B).

Variations upon these methods of examination are successfully used.

Enlargement of the gland into the thoracic inlet may prevent palpation of the lower poles. An enlarged gland which has descended into the thoracic inlet occasionally may be made to rise into the neck and become visible when the patient performs the Valsalva maneuver. Percussible retrosternal dullness can also help in delineating such enlargement.

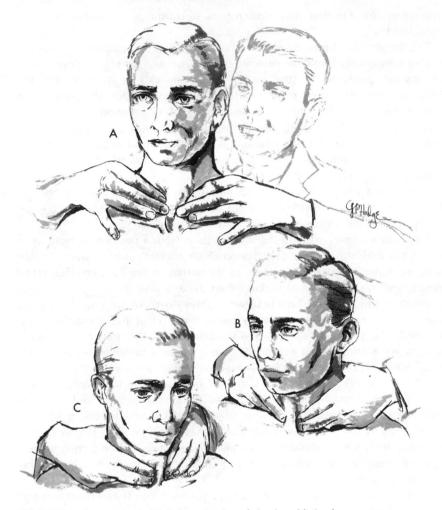

FIGURE 4-1. Posterior approach for palpation of the thyroid gland.

CARDINAL SYMPTOMS AND ABNORMAL FINDINGS

Headache As a symptom caused by a pituitary tumor, it is frequently frontal or bitemporal in location.

Impairment of peripheral vision A symptom produced when the optic chiasm is compressed by a pituitary tumor.

Increased perspiration May be associated with excess growth, thyroid, or adrenal medullary (epinephrine) hormone and with excess insulin activity. *Decreased perspiration* is associated with deficient thyroid activity.

Hand tremor and/or sensation of tremulousness Associated with excess thyroid hormone or adrenal medullary hormone (epinephrine) or excess insulin activity.

FIGURE 4-2. Anterior approach for palpation of the thyroid gland.

Excess dark body hair May be associated with disorders of ovary and/or adrenal cortex in which there results excessive androgenic hormone production.

Temporal and vertex hair loss May be associated with more profound degrees of excess androgenic activity in the female.

Loss of axillary and pubic hair May be associated with decreased production of androgenic hormone classically associated with panhypopituitarism in either sex.

Excess pigmentation (darkening) of the skin Frequently a complaint in Addison's disease.

Fatigue and weakness Associated with a large number of endocrine disorders, including Addison's and Cushing's diseases, primary aldosteronism, hyperparathyroidism, hyper- and hypothyroidism, and testicular and pituitary failure.

Lightheadedness or faintness, especially upon sudden standing A frequent complaint in Addison's disease.

Muscle cramps, spasm May be associated with hypoparathyroidism and primary aldosteronism.

Intolerance to heat or cold The former is associated with hyperthyroidism and the latter with hypothyroidism.

Excessive volume of urination Associated with hyperparathyroidism, primary aldosteronism, diabetes mellitus, and diabetes insipidus.

Protrusion of the eyes Often associated with thyrotoxicosis.

The Anterior Pituitary

Gigantism is defined as body height above the range of normal for the patient's age and race. Gigantism may be caused by excess production of growth hormone by the anterior pituitary, in which case features of acromegaly may also be apparent (acromegalic gigantism), or it may be associated with eunuchoidal body proportions and hypogonadism (eunuchoid gigantism), or it may be present in individuals who are otherwise normal (normal gigantism). In acromegalic and normal gigantism, growth is increased but body proportions are usually normal. The former is the result of excessive production of growth hormone before the epiphyses are closed. A pituitary adenoma may be identifiable as the source of this growth-hormone excess.

Acromegaly is a disorder characterized by generalized overgrowth of bony, cartilaginous, and soft tissue and results from an excess production and action of growth hormone in the adult. The presence of a pituitary adenoma (predominately eosinophilic and/or chromophobic cells) which secretes excessive amounts of growth hormone usually can be shown to be present. The earlier and more subtle physical findings include mild overgrowth of soft tissue and excessive sweating. Comparison of the patient with his earlier photographs helps in detecting subtle evidences of coarsening of the facial features. Palpation of thenar and hypothenar eminences of the hand may show these soft tissues to be prominent, warm, moist, and spongy. Later the facial and frontal bones become prominent and the nose is enlarged and broadened. Further soft tissue overgrowth causes redundancy of the scalp, longitudinal ridging of the forehead, and accentuation of nasolabial folds. At more advanced stages the overdevelopment of the frontal sinuses produces prominence of the forehead. The mandible elongates, causing overbite of the lower incisors, which may be abnormally separated from one another. The thyroid may be palpably enlarged. There is kyphosis of the dorsal spine. Degenerative arthritic changes may develop, especially in the knees. The hands and feet become enlarged and broadened, with thickening especially of thenar and hypothenar eminences and heel pad. The voice is husky and speech dysarthric due to enlargement of the tongue, vocal cords, and laryngeal cartilages. There is generalized enlargement of all the viscera, particularly the heart, liver, and spleen. Vision is reduced in the temporal fields when the optic nerves are compressed at the optic chiasm by a pituitary adenoma.

Dwarfism refers to body height that is below the limits of normal for the individual's age and race. Pituitary dwarfism due to hypofunction of the pituitary gland is uncommon but recently has begun to be recognized more frequently. The body proportions are normal, but the features are childlike, and sexual development fails to occur or is incomplete. In such cases there usually are varying degrees of deficiency of gonadotrophic, thyrotrophic, melano-trophic, adrenocorticotrophic, and growth hormones. Many constitutional and nutritional disorders during childhood may stunt growth and produce dwarfism.

The cause of short stature in a large number of patients who complain of it, however, is traceable to short normal parents.

Hypopituitarism in the adult results in decreased hormone secretion by the thyroid, adrenals, and gonads. In addition, there may be diminished production of growth hormone and melanocyte-stimulating hormone (MSH). When the activities of all of these hormones are reduced, the patient is said to have panhypopituitarism. The physical features of myxedema are present but usually are not as profound as in the more severe degrees of myxedema noted in primary hypothyroidism. Hypoadrenocortical function and hypogonadism result in loss of axillary and pubic hair as well as body hair. Hypoadrenocortical function results in postural hypotension, weight loss, and asthenia. Decreased MSH production results in skin, areolar, and genital depigmentation and a reduced ability to suntan. The combined effect of decreased production of thyroidal, adrenocortical, and gonadal hormones and MSH results in skin which is pallid, hairless, smooth, and dry, and which has been called "alabaster" skin. Typically, the nutritional state of the patient appears to be normal, although in advanced stages there may be weight loss or even emaciation.

The Thyroid

The thyroid gland may enlarge as a result of iodine deficiency or as a result of the action of certain goitrogens with or without overt clinical evidence of thyroidal dysfunction. The gland may be diffusely enlarged, nodular, and occasionally cystic. Neoplastic and inflammatory disease may also produce enlargement of the thyroid.

Thyrotoxicosis, the metabolic expression of overproduction of thyroid hormone(s), usually is accompanied by clinical thyroidal enlargement. Because of the increased blood flow which occurs with this condition a hum or systolic bruit sometimes accompanied by a thrill may be detectable over the gland. As a result of hypermetabolism, there may be weight loss, with disappearance of subcutaneous fat. The skin is warm and moist and has a fine texture suggesting velvet. There is a fine tremor of the hands, and the tendon reflexes may be exaggerated. The eyes may be protuberant. The ocular signs of thyrotoxicosis result from (a) retraction of the lids, producing widening of the palpebral fissures, a staring expression, and lid lag; (b) swelling of the orbital contents, producing forward displacement of the globe; (c) swelling of the conjunctiva (chemosis), due in part to (b) and in part to the effects of trauma to the exposed globe; and (d) weakness of the extraocular muscles with limitation of upward gaze; later, convergence and lateral movement are impaired.

The pulse rate almost invariably is elevated, and the pulse has a bounding quality. The pulse pressure frequently is widened due to an elevation of the systolic pressure, the result of increased cardiac stroke volume. Cardiac arrhythmias are common. The heart sounds are loud and hyperactive, and a functional systolic murmur may be present.

Myxedema, the result of severe and often prolonged *hypothyroidism,* presents a characteristic appearance. The face is puffy. This is particularly noticeable in the eyelids. The lips and tongue may be thickened. The speech is slow and the voice deep. The skin is thick and dry with rough, scaly texture and appearance. There may be thinning of scalp hair and of the lateral aspect of the eyebrows. The hair generally is coarse and brittle. If there is sufficient myxedematous involvement of the pituitary gland, sex hair may be diminished. The body temperature usually is subnormal. The thyroid gland usually is not palpable, but occasionally it may be enlarged and palpable. The pulse rate is slow. Blood pressure usually is normal; however, hypertension occurs at a greater frequency than in nonmyxedematous persons. Heart sounds usually are soft and muffled. The deep tendon reflexes are characteristically hypoactive, with a slow recovery phase.

It should be appreciated that these are descriptions of classic or fully developed states of thyroid dysfunction in the adult. Hypothyroidism in the child may result in dwarfism, in which case body skeletal proportions tend toward infantile proportions. The physician will wish to detect these conditions earlier when signs are more subtle. The same may be said for the descriptions which follow.

The Adrenals

Adrenal medullary hyperfunction (pheochromocytoma) is characterized by persistent or paroxysmal hypertension. As a result of increased circulating epinephrine and norepinephrine there may be tachycardia or bradycardia, sweating, blanching of the skin, tremor of the hands, and dilatation of the pupils, along with an elevation of the blood pressure. These physical changes are manifest primarily during "attacks" of this disorder.

Chronic adrenal cortical insufficiency (Addison's disease) results in increased pigmentation of the skin and mucous membranes. The pigmentation usually is brown but may be tan, smoky, and occasionally brown-gray or blue-gray, especially when it is present on the buccal mucosa, tongue, and gums. The pigmentation is maximal on the exposed surfaces, generally becoming accentuated toward the distal parts of the extremities. It is accentuated over pressure points (wrists, knees, and elbows, and where tight-fitting garments exert pressure), in scars, in body folds, on the areolae and nipples, on the external genitalia, and in the creases of the palms. Areas of vitiligo may also develop and are particularly noticeable in contrast with the pigmented areas. There is marked muscular weakness, asthenia, and anorexia, usually associated with weight loss. The blood pressure is low, the pulse pressure narrow, and the systolic pressure commonly under 100 mm. Hg. Orthostatic hypotension and tachycardia are characteristic. Axillary and pubic hair is decreased in both sexes. It is decreased especially in women, a large proportion of whose total androgen production is from the adrenal cortex. Acute adrenal cortical failure results in nausea and vomiting, dehydration, prostration, and shock, an alarming clinical picture.

Adrenal cortical hyperfunction may result in an excess secretion of glucocorticoids, mineralocorticoids, and androgenic or estrogenic steroids:

a. *Glucocorticoid excess* (Cushing's syndrome) results from chronic overproduction of cortisol and is characterized by accumulations of fat in the neck and upper trunk, loss of protein (thin skin, easy bruising, striae, bone tenderness and pain, and muscle atrophy), and hypertension. The face becomes rounded because of preauricular fullness and bulging cheeks which are frequently plethoric in appearance. The general appearance of the face is often described as "moon facies." The mouth pouts like that of a sunfish. Characteristically the supraclavicular fossae become obliterated by the deposition of fat in this area (supraclavicular fat pads). Fat may also accumulate anteriorly over the manubrium and posteriorly in the region of the cervicodorsal spine (cervicodorsal or buffalo hump). Wasted abdominal wall musculature contributes to the characteristic abdominal protuberance.

In contrast to the face and torso, the limbs are thin and wasted, with atrophy most evident in the forearm and quadriceps femoris muscles. The appearance of some such patients brings to mind that of the well-known "Humpty Dumpty" of children's verses. Hirsutism consisting of a light-colored, soft, fine, downy type of hair (lanugo hair) is common and more easily observable in women. The skin is thin, and there are characteristic purplish striae located on the anterior and posterior axillary folds, breasts, abdomen, and lateral aspects of the buttocks and thighs. There is a livid, plethoric appearance to the skin resulting from the thinning and from associated polycythemia; bruises are common, particularly on the extremities and at needle-puncture sites. Facial acne and the faun-colored lesions of tinea versicolor on the anterior chest are common. If Cushing's syndrome (glucocorticoid excess) is accompanied by 17-ketosteroid excess of sufficient androgenicity, protein loss and the signs thereof will be minimized, while some degree of virilization (clitoral enlargement and dark coarse facial hair) may be present.

b. *Mineralocorticoid excess* (primary aldosteronism) is characterized by hypertension and signs and symptoms of potassium depletion. The latter condition results in muscular weakness and sometimes paralysis, particularly of the lower extremities. Alkalosis associated with potassium deficiency may result in symptoms and signs of tetany. More uncommonly the mineralcorticoid produced in excess is desoxycorticosterone, in which case the signs and symptoms are similar to those of primary aldosteronism.

c. *Adrenal androgen excess* (adrenogenital syndrome) is the result of excessive androgen production by the adrenal cortex, the effects of which are easily recognized in infants, children, and females. Whenever the biologic effect of the secreted androgen is intense, females become virilized, displaying dark coarse androgenic-type hair on face and neck (beard), on the chest, breasts, and about the areolae, and on arms and legs. The pubic hair extends upon the abdomen toward the umbilicus (male escutcheon). Concomitantly, there frequently is a degree of male pattern of baldness with recession of the hairline in the temporal region and vertex thinning. The clitoris enlarges, the breasts decrease in size, the muscles increase in size, and the figure becomes masculine. The voice deepens. In some patients plethora, obesity, and hypertension suggest a mixed picture of virilism and Cushing's syndrome. A milder form of the adrenogenital syndrome in the female produces hirsutism without signs of virilism (balding, clitoral enlargement, deepening of the voice). Family and/or race may be associated with a normal propensity toward lesser or greater amounts and varying color of body hair. Such factors unaccompanied by any demonstrable hormonal disturbance apparently account for many cases of hirsutism.

The Parathyroids

Hyperparathyroidism is accompanied by physical signs that result from the accompanying hypercalcemia and bone disease. In addition to muscular hypotonia and weakness there may also be mental confusion, obtundation, or coma due to the hypercalcemia. Calcium deposits, another consequence of hypercalcemia, may be detected in the cornea as a faint white band, principally at the 3 and 9 o'clock positions and separated from the limbus by a narrow uninvolved area (band keratopathy). Calcium deposits also may be detected as white flecks on the conjunctival surface of the tarsal plates of the eyelids and as calcifications in the eardrum. Demineralization of bone may result in fracture of the long bones or loss of height due to vertebral collapse. Bone cysts and tumors may be found as palpable enlargements, particularly at the ends of long bones and in the jaw (epulides).

   Hypoparathyroidism results in hypocalcemia and consequent tetany. Carpopedal spasm, the principal physical sign of tetany, may occur spontaneously in overt tetany, resulting in flexion at the elbows, wrists, and metacarpophalangeal joints; extension of the interphalangeal joints and adduction of the thumbs; turning down of the toes; and arching of the plantar surface of the foot. These signs in the arm and hand, which are elicited by interrupting the circulation to the forearm for three minutes by applying a blood pressure cuff and inflating it above systolic pressure, are denoted Trousseau's sign. In the absence of overt tetany, this sign is diagnostic of latent tetany. Chvostek's sign carries the same connotation but may be elicited in 10 percent of normal adults. It is elicited by tapping over the facial nerve in front of the ear with the finger. The sign is the contraction of the facial muscles and orbicularis oculi on the side tapped. Additional signs of hypoparathyroidism are cataracts and, in severe, long-standing disease, papilledema. When hypoparathyroidism is chronic and early in onset, dental hypoplasia may result. The skin is characteristically dry and scaly, the body hair thin and patchy, and the nails brittle.

The Gonads

The complex syndromes of aberrant sexual development and gonadal dysgenesis are beyond the scope of this discussion. Only the effects of prepubertal and postpubertal gonadal failure on the adult male and female will be considered here.

In the adult male, destruction or removal of the testes or failure of the secretion of gonadotrophic substances results in only partial regression of the secondary sex characteristics. Facial and body hair usually remains but may be diminished in amount and rate of growth. Penile size usually is unchanged, but erection becomes difficult, and both libido and potency may fail. There is usually some loss of muscle strength due to diminished androgen production, and the skin becomes less rough.

When primary or secondary testicular failure occurs early so that puberty fails to be initiated, the genitalia remain underdeveloped. Scalp hair may be luxuriant, but body hair is sparse. The face remains juvenile and beardless, and the skin maintains a fine texture. Such individuals generally become tall, with long fingers and narrow hands and feet. The arms and legs are abnormally long, the symphysis pubis—to—heel distance exceeding the symphysis pubis—to—crown distance and the arm span exceeding the height. There may be a female-type distribution of body fat. This general appearance is called *eunuchoidism* and these skeletal proportions are called *eunuchoidal proportions.*

In the adult female, primary failure of ovarian function or isolated failure of the secretion of gonadotrophic substances results in amenorrhea together with some decrease in breast size and atrophy of the external genitalia. Females in whom gonadal failure occurs early and in whom therefore puberty fails to be initiated usually are taller than normal, with eunuchoidal proportions. The general appearance is immature, with absent breast development and with infantile external and internal genitalia. Hair growth is scant, particularly in the pubic and axillary region. The hands and feet are narrow.

The Pancreas

The islets of Langerhans constitute the endocrine portion of the pancreas. The hormones produced and secreted by the islets are insulin and glucagon, both having manifold and widespread effects upon metabolism. Excessive (islet cell tumor) and deficient (diabetes mellitus) production of insulin both result in readily discernible physical manifestations.

Excessive insulin action lowers blood glucose into the subnormal range (hypoglycemia). This results in the release of catecholamines and in attendant hunger, tremulousness, anxiety, perspiration, pupillary dilation, and tachycardia. When hypoglycemia is profound or prolonged, there is a more direct consequence to the nervous system, evidenced by headache, double vision, confused mentation, obtundation, various paralyses, and/or convulsions and coma.

Deficient insulin action, when it is profound, results in elevated blood glucose (hyperglycemia) and in ketoacidosis. The symptoms and physical signs proceed from the disordered metabolism. Early symptoms include increased urination, thirst, and fatigue; later symptoms are nausea, vomiting, abdominal pain, weakness, and malaise. Physical signs begin with tachypnea and fruity breath and progress to signs of dehydration (loss of skin turgor, softening of the globes, decreased tongue volume, dry mucous membranes, tachycardia, hypotension), deep respirations, drowsiness, stupor, and coma.

The signs and symptoms attendant upon the chronic state of diabetes mellitus are reflections of abnormalities principally in nerve tissue and in the macro- and microvasculature. They are considered in the chapters on the Eye, Circulatory System, Genitourinary System, and Nervous System.

SPECIAL TECHNIQUES

A variety of laboratory tests is employed to confirm the diagnosis of endocrine disorders. The tests include radiographic and radioisotope studies, biochemical determinations, and bioassay and radioimmunoassay techniques.

REFERENCES

1. Martin, L. *Clinical Endocrinology for Practitioners and Students* (3rd ed.). Boston: Little, Brown, 1961.
2. Turner, C. D., and Bagnara, J. T. *General Endocrinology* (5th ed.). Philadelphia: Saunders, 1971.
3. Wilkins, L. *The Diagnosis and Treatment of Endocrine Disorders in Childhood and Adolescence* (3rd ed.). Springfield: Thomas, 1966.
4. Williams, R. H. (Ed.). *Textbook of Endocrinology* (4th ed.). Philadelphia: Saunders, 1968.

SKIN 5

Peter J. Lynch

GLOSSARY

Albinism A disease of generalized hypopigmentation in which little or no melanin is formed.

Bulla, pl. bullae Loculated fluid in the skin; a large blister.

Café au lait spots. Sharply marginated brown patches. The presence of one or two such spots is normal.

Carotene A yellow-orange pigment found in many foods.

Crusts Dried or hardened serum proteins on the surface of the skin.

Erosion A superficial loss of skin.

Erythema Redness.

Freckles Sharply circumscribed small brown macules.

Keratin The protein product of skin metabolism. Scale is formed by the macroscopic accumulation of keratin.

Lentigo, pl. lentigines Sharply circumscribed small brown macules which appear late in life following years of sun exposure.

Macule A circumscribed area of color change.

Melanin The brown pigment of the skin which is made by melanocytes.

Nodule A large palpable mass, usually elevated above the skin surface.

Papule A small palpable mass, usually elevated above the skin surface.

Plaque A flat elevated mass, the confluence of papules.

Pruritus Itching.

Pustule A cloudy or white vesicle, the color of which is due to the presence of polymorphonuclear leukocytes.

Scale The macroscopic accumulation of keratin.

Scleroderma A disease in which the dermal component of the skin is thickened.

Seborrhea Oiliness of the skin due to lipids that originate in the sebaceous gland.

Ulcer As applied to the skin, a deep loss of tissue or large erosion.

Urticaria The presence of edema in the skin secondary to histamine release.

Vesicle A small loculation of fluid in the skin; a small blister.

Vitiligo Circumscribed areas of pigment loss.

Wheal An edematous papule, the primary lesion of urticaria, a "hive."

Xerosis Dryness of the skin.

What portion of the physical examination could be easier than the examination of the skin? The skin is directly visible, both in color and in three dimensions; it is directly palpable; and it is thin enough so that even the deepest cutaneous pathology is only a matter of millimeters away from the examiner. But this easy accessibility also creates some problems — it provides a great quantity of information, much of which is unimportant or unrelated to the patient's presenting problem. This is complicated by the fact that we are so accustomed to viewing the skin we really do not "see" it at all. As a result most of us, in performing the physical examination, ignore all the information provided by the skin rather than make the effort to sort the important from the unimportant. This chapter is designed to assist the examiner in the sorting process.

TECHNIQUE OF EXAMINATION AND NORMAL FINDINGS

Examination of the skin is carried out through inspection and palpation. Inspection requires both adequate lighting and adequate exposure of the patient's skin. Bright, diffuse, overhead fluorescent light is the best way to achieve adequate lighting. Daylight and incandescent light can be used, but light from these sources is usually insufficiently bright and is too directional.

Proper exposure of the patient's skin for a general clinical examination requires the removal of clothing down to the underwear, over which may be worn a hospital examining gown of the type that opens to the rear. I cannot emphasize too strongly that inadequate exposure is almost always the fault of the examiner rather than that of the patient. In my professional lifetime I have met only a handful of patients who, because of modesty, were reluctant to disrobe sufficiently for proper examination. On the other hand, I have watched countless examinations in which the examiner, having failed to request proper exposure, attempted the shortcut of merely lifting the clothing for a quick peek underneath. The inadequacy of this approach is apparent. Unfortunately, I have also seen examiners, brusque and unthinking in their disregard for a patient's modesty, literally force a patient into a protective, covering response. This latter situation is most likely to develop in the handling of the examining gown. The following approach, which requires the use of a second sheet, is suggested.

First, with the patient sitting on the edge of the examining table, examine the exposed areas including the scalp, hair, face, mouth, neck, arms, hands, and fingernails. Second, ask the patient to assume a supine position on the examining table. Then place the folded second sheet such that it covers the lower half of the patient from the waist to below the knees. Third, ask the patient to pull the hospital gown out from under the covering sheet so that the gown is bunched over the breasts, allowing examination of the abdomen, lower chest, anterior surface of the lower legs, feet, and toenails. Fourth, ask the patient to sit up with the legs still extended on the examining table. Undo the snaps or ties at the back of the gown and examine the patient's back. Finally, uncover and examine the shoulders and upper chest. If it is necessary to examine the breasts, this may be accomplished either in the third step while the patient is lying down or in the last step. Finally, I would like to emphasize that this examination includes the all-too-often-overlooked hair, nails, and oral mucous membranes.

Examination in these steps allows all but the posterior legs, genitalia, and buttocks to be examined and is sufficient in the absence of known cutaneous disease in the unexamined areas. Obviously not all of the steps listed above have to be covered consecutively; the various positions can be utilized as they occur during other parts of the examination. Unfortunately, nonconsecutive viewing of the skin in a piecemeal fashion during the rest of the physical examination often leads to incomplete and inadequate information regarding the skin.

Most of us have acquired, on a nonmedical basis, significant knowledge regarding the appearance of normal skin, but this knowledge is probably insufficiently organized to be useful. This section will attempt to put that knowledge into appropriate perspective.

Color

The range of color which can be considered normal is great and depends on many variables such as race, nationality, and degree of sun exposure. Physiologically, skin color is derived from three major sources: (1) erythematous hues that come from oxygenated hemoglobin contained in the cutaneous vasculature, (2) brown hues that come from melanin pigment produced by the melanocytes of the epidermis, and (3) yellow hues that come from the natural color of nonvascularized collagen and from bile and carotene pigments. Cutaneous color changes, whether pathologic or physiologic, are related to changes in the balance of these three hues.

Texture

The characteristic "feel" of skin depends on a number of physiologic processes. These include *softness* as provided by the layer of fat cells which abuts the lower portion of the dermis; *moisture* as provided by water diffusion through the skin and by sweating onto the surface of the skin; *lubrication* as provided by the sebaceous glands; *warmth* as provided by the circulation of internally warmed blood; and the presence or absence of *roughness* depending on the amount of scale (keratin) produced by the epidermal cells. Balance among these factors depends on the patient's age and sex and, of course, on the region of the skin being examined.

Mucous Membranes

Mucous membranes are characteristically pink in color and moist to palpation. Mottled brown or black melanin pigmentation may be present on the oral mucous membranes of black patients.

Hair

Normal hair distribution is well appreciated by most examiners and need not be considered here. However, it should be remembered that facial, axillary, and pubic hair depend on the presence of sex hormones and thus on both the sex and the age of the patient. The scalp hair should be specifically examined for length, texture, fragility, sheen, and the ease with which hairs can be manually removed from their follicles. Scalp hair normally grows about 0.3 mm per day or, in more practical terms, about one-half inch a month.

Nails

The fingernails should be smooth, translucent, and evenly attached to the underlying nail bed. Paronychial tissue should be intact. Fingernails and toenails normally grow about 0.1 mm. per day.

CARDINAL SYMPTOMS AND ABNORMAL FINDINGS

Symptoms

Two important symptoms relate to cutaneous pathology: *pruritus* and *pain.* These two symptoms are related insofar as the respective sensations are both carried by the peripheral nervous system via its cutaneous branches. Pain occurs under several sets of circumstances. Most commonly it appears when the skin around the nerves is no longer intact, exposing the sensitive nerve endings to the dry hostile environment outside the body. Pain may also occur as a direct effect of biochemical mediators in some kinds of cutaneous inflammatory reactions. Finally, pain may be the result of simple nonpenetrating external trauma. Usually these situations are clinically apparent; thus pain is more often a therapeutic problem than a diagnostic one.

Little is known regarding the pathophysiology of pruritus, but for the purpose of this chapter it suffices for us to know that itching is carried by the small nerve fibers of the skin. Generalized itching, and subsequent scratching, may occur in a variety of systemic diseases, notably in association with chronic disease of the thyroid, pancreas, liver, and kidneys. It may also occur in patients with diseases of the hematopoetic system, especially Hodgkin's disease and polycythemia vera. But most often, even after thorough examination, no explanation for generalized pruritus can be found. In these instances the functional disability often present may suggest a psychiatric cause.

Localized areas of itching are exceedingly common and, when unassociated with visible cutaneous disease, probably have no pathologic meaning. Many, if not most, kinds of visible cutaneous disease are associated with more or less pruritus. This pruritus causes a dual problem. First, the itching almost literally may "drive a patient crazy," causing serious interference with ability to function. Second, the concomitant scratching intensifies and continues the skin disease which was responsible for the itching in the first place. This itch-scratch cycle is one of the most vexing therapeutic problems in all of dermatologic disease.

Change in Color

As was pointed out earlier, the normal range of skin color is great. The simple presence of one color or another is not necessarily significant, but the fact that the presenting color represents a *change* from what existed before is important. Changes in color may be generalized or localized. Small, localized areas of color change are called macules; larger areas are called patches. The major kinds of color change include the following.

Brown. Generalized darkening of melanin pigmentation is an important clue to some types of pituitary, adrenal, and liver diseases. Localized increase in melanin pigmentation may be seen in the brown macules or patches of café au lait spots, freckles, lentigines, nevi, and areas of postinflammatory hyperpigmentation.

White. Absence of melanin gives the skin a white color. Generalized hypopigmentation may be seen in albinism, and localized areas of hypopigmentation may be seen in the macules or patches of vitiligo, scars, postinflammatory hypopigmentation, and a variety of other cutaneous diseases.

Yellow. Generalized yellowness of the skin due to an increase in cutaneous bile pigment may be seen in liver failure, in which case it is known as jaundice. More rarely, diffuse yellowness occurs in diabetics and vegetarian food fadists as a result of increased carotene pigmentation. Finally, a pale yellow color may be seen in anemia, in which the contribution of the red, oxygenated blood decreases, allowing accentuation of the normal yellow color of collagen. This latter phenomenon is particularly prominent in pernicious anemia and in anemia of chronic renal disease.

Erythema. Increased cutaneous blood flow, most commonly as a component of inflammation, leads to increasing redness of the skin. Thus generalized erythema may occur with drug eruptions, viral exanthems, and urticaria. Localized inflammation and redness occur nonspecifically in a vast array of cutaneous disease. Noninflammatory redness occasionally occurs due to an increased number of intravascular red blood cells (polycythemia) or due to extravascular presence of red blood cells such as occurs in petechiae and purpura.

Other Colors. Rarely, some medications injected or ingested contribute their own color to the skin. Examples include the slate-gray color due to silver salts and the yellow color due to quinacrine.

Change in Texture

After puberty the scalp and face of most patients will feel oily (a condition known as seborrhea), but there are no pathologic states specifically associated with an increase in sebaceous secretion. On the other hand, decreased lubrication (dryness, chapping, or xerosis) is common after the age of 60 and occasionally occurs in younger people as a result of too-frequent bathing. Rarely, xerosis reflects a deficiency of thyroid or sex hormones.

As a result of thermal stimuli (such as fever or an overly warm examining room) patients may develop a palpable moistness of the skin associated with generalized sweating. Under emotional stimuli localized sweating occurs on the forehead, palms, soles, axilla, and groin. Rarely, moist skin may be a reflection of the increased metabolic rate which occurs in hyperthyroidism.

Increased warmth of the skin occurs when an increase in cutaneous blood flow delivers body heat to the surface of the skin, where it is then lost by convection, conduction, and radiation. This may occur with fever or following exercise. Localized areas of increased warmth may accompany the increased blood flow seen with cutaneous inflammation. Coolness of the skin reflects decreased blood flow such as is seen in the lower legs of patients with peripheral arteriovascular disease.

Finally, the skin may lose its elasticity, or feel tough, when it is distended by edematous fluid, when the cutaneous fat is replaced by collagen as in scleroderma, or when normal collagen is replaced by scar tissue.

Specific Cutaneous Lesions

Thus far we have talked primarily about functional changes that occur in the normal components of the skin. In this section we will discuss structural changes, that is, the development of lesions which, strictly speaking, always represent cutaneous pathology. In some instances the pathology has little significance (e.g., nevi and senile angiomas), but the inexperienced examiner should consider the presence of any of these structural changes as potentially important until he has a chance to learn which can be safely ignored. The most common structural changes are shown in Figure 5-1 and are discussed below.

Palpable Lesions. Localized lesions which have substance or mass are always palpable and are usually elevated above the surface of the skin. Small palpable lesions are called *papules* and large papules are called *nodules.* A confluence of papules or nodules, resulting in a large flat-topped lesion, is called a *plaque.* The palpable substance of papules, nodules, and plaques occurs as a result of one or more of the following processes: (1) proliferation of the various cells normally found in the skin (e.g., inflammatory cells, metastatic tumor cells, leukemic cells, etc.); and (2) the accumulation of fluid within the skin. Fluid accumulating within the skin may be present in a diffuse fashion (a *hive* or *wheal*) or in a loculated fashion (as in a blister). Small blisters are known as *vesicles* and large blisters are known as *bullae.* Vesicles or bullae which contain many polymorphonuclear leukocytes appear cloudy or white and are called *pustules.*

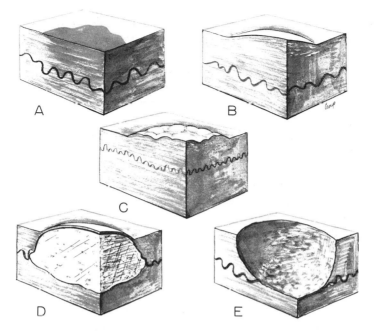

FIGURE 5-1. Primary lesions of the skin: A. Macule. B. Papule. C. Plaque. D. Vesicle or bulla. E. Erosion or ulcer.

Erosions and Ulcers. Superficial loss of skin is called an erosion, whereas deep loss is termed an ulcer. In both situations the barrier function of the skin is lost, and serum together with inflammatory cells exudes to the surface as "weeping" or "oozing." When this exudate dries, it forms *crusts.* It is important to distinguish between the light gray flakes of scale and the yellow-brown friable granules of crusts, inasmuch as the latter represents epithelial loss and the former represents epithelial proliferation.

Erosions and ulcers arise in three major ways: (1) as a result of external trauma, most commonly scratching; (2) from the unroofing of vesicular or bullous lesions; or (3) from the necrotic effect of vascular ischemia. Ability to determine which of these three mechanisms is responsible for an ulcer greatly simplifies the preparation of a differential diagnosis.

Once examiners have learned to recognize the kind of pathological conditions described above and can use appropriate terms in their written or verbal description, they are ready to formulate differential diagnoses. To assist the student in this task, many dermatology textbooks are arranged or organized according to the different patterns of cutaneous pathology; the student then may go directly from his own descriptions of the lesions to the appropriate textbook chapter.

LABORATORY TECHNIQUES

There are four laboratory procedures that are so common as to be a part of many routine examinations of the skin. The first of these is the biopsy. A small circular or elliptical piece of skin is surgically removed under local anesthesia. After formalin fixation, paraffin embedding, and suitable staining, thin sections of this skin are examined under the light microscope. The histologic appearance of the skin, together with the clinical appearance of the patient, is extremely helpful in confirming a suspected diagnosis.

The second common laboratory test is the potassium hydroxide (KOH) examination. In this test, scale or hair from a suspected fungal infection is placed on a microscope slide, and a drop of 10 percent potassium hydroxide solution is added. The potassium hydroxide causes separation of the epithelial cells and allows easier visualization of the fungal hyphae.

The third test is the microbial culture for pathologic bacteria, fungi, and viruses. In this test, suitable material for culture is removed from the cutaneous lesion by the clinician. This material is then cultured and identified by technicians specially trained in these techniques.

The fourth test is the patch test. This test is useful in confirming a suspected diagnosis of allergic contact dermatitis. The suspected allergen (diluted if appropriate) is placed on the normal skin in a semi-occluded fashion. After 48 hours the allergen is removed and, if the patch test is positive, the site of allergen application will exhibit an inflammatory reaction similar to the patient's original disease.

EYE 6

John W. Henderson

GLOSSARY

Amblyopia Decreased vision of an eye from any cause.

Anisocoria Inequality in size of the pupils.

Anterior segment Anterior portion of the eye.

Arcus senilis A white ring around the limbus of the cornea occurring in elderly persons.

Asthenopia Discomfort related to use of the eyes.

A-V notching The indentation of a retinal vein by an overlying retinal artery in arteriosclerosis.

Cataract Opacity in the crystalline lens of the eye.

Conjunctivitis Inflammation of the conjunctiva.

Diplopia Double vision.

Ectropion Eversion of the lid border.

Entropion Inversion of the lid border.

Epiphora The overflow of tears down the cheek.

Exophthalmos Protrusion of the eyeball.

Funduscopy Examination of the interior of the eyeball using an ophthalmoscope.

Glaucoma Increased intraocular pressure as well as the disease resulting from such pressure.

Hyphemia Presence of blood in the anterior chamber of the eye.

Hypopyon Presence of pus in the anterior chamber, often with a horizontal fluid level.

Iritis Inflammation of the iris.

Miosis Constriction of the pupil; a drug which constricts the pupil is called a *miotic*.

Mydriasis Dilation of the pupil; a drug which dilates the pupil is called a *mydriatic*.

Nystagmus Irregular jerking movement of the eyes.

O.D. Abbreviation for the right eye (*oculus dexter*).

O.S. Abbreviation for the left eye (*oculus sinister*).

Optic atrophy A loss of tissue of the optic nerve due to prior disease, usually making the optic disc appear whiter in color under the ophthalmoscope.

Papilledema Swelling of the optic nerve head.

Phoria Latent tendency to deviation of the visual axes which is held in check by the fusion reflex.

Photophobia Sensitivity to light which is usually associated with corneal disease.

Posterior synechiae Adhesions between the iris and the lens.

Presbyopia Diminution of the power of accommodation of the eye due to the aging process.

Proptosis Forward displacement of the eyeball.

Ptosis Drooping of the upper lid (*blepharoptosis*).

Tropia Deviation of the visual axes from parallelism which is not overcome by the fusion reflex.

Visual field Area of vision of each eye measured while the eye is directed straight ahead.

Yoke muscles Muscle pairs, one on each eye, which lead in a specific diagnostic position; for example, the right lateral rectus and the left medial rectus in gaze to the right side.

TECHNIQUE OF EXAMINATION AND NORMAL FINDINGS

A complete routine ocular examination includes measurement of visual acuity, assessment of ocular movements, inspection of external structures, testing of the pupillary reactions, evaluation of the ocular fundus, and estimation of the intraocular pressure.

Visual Acuity

Accurate estimation of the patient's visual acuity is often mandatory. In cases involving head trauma or injury of the face or eyes, the vision should be measured in case of future compensation or legal action. For this purpose a Snellen chart is utilized. Seat the patient a measured twenty feet from the chart, cover one eye at a time, and ask him to read the letters starting from the top down. The last row in which he is able to read the majority of the letters should be recorded.

The Snellen chart is constructed so that the visual angle subtended by each letter will occupy an angle of five minutes of arc at the eye (Figure 6-1). The letters are blocked out in a grid pattern occupying one minute for each block.

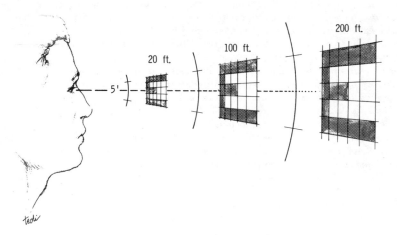

FIGURE 6-1. Use of the Snellen chart.

They are then made of proper size so that the 20/200 letter at the top of the chart occupies the five minutes of arc at a distance of 200 feet from the patient, the 20/100 letters the same angle at 100 feet, and the 20/20 letters the five-minute arc at 20 feet. Therefore, the lower figure in the designation of visual acuity represents the distance at which the normal eye would see the letter, and the upper figure always refers to the distance the patient is seated from the chart. For example, the notation of 20/40 vision for an eye indicates the standard chart distance from the patient and the fact that the subject cannot read past the line which would ordinarily subtend the five-minute arc 40 feet away. The numbers are in no sense a fraction.

Ask the patient to cover one eye and read down the chart, preferably with his distance glasses on. If he reaches the 20/30 row and misses two letters, the record is made: 20/30—2. The cover is shifted to the other eye, and if he reads through the 20/40 row and gets only two letters in the next row the record is: 20/40ı2. The result is written "O.D. (right eye) 20/30—2, O.S. (left eye) 20/40+2." If glasses are worn, note "with correction"; if not, "without correction."

If the patient cannot see any of the letters on the chart, move him closer to it until he can read the large letter at the top. If this is five feet from the chart, write "5/200" (since the upper figure designates the distance of the observer from the letter). If no letters can be read, hold up your fingers and ask him to count them, recording the distance at which this occurs: "Counts fingers at 2 feet," for example. If this is impossible, check him for the ability to see moving objects or the direction from which your flashlight beam is shining on his eye; record "moving objects" or "light projection." "Light perception" may be recorded if he is unable to recognize direction. Do not record the eye as blind unless no light is perceived.

In some cases it may be desirable to record the near vision. Special, graded reading cards are available for this, but in a general examination ordinary newsprint will suffice for your records.

For general purposes a vision of 20/30 or better in each eye can be accepted as normal. Be sure that the patient is wearing his glasses if a refractive error is present. In normal individuals there is not usually more than a one-line difference between the vision of the two eyes. It is normal for persons over 40 to begin to have near-vision difficulty because of presbyopia and to require glasses for reading.

Ocular Movements

Seat yourself facing the patient and hold your flashlight in the midline between yourself and the patient, asking him to look directly at the light. Observe the position of the reflection of the light on each of his corneas with respect to the location of the pupils (corneal light reflex). Then cover one eye while he looks at the light. Remove the cover quickly and notice whether the eye moves in regaining fixation on the light. This may indicate a drift of the eye behind the cover which can be indicative of muscle imbalance. Then shift the cover across from one eye to the other and back again, always observing the movement the uncovered eye makes to regain fixation (alternate cover test). If weakness or paralysis of one of the horizontal rectus muscles is present, a horizontal shift will occur. If one of the elevator or depressor muscles is involved, a vertical movement will be noted.

The maintenance of parallel eyes (and therefore of corneal light reflexes which are equally centered) occurs in most patients because of the fusion reflex which makes binocular vision possible. If a deviation of the eyes occurs behind the cover which then recovers when both eyes are open, you are dealing with a *phoria*, a latent tendency to deviation which is held in check by fusion. If, on the other hand, the alternate cover test demonstrates shift which does not regain parallelism with removal of the cover, the muscle imbalance is called a *tropia.*

In evaluating possible weaknesses of individual extraocular muscles and consequently of their innervation in most cases, it is helpful to use a system of "diagnostic positions of gaze" to unravel the complexities of movement of twelve muscles (six on each eye) acting in concert (Fig. 6-2).

In each of the diagnostic positions indicated in Figure 6-2A you will note that two muscles are listed. These are the leading muscles in this position, one on each eye, and the diagram is arranged as if the patient were facing you. The muscle pairs indicated are also known as "yoke muscles," since they are yoked together by innervation to act in unison.

If the patient's eyes are turned to the right, the right lateral rectus muscle

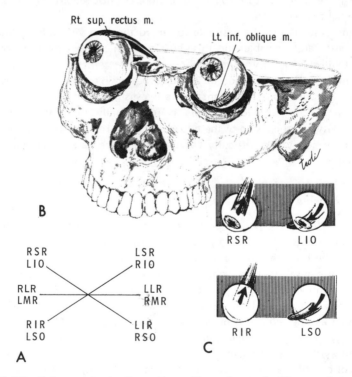

FIGURE 6-2. Yoke muscles in diagnostic positions of gaze. A. Diagram of muscle pairs, patient facing examiner. B. Right-hand elevators. C. Right-hand depressors. Note: Vertical recti evaluated in out-turned eye, obliques in in-turned eye.

(RLR) is leading the right eye, the left medial rectus muscle (LMR) the left eye. Conversely, when the patient looks to the left the muscle pair considered is the left lateral rectus (LLR) and the right medial rectus (RMR).

Since the vertically acting muscles (the superior and inferior recti, the superior and inferior obliques) have a more complex action if considered from the straight-ahead position, it is more convenient to turn the eye into a position where a single action of each muscle predominates. Since the insertion of the superior and inferior rectus muscles is lateral to the origin by about 25 degrees, turning the eye outward places these muscles in position to either raise the eye (superior rectus) or lower it (inferior rectus). Subsidiary actions can then be ignored. Since the origin of the inferior oblique and the trochlea of the superior oblique are medial and forward of their insertions, turning the eyeball inward toward the nose (about 51 degrees) places these muscles in a more favorable position either to raise the eye (inferior oblique) or to lower it (superior oblique). Therefore, the out-turned eye is in position to allow observation of superior or inferior rectus function, and the in-turned eye is in position to observe superior or inferior oblique action.

To take an example, if the eyes are held in an up-and-right position (see Figure 6-2) one is comparing the right superior rectus (out-turned eye; rectus muscle) with the left inferior oblique (in-turned eye; oblique muscle). The other muscle pairs can be worked out from the diagram in the same way.

Ask the patient to follow your light with his eyes and observe him in each of these diagnostic positions: up to the right, directly to the right, down to the right, down to the left, directly to the left, and up to the left. Notice the position of the corneal light reflexes in each eye to make sure they are parallel. If the eyes go out of parallel in a certain position, the lagging muscle can be designated directly. In addition, ask the patient whether he sees two lights in any of the positions (diplopia, or double vision). This will give you a clue as to the muscle pair to be analyzed further.

It is normal to find a small amount of horizontal shift of the visual axes with the alternate cover test, provided the eyes regain a parallel position as soon as the cover is removed (phoria). It is more common to find a small outward deviation (exophoria) than a minor inward deviation (esophoria). In both instances parallelism is regained. Usually any vertical shift behind the cover is abnormal. Any lack of parallelism in the diagnostic positions of gaze is beyond the range of normal. Any tropia (constant deviation from parallel) is abnormal. A transient jerking movement of the eyes in the extremes of horizontal gaze ordinarily falls within normal limits.

External Examination

Even before checking the visual acuity and the ocular motility you should observe the patient's face and eyes along with his general physical characteristics. Clues as to general disease may be evident in the facial and ocular expression, the state of the face and eyelids, the prominence of the eyes, the alert or dull expression they may convey.

Always remember the order in which observation of external findings should be made: lids, conjunctiva, cornea, anterior chamber, iris, pupils, and lacrimal ducts. If each of these is thought of in turn, findings will not be overlooked.

Note the appearance of the *eyelids*. There may be edema, changes in the skin, and differences in the height of the palpebral fissures; a staring expression may be evident; or the upper lids may be pulled upward in a retracted position. Many structural variations may fall within the normal range. A slight difference in the palpebral fissures is not significant unless associated with unequal pupils. Differences in the depth of the upper lid fold may occur. Normal racial differences can be expected.

The *conjunctiva* covers the anterior eyeball aside from the cornea and is reflected back onto the posterior lid surfaces. Its appearance (edema, pallor, vascular injection) should be noted. The palpebral conjunctiva and the fornices must be seen, since here foreign bodies may lodge. To examine the conjunctiva of the lower lid place your index finger firmly over the midpoint of the lid just above the bone of the lower orbital rim and pull downward. This everts the lower fornix, and changing the position of your finger will expose different areas.

In order to see the superior fornix and the conjunctiva of the upper lid, the eyelid must be everted (Figure 6-3). Ask the patient to look downward, grasp the upper lashes gently with the thumb and forefinger of one hand, and use a pencil tip or side of a tongue blade to form a fulcrum just at the upper border of the tarsal plate. By pushing down at this point and pulling upward on the lashes, you will evert the upper lid, exposing the posterior surface. As long as the patient keeps looking downward, the lid will remain in this position. After inspection, ask the patient to look upward, and the lid will flip over into its normal position.

There is considerable normal variation in the degree of vascularity of the conjunctiva. In general, the more florid the patient's complexion, the more he is apt to have reddened eyes. A small amount of discharge present in the inner canthus merely reflects the "wastebasket" function of the tear drainage mechanism and is to be expected, more so in patients with oily skin and greater meibomian secretions.

Observe the *cornea*. It should be shiny and bright when illuminated by your flashlight. Any break in this clear continuity, such as scarring, vascularization, or ulceration, will dull the reflection. Photophobia may be a clue to corneal disease. The size (horizontal diameter 9 to 11 mm.) and curvature should be noted. In elderly patients a partial or complete white ring about the periphery of the cornea (arcus senilis) may be expected.

Observe the depth of the *anterior chamber*. There should be adequate clearance between the cornea and iris with no irregularities in depth. The depth varies somewhat among normal persons, but if the iris appears to bulge forward

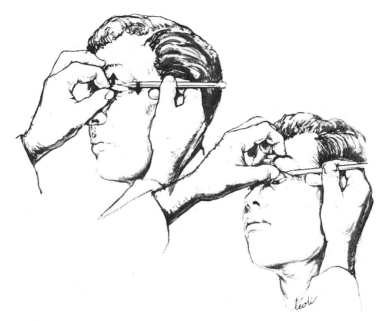

FIGURE 6-3. Technique of visualizing the superior fornix and conjunctiva of the upper lid.

and the space is very shallow you should immediately think of the possibility of glaucoma. Anything other than clear aqueous occupying the anterior chamber is abnormal.

Note the color and consistency of the *iris*. A difference in color between various parts of the same iris or between the two eyes may be important. Abnormalities of the position of the pupils and irregularities in shape or difference in size between the two (anisocoria) should be recorded. The size of the pupils varies in normal individuals. Large pupils are generally found in myopic eyes, smaller ones in patients with hyperopia. A small difference in pupil size between the two eyes is usually not significant unless accompanied by abnormal reflexes.

Next test the *pupillary reflex*. Shine your light quickly into one eye and observe the responses of both eyes. The illuminated eye shows constriction of the pupil (direct light reaction), and the pupil of the other eye also constricts an equal amount (consensual light reaction). Then shine the light into the other eye and observe the same reactions. Then ask the patient to look at your finger and move it directly toward his nose. With focusing for this near object the pupils will both constrict (near reaction, reaction in accommodation).

The rapidity of response of the pupillary reflexes varies considerably in normal patients. Usually the presence of the response is sufficient — as long as it is equal in the two eyes. Older patients who do not accommodate well often show a normal slowing of the near reaction.

Remember that when you are looking at the pupillary opening you are also inspecting the *lens* of the eye, which is normally transparent.

Inspect the *lacrimal drainage system*. Observe the position and patency of the tear point at the inner end of each lid. The puncta should be turned backward slightly to contact the pool of tears in the inner canthus. Make gentle pressure over the position of each lacrimal sac and note whether any material regurgitates back into the eye. Tears should not spill over onto the cheek (epiphora), and no elevation should be palpable over the lacrimal sac.

Ophthalmoscopic Examination

Evaluation of the ocular fundus offers you a unique "window" through which your ophthalmoscope will often aid in general physical evaluation of the patient. Constant practice is necessary and most rewarding.

Seat the patient comfortably and ask him to look straight ahead. Remember the basic rule for use of the ophthalmoscope: right hand, right eye, right side of the patient — left hand, left eye, left side of the patient (Figure 6-4). Hold the instrument in your right hand, and stand on the patient's right side. Examine his right eye with your right eye; the reverse holds for his left eye. A +8 lens (black numbers on the dial) is placed in the instrument aperture and the pupil observed from a distance of 6 to 8 inches. This illuminates the retina, and any opacity or obstruction to the emerging light will be seen as a dark spot or shadow against an

FIGURE 6-4. The initial position for examination of anterior segment, lens, and vitreous. The examiner then moves as close as possible to visualize the retina.

orange-red background. Gradually move closer to the patient, holding his upper lid open gently with the thumb of your free hand, resting your hand on his forehead. The lens wheel of the ophthalmoscope is turned (toward zero on the dial) until the lighter color of the optic disc is seen just nasal to the center of the retina. Then refocus your instrument until details of the optic nerve head are seen clearly. Note the size, shape, color, margins, and physiologic depression of the optic nerve. Any elevation can be measured by focusing on the highest part of the disc. Then throw the light beam about two disc diameters nasally to the nerve head and refocus on the retina. Note the number of clicks as the lens wheel is turned to refocus; then read the difference directly from the dial.

If you are one of a considerable number of students with one weak eye, change the technique. Do your ophthalmoscopic examination with the patient on a stretcher, and stand at his head, looking at each of his eyes in turn with your good eye.

You must follow a definite order of examination of the retina, always reserving the central or macular area until last. This avoids dazzling the patient early in the examination and will insure better cooperation for prolonged viewing.

The blood vessels of the retina emerge from and enter the disc in four main pairs (Figure 6-5). Examine the superior nasal vessels first, following them out as far as possible without having the patient turn his eye. This is followed in order by the inferior nasal, inferior temporal, and superior temporal pairs. The retina adjoining each of these pairs is inspected at the same time. Then ask the patient

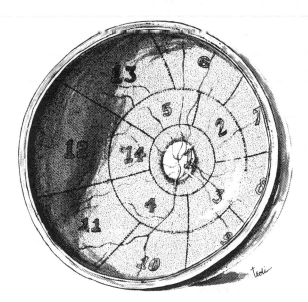

FIGURE 6-5. Diagrammatic representation of retina, showing order in which areas are examined. Note that macular area, position 14, is examined last.

to look up, up and in, directly toward the nose, down and in, straight down, down and out, directly outward, and then up and out. This covers eight successive overlapping zones of the peripheral retina, and each zone joins the margins of the more central retinal areas seen along with inspection of the vessels. The central retina and macular area are visualized last, the patient being asked to look directly at the light if necessary. The foveal pit is seen as a small, bright dot produced by light reflection from the indentation.

Complete observation of the ocular fundus with the ophthalmoscope is best done with a dilated pupil. In the vast majority of patients a mydriatic drop may be used safely, but you must be aware of the possibility of inducing a rise in intraocular pressure (glaucoma) in a predisposed eye. This is more often the case in patients over 40 years of age. Always question the patient as to the previous diagnosis of glaucoma, observe the optic nerve head for abnormality, evaluate the tactile tension, and if there is any question, measure the intraocular pressure with a tonometer. The presence of a small corneal diameter with a shallow anterior chamber makes the possibility of glaucoma greater. After checking for these points, instill one drop of a mydriatic (among others, 0.5% Mydriacyl or 10% Neo-Synephrine) into each eye. If this is done early in the physical examination, you may return to the funduscopic examination later. When you are through, measure the tonometer tension of the eyes to screen for possible glaucoma.

NOTE: If there is acute neurologic disease, especially a recent brain hemorrhage or head trauma, the observation of serial changes in the pupils is crucial in planning treatment. In order to avoid masking important changes, *do not* dilate the pupils.

It is impossible to enumerate the myriad normal variations encountered in the funduscopic examination, since they often merge into the abnormal. Considerable variation in the size and shape of the nerve head may occur. If the disc is small the nasal border may be blurred. If the eye is myopic there may be an arc of retraction along the temporal side. The physiologic depression may vary from highly prominent to nonexistent. Pigment may be noted along the disc border. The position of the vessels on the nerve head may vary. The retinal vessels are usually gently sinuous in their courses, with approximate right angles at division points. They usually cross each other without noticeable indentations. The veins are somewhat darker than the arterioles, and are about one-third wider. Both can be traced almost to the visible periphery in the eye with a widely dilated pupil. The background color of the fundus may be darker in brunets, very light in blonds, and in the latter the deeper choroidal vessels may shine through. The healthy retina is transparent and produces no visible findings except for highlights which often follow the expected pattern of the retinal nerve fibers.

Measurement of Intraocular Pressure

The intraocular pressure should be recorded in every complete physical examination. The Schiotz tonometer measures the indentation pressure of the cornea by means of a central plunger fitted into a curved footplate, with movement of the plunger transmitted to a lever scale (Figure 6-6). The reading is translated into millimeters of mercury and the range of normal is from 12 to 22 mg. Hg. The patient is made comfortable in a reclining position and one drop of a local anesthetic such as 0.5% tetracaine is instilled into each eye. When the

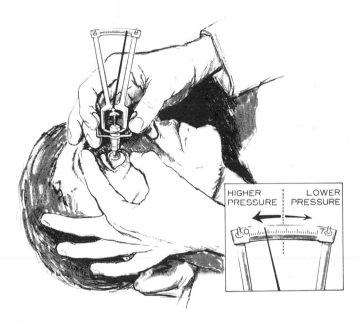

FIGURE 6-6. Measurement of intraocular pressure by tonometry.

eyes are anesthetized, the patient is directed to look straight above him, fixating either his extended finger or a spot on the ceiling. The instrument is held lightly in one hand, and the lids are separated gently with the fingers of the other hand. The footplate is rested on the cornea and the reading taken from the dial. A small rhythmic deviation in the pointer should be observed (transmitted pulse pressure) to ensure a proper reading. The procedure is repeated for the other eye. Care must be taken not to exert pressure on the lids, and the instrument must be kept meticulously clean.

As a means of rapidly checking for a potentially high intraocular pressure in the differential diagnosis of an acutely red eye, tension may be estimated tactilely (Figure 6-7). Ask the patient to look downward, and place the tips of both index fingers on the upper lid above the position of the downturned cornea. While one index finger rests firmly on the globe, the other is pressed inward to indent the eyeball. This indentation is then alternated to the other

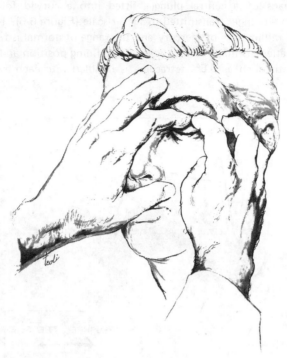

FIGURE 6-7. Tactile test for intraocular pressure.

index finger and repeated several times with a reciprocal, pistonlike movement, pressing directly toward the center of the globe.

A grossly soft eyeball or a marked rise in pressure may be determined in this way, but in most instances the method is no substitute for tonometry.

CARDINAL SYMPTOMS AND ABNORMAL FINDINGS

Pain. The complaint of pain in or about the eyes must be evaluated as to location, duration, type, and mode of onset. It may be related to a specific incident, as in the case of an injury to the eye. Pain in the eyes and forehead may be the result of uncorrected refractive error or ocular muscle imbalance, in which case it will often follow use of the eyes. Almost all patients with headache will ask whether the eyes are involved, but few headaches will be found due to eye disease; a typical unilateral migraine headache, for example, will often include ocular pain on that side.

Very severe localized pain in one eye usually suggests surface disease such as a foreign body or corneal ulcer. A dull throbbing pain is more typical of iritis or of acute glaucoma, and may be worse when the patient is lying down.

Visual Loss. This may range from blurring of vision to complete blindness in one or both eyes. The earliest signs of double vision may be recognized by the patient only as a visual blur related to the overlapping images. A reduction in visual acuity can reflect systemic disease, such as early diabetes where a shift to myopia may occur. The early symptoms of senile cataract may include visual blurring, especially for distance.

Sudden loss of vision usually signals retinal or optic nerve disease and may be related either to inflammation or to vascular embarrassment. Central visual loss must be distinguished from that of the peripheral areas; the patient is usually much more aware of loss of central reading acuity than of a peripheral narrowing of the visual field.

Double Vision. Visual confusion is characteristic of the patient with double vision (diplopia). He will often close or cover one eye for relief. In certain cases of incomplete paralysis of an eye muscle he may assume an unusual head position in order to maintain single vision, for example, head turning to the left with the eyes directed to the right in the case of a partial paralysis of the left lateral rectus muscle. Diplopia usually signifies either muscular or neurologic disease. A lack of parallelism of the eyes may not be associated with double vision when it is the result of a "lazy" or amblyopic eye dating from a muscle imbalance in childhood.

Photophobia. Sensitivity to light varies considerably among normal individuals, but where it is significant there is usually disease of the cornea or of the anterior segment of the globe. In acute cases a corneal foreign body, corneal ulcer, or iritis may be suspected, although old corneal scarring and vascularization may also produce a hypersensitivity to light.

Discharge. Material emerging from the eye as either watery or more viscid discharge usually suggests either conjunctival disease or difficulty in the lacrimal drainage system. In acute infections it may be more purulent and may collect on the lid borders. Where the lacrimal system is occluded, a flow of tears may be continuous and mucus may collect in the tear sac so that pressure over this structure causes regurgitation of material into the eye. In allergic conjunctival disease the consistency of the discharge is stringy and tenacious. Tears may also be a part of the light-sensitivity response of a diseased eye.

Redness. Vascular congestion or redness of the eyes must be evaluated in terms of other findings. If it is greater toward the fornices of the conjunctiva, there is usually associated discharge and conjunctivitis. If it is greater around the margin of the cornea, there may be associated corneal or anterior segment disease such as iritis. It may be generalized and related to surface irritation or sensitivity. In certain florid individuals it may not be significant unless other findings support the diagnosis of disease.

"Eye Strain." This commonly used term is not sound from the medical standpoint and should not be used to denote the symptom of fatigue. "Strain" connotes damage, but eye use in itself never produces irreversible change. If there is discomfort with use of the eyes (asthenopia), a search should be made for refractive error or ocular muscular imbalance.

Visual Acuity

Reduced vision always suggests either local or general disease, especially when loss has been sudden or when relatively recent glasses no longer aid. A reason must be sought for any instance of reduced vision. Causes can range from local ocular pathology, such as cataracts, retinal detachment, and vitreous or retinal hemorrhage, through optic nerve involvement as a part of neurologic disease. Sudden changes may suggest circulatory insufficiencies of various types, such as cerebral hemorrhage, major carotid occlusions, and the like. Poor vision in one eye may be noted in many patients who had a strabismus in childhood, with resulting suppression of central vision. Visual-field defects of intracranial disease may be associated with complaints of visual loss, but the central acuity often may be normal and visual-field examination necessary.

Any case of reduced vision, either unilateral or bilateral, *demands an explanation.*

Ocular Motility

Paralysis of specific nerve supply to the extraocular muscles produces characteristic findings. The patient with an oculomotor paralysis will have his eye turned down and out, with ptosis of the upper lid. Diplopia will not occur because of the closure of the lid. With an abducens paralysis the involved eye will be turned in toward the nose because of unopposed action of the normal medial rectus, and the esotropia will be greater when looking in the direction of normal action of the paralyzed muscle. The patient with a trochlear palsy will complain of difficulty with vision in the lower field, as in reading, and if he has been able to maintain single binocular vision his head will be tilted toward the shoulder opposite the side of the paralysis. It is useful to be able to test for intact trochlear function in the presence of a complete oculomotor paralysis, since in some conditions both nerves may be involved together. With the eye in a down-and-out position, the superior oblique muscle is in a more favorable position for torsional movement, and when the patient is asked to look downward, an intact superior oblique will produce a twisting movement of the eye, with the superior pole of the cornea intorted toward the nose.

Isolated involvement of ocular muscles may occur with certain neuromuscular diseases, may follow orbital or facial fractures, or may be part of the findings of the endocrine exophthalmos of thyroid disease.

Nystagmus (irregular jerking movements of the eyes) may indicate either a congenital abnormality or a disease of the vestibular or central nervous system.

External Examination

The *general appearance* of the eyes and face may reveal a staring expression with retraction of the upper lids and prominence of the eyeballs in hyperthyroidism. Unilateral prominence of an eye (proptosis) indicates a space-occupying lesion of some type in the orbit, and bilateral prominence may be associated with thyroid abnormalities or with diseases of the blood-forming organs. Drooping upper lids may indicate extreme debility or neuromuscular disease. Partial ptosis may be a part of Horner's syndrome, with involvement of the cervical sympathetics. More marked ptosis of an upper lid associated with decreased pupillary light reaction may be the earliest indication of an oculomotor paralysis.

The appearance of the eyelids themselves may reveal systemic edema in advance of more noticeable changes elsewhere. The skin may be involved in either local or generalized dermatologic conditions.

An inturning of the lid border (entropion) causes irritation by abrasion of the lashes against the cornea, and may be due to lid spasm or to contraction of scar tissue. Eversion of the lid border (ectropion) may result from scar tissue or from senile laxity, and is associated with overflow of tears (epiphora).

Almost all diseases affecting the *conjunctivas* produce injection of the vessels together with discharge. In conjunctivitis of infectious origin the injection usually increases in the fornices, and secretions are present. More severe involvement produces small hemorrhages beneath the conjunctivas. Spontaneous ecchymoses may occur beneath the conjunctivas in otherwise healthy patients, and usually have no significance unless signs of a bleeding tendency are present elsewhere.

Edema may produce a boggy conjunctiva in patients with systemic fluid retention, or it may be a manifestation of endocrine exophthalmos or of local inflammation or vascular stasis. In patients with jaundice, a yellow color of the sclera visible through the conjunctiva may antedate other physical evidence.

Pain in the eye is usually severe in acute disease of the cornea. Photophobia also occurs in most corneal diseases. In abrasions or ulcers of the cornea there is often increased redness of the globe around the corneal limbus. Loss of the bright surface reflection occurs in surface lesions, and at times a surface area of involvement will cast a shadow on the underlying iris if the flashlight is directed at the proper angle. Enlargement of the cornea is the commonest finding in infantile glaucoma and is usually associated with clouding and photophobia. Edema of the cornea may be a part of local disease or of acute glaucoma.

Any vascularization or visible white scarring of the cornea is indicative of disease. Where a white arcus occurs around the limbus in younger individuals, it may indicate abnormality in the lipid metabolism. A brown ring of pigment occurs around the corneal limbus in hepatolenticular degeneration (Wilson's disease), and is known as a Kayser-Fleischer ring.

Any material visible in the *anterior chamber* is abnormal. Blood may be present after injury (hyphemia), or pus may level out in the lower chamber in association with corneal infection (hypopyon). The chamber may be lost in perforating wounds with aqueous leakage.

Findings of the *iris* and *pupil* often are associated. An irregular pupil may be due to adhesions of the iris to the lens as the result of prior iritis. Congenital abnormalities such as multiple or displaced pupils must be differentiated from tears of the iris base due to prior trauma (iridodialysis). Any localized elevation of the iris is immediately suspect of possible tumor, especially if darkly pigmented.

Iritis is a nonspecific inflammation of the iris. The patient notes throbbing pain and visual blurring. There is circumcorneal injection with a small pupil. The eye is usually quite soft to palpation and tender. A tendency to adhesion of the pupillary border to the lens may produce an irregular pupil due to posterior synechiae.

Pupillary Reflexes

Loss of the pupillary light reflex is always important. Where it is unilateral due to blindness, neither a direct reflex nor a consensual reflex will occur when the blind eye is tested. Where loss is bilateral in the presence of sight, neurologic disease is usually present. One example is the Argyll Robertson pupil of central nervous system syphilis, in which the light reflexes are gone but the near reflex in accommodation is present. A unilaterally fixed, dilated pupil may be the result of local trauma to the eye, but more often it is a serious sign in a patient with recent head injury, since it indicates beginning involvement of the oculomotor nerve. A miotic pupil associated with partial drooping of the upper lid may indicate disease of the cervical sympathetic on that side and is a part of Horner's syndrome.

Any visible clouding of the *lens* as seen through the pupil is indicative of cataract formation and may also be seen as a dark shadow against the light of the fundus in the beginning of the ophthalmoscopic examination. If the lens has been removed or is dislocated, the normal support of the iris will not be present, and the iris will "flutter" with quick movements of the eye (iridodonesis).

The patient with disease of the *lacrimal drainage system* usually has overflow of tears from the eye (epiphora). When the nasolacrimal duct is obstructed, secretions accumulate in the tear sac and regurgitate back into the eye. Pressure over a palpable lacrimal sac will enhance the regurgitation. In some patients, associated redness and tenderness indicate secondary tear-sac infection.

Ophthalmoscopic Examination

A blurred image of the fundus that will not clear with proper focus of the ophthalmoscope is usually due to clouding in the *media* of the eye — the cornea, lens, or vitreous. Vitreous haziness is present in intraocular inflammation, whereas cataract in the lens or corneal scarring may be more evident. In an occasional patient with a high refractive error the fundus may be difficult to see, and in this case you should attempt funduscopy through his correcting lenses. Specific localized opacities in the media will be evident with use of the +8 lens noted above.

Major abnormalities of the optic nerve are *optic atrophy, cupping of the disc in glaucoma,* and *papilledema.* In optic atrophy the color of the disc is paler than normal and may be chalky white in advanced cases. There may be associated superficial scar tissue or loss of substance, and usually there is resultant decrease in vision. The cupping of the disc that occurs in glaucoma consists of an exaggeration of the physiologic depression on the temporal side which extends to the temporal border. The cup may be deep and is a bluish-white in color, and the emerging retinal vessels may disappear behind the shelf at the edge of the cup, then emerge over the edge to reach the retina. Cupping of the disc is expressed as a ratio of the cup diameter to the horizontal disc diameter (*C/D ratio*). While no exact C/D value differentiates "physiologic" from abnormal, a ratio greater than 0.3 or an asymmetry between the two eyes should be viewed with suspicion. Swelling of the nerve head (papilledema) may be unilateral in localized optic nerve disease or bilateral in the choked disc of increased intracranial pressure. The amount of elevation should be measured with the ophthalmoscope for estimation of future change. Early papilledema is difficult to determine, but filling in of the physiologic depression, blurring of the margins of the disc, fullness of the retinal veins, and loss of spontaneous venous pulsation on the disc are early signs. Where the process is more advanced, superficial hemorrhages and exudates around the disc make diagnosis more certain.

Evaluation of the *retinal vessels* is helpful in assessing the patient with arteriosclerosis or hypertensive disease. The normal retinal arteriole is seen only as a blood column. Where thickening of the wall occurs, a shinier reflection is noted and may be copper-colored in less advanced sclerosis, but will appear like a silver wire in advanced disease. This change is due to visibility of the vessel wall, and at times the wall itself may be seen along the edge of the blood column. In addition, where an involved artery crosses a retinal vein it may indent the vein and even cause evidence of back pressure in the vein distal to the crossing (A-V notching). The disappearance of the vein on both sides of the blood column representing the artery is a measure of thickening of the arteriolar wall.

With hypertension there is localized or generalized narrowing of the arteriolar blood column. This is identified by a change in the ratio of arteriolar to venule size. In tracing out the smaller branches toward the periphery, the arterioles will disappear earlier than the veins. In the more severe degrees of hypertension, leakage of the vascular walls will lead to hemorrhages and deposits in the retina, and papilledema may ensue.

Other changes in the blood vessels include the full, tortuous dark vessels associated with polycythemia, the whitish sheathing in the perivascular spaces occurring with severe systemic lipemia, and the prominent arterial pulsation seen in aortic valvular insufficiency.

Hemorrhages and *exudates* in the retina are always important signs of disease. The shape of the hemorrhage indicates the depth of the retina at which it occurs. Superficial hemorrhages are flame-shaped or splinterlike in contour, and lie in the nerve fiber layer of the retina. They usually occur where venous back pressure is present, such as in occlusion of the central retinal vein or one of its branches, or in association with papilledema. In contrast, closure of the central retinal artery or a branch results in ischemic edema of the area involved, and where the artery is completely closed the retina is pale and edematous, with the thinner macular area shining through as a "cherry-red spot." Hemorrhages which are located in the deeper retinal layers are more round or blotchy in contour and are often associated with exudates. These latter deposits represent residues of edema and of blood substances which are incompletely absorbed due to poor retinal circulation. Sharply defined yellow or white deposits should be distinguished from more fuzzy *cotton-wool* patches, which are small ischemic infarcts in the nerve fiber layer. The association of hemorrhages and exudates occurs in advanced hypertension, severe renal disease, certain of the collagen diseases, advanced diabetes, the blood dyscrasias, and severe retinal venous occlusions. The early hemorrhages of diabetic retinopathy may be punctate or clusterlike, since they are really venous microaneurysms. They are usually present only in the macular area and surrounding posterior pole of the eye.

When round, blotchy hemorrhages occur that are noted to have whiter centers, one must think of blood dyscrasia or of the metastatic lesions of subacute bacterial endocarditis. When blood emerges in front of the retina (preretinal hemorrhage) it will obscure the underlying details and may show a gravitational fluid level. Such bleeding is associated with sudden intracranial hemorrhage.

Any area of *retinal elevation* is significant. A solid mass indicates tumor growth, usually arising in the choroid. If it has a dark color it is likely to be melanoma, and if lighter, one must think of metastatic malignancy. When the elevation is transparent and wrinkled, retinal detachment must be considered.

Areas of localized *chorioretinal scarring* are recognized by irregular pigment deposition around a paler center. If they are clear-cut and sharp in outline no current inflammatory activity is suspected, but if they have fuzzy borders, hemorrhages along the margin, or associated clouding in the vitreous, an active process may be suspected. Irregular mottling of pigment and scar-tissue change in the macular areas may be seen in older patients who are suffering from senile macular degeneration, and associated decrease in central vision will be found.

Intraocular Pressure

Any elevation in the intraocular tension is abnormal and should be investigated further. Glaucoma is the greatest single cause of blindness in patients over 40 and can be controlled by proper treatment if discovered early. Acute glaucoma is rare but is associated with pain in the eye, a steamy cornea, moderately dilated pupil, and systemic symptoms such as severe headache, nausea and vomiting, and prostration. Abdominal symptoms have led to misdiagnosis of an acute abdomen in rare instances. The more common chronic glaucoma may cause only variable visual blur, minor headache, and peripheral visual loss, or may be entirely silent. Clues here are indications of moderately elevated intraocular tension and early cupping in the optic nerve head. Tonometry and evaluation of the visual fields must be done. Many authorities feel that a complete physical examination of any patient over 40 years of age must include use of the tonometer, in order that suspected cases may be referred promptly for ophthalmologic care.

Ocular Pain

Pain in the eye is always important. When associated with uncorrected refractive error it usually follows use of the eyes, may be associated with reduced acuity, and often is referred to the forehead. Following trauma there may be a foreign-body sensation due either to a surface abrasion of the cornea or an actual foreign particle on the cornea or lids. Other signs of injury to the anterior segment may be noted in more severe cases. With associated redness one must think of acute conjunctivitis. In acute iritis the pain is throbbing, and more injection of the globe occurs around the periphery of the cornea (circumcorneal injection). In addition, there may be a small pupil and photophobia. Acute glaucoma causes more diffuse redness, the cornea is steamy, and the pupil moderately dilated.

Deep pain in the orbit may not be associated with local signs but may indicate neighboring disease such as sinus disease or intracranial sensory nerve involvement. Irritation of the meninges or increased intracranial pressure may produce orbital pain.

SPECIAL TECHNIQUES

Examination of the Visual Fields. In any case of suspected disease of the central nervous system, localization of a possible lesion requires measurement of the fields of vision. This can be done grossly by confrontation testing, but using a perimeter to record the edge of the peripheral field, or a tangent screen to test the central 30 degrees of the visual area, is more accurate. Since visual-field defects occur early in cases of uncontrolled chronic glaucoma, serial examinations are necessary to assess the response of the disease to treatment. When central vision is impaired, search for a central area of blindness is done by visual-field techniques.

Refraction. Measurement of the error of refraction must be done in every case of subnormal vision. If the visual decrease is due only to need for glasses, this will become evident. When glasses do not improve the vision, a further organic cause must be sought.

X-ray Studies. X-rays of the orbit and of the optic foramina are necessary in any case of orbital disease or of optic nerve involvement. Any patient with the possibility of an intraocular or intraorbital foreign body after trauma must have x-ray examination.

Slitlamp Biomicroscopy. In order to visualize fine detail in the anterior segment of the eye, a special binocular microscope with a fine slitlight beam must be used. In particular, the presence of cells and protein material in the anterior chamber aids in the diagnosis of iritis. With special prismatic contact lenses, the angle of the anterior chamber as well as the retina may be studied in this way.

Exophthalmometry. Measurement of the exact degree of protrusion of the eyeball is needed in patients with orbital disease. The exophthalmometer is placed on the bone of the lateral orbital rims, and by means of a mirror system the distance of the corneal apex from this fixed bony point may be recorded in millimeters. Serial measurements will indicate progression or regression of the pathologic process.

Ophthalmodynamometry. Measurement of the pressure of the central retinal artery is helpful in patients with circulatory disease such as carotid insufficiency. The ophthalmodynamometer, which has a smooth, round footplate extending into a spring-tension cylinder, is placed on the sclera in the outer canthus. Pressure is exerted on the instrument while the observer watches the central retinal artery. When arterial pulsation begins, a reading which represents the diastolic level is taken; with greater pressure, collapse of the artery indicates the systolic level. A difference between the readings of the two eyes or an abnormally low diastolic level may be significant.

REFERENCES

1. Allen, J. H. (Ed.) *Manual of Diseases of the Eye: For Students and General Practitioners* (24th ed.). Baltimore: Williams & Wilkins, 1968.
2. Havener, W. H. *Synopsis of Ophthalmology* (3rd ed.). St. Louis: Mosby, 1971.
2a. Newell, F. W., and Ernest, J. T. *Ophthalmology, Principles and Concepts* (3rd ed.). St. Louis: Mosby.
3. Scheie, H. G., and Albert, D. M. *Adler's Textbook of Ophthalmology* (8th ed.). Philadelphia: Saunders, 1969.
4. Vaughan, D., Cook, R., and Asbury, T. *General Ophthalmology* (6th ed.). Los Altos, Calif.: Lange Medical Publications, 1971.

MOUTH AND JAWS 7

James R. Hayward

The oral cavity is the most accessible body orifice and may reveal significant local diseases as well as signs of systemic disease. Few areas of the body are exposed to the degree of continuous insult to which the oral tissues are subjected. The constant exposure to mechanical, thermal, chemical, and microbiologic stress makes the tissues of the oral region a significant index of tissue tolerance and systemic defense. Therefore, systemic diseases that reduce tissue tolerance often have oral manifestations. Reactive lesions to local injury also are common in the mouth.

The guidance of the patient's history is as important to the examination of the mouth as to that of any other part of the body. However, minor discomfort in the oral region is apt to be a common and transient experience. The fact that significant oral pathology may present only slight discomfort should stimulate the examiner to inspect this region with critical interest. Although the mouth and jaw region is commonly examined and treated by the dentist on a periodic basis, it is important for the physician to recognize the normal spectrum, the common abnormalities, and the oral manifestations of systemic diseases. Many conditions will prompt the referral of the patient to a dentist for appropriate treatment.

TECHNIQUE OF EXAMINATION

The examination of the mouth and jaws is carried out by inspection, palpation, percussion, and transillumination.

Examination equipment includes mouth mirror or tongue blades, bright mobile light source (pen flashlight), gauze sponges, rubber finger cots, lubricant for lips, and a fine metal probe of dental or lacrimal design.

Seat the patient comfortably and, if possible, stabilize the head with back support. Observe the symmetry in form and function of the lips in pursing action. Ask the patient to remove any dental appliances. With the patient's mouth only slightly open, retract the lips and cheeks with the tongue blade, and with direct light inspect the inner lip and cheek surfaces and all recesses of the gingivobuccal fornices and gums. Next, ask the patient to open his mouth wide and tilt the head back so that the hard and soft palates can be seen. Depress the dorsum of the tongue with a blade and request "ah" phonation to observe midline uvula elevation and coordinated pharynx constriction (Figure 7-1A). Have the patient resume the original head position and protrude the tongue. Note here the symmetry and muscle coordination of midline protrusion as well

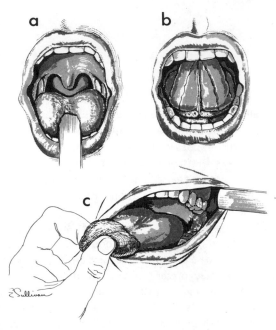

FIGURE 7-1. Positions for mouth examination. In (a) tongue is depressed for view of dorsum, palate, and pharynx. In (b) mouth is open and patient is attempting tongue-tip contact with palate for inspection of ventral surface and floor of the mouth. In (c) tongue is grasped in gauze sponge and retracted to inspect and palpate lateral posterior tongue border with cheek retracted.

as the dorsal surface characteristics of the tongue. Complete the inspection of the oral cavity by retracting the patient's tongue laterally, to view this posterior surface and the floor of the mouth (Figure 7-1C). The importance of this maneuver lies in the frequency of malignancy in this area. Ask him to touch the hard palate with the tip of the tongue in open-mouth position. Observe the ventral surface of the tongue and structures in the anterior floor of the mouth (Figure 7-1B).

Begin palpation with a request for tongue protrusion, which is maintained by grasping the tongue between layers of gauze. With this control, use the index finger of the opposite hand, covered with a finger cot, to palpate gently but firmly the soft, smooth tongue surfaces. Release tongue and continue palpation to the floor of the mouth. Palpate these sublingual structures with the opposite hand supporting the submental and submandibular tissues. Continue bimanual palpation between oral mucosa and facial skin in the cheek and lip regions. Conclude palpation with the hard and soft palate areas. Note the flow and secretion quality from the orifices of the submaxillary and parotid ducts.

Examine the teeth for their form, function, and support in the jaws. Light percussion with a mirror handle may be helpful in localizing painful dental conditions. Finally, observe the excursions of the mandible and occlusion of the teeth that determine the functional potentials of the masticating system. Palpate the condyles of the temporomandibular joint in motion by placing your fifth fingers in the external auditory canals during jaw excursions.

The examination of the oral region concludes with evaluation of the regional lymph nodes for their relation to oral pathology.

REGIONAL FINDINGS

Oral Mucous Membranes

Although the region of the mouth and jaws is approached in order of anatomic areas, certain basic considerations pertain to the majority of mucosal surfaces. The oral mucosa has a rich vascular supply and a resilient, flexible epithelial surface. Except for the vermilion tissue of the lips, the oral mucosal surfaces are kept moist by numerous submucosal accessory salivary glands adding to the major salivary gland secretions.

Normal mucosal surfaces are pale coral pink. Bright red surfaces generally indicate the erythema of inflammation, while pallor indicates localized ischemia or generalized anemia. Cyanotic color changes may indicate local congestion or many systemic states which produce hypoxemia. The oral mucous membranes are normally pigmented with generalized and local melanin on a racial basis. Local deposits of brown pigment in the mucosa also are seen in some metabolic disturbances such as hypoadrenalism (Addison's disease). Linear pigmentation of the terminal capillary beds of the free gingival margin may indicate heavy-metal absorption which may correlate with toxic symptoms. Localized bluish pigmentation in the gingival areas that is not related to underlying vascular abnormality may be due to the accidental implantation of metal dental filling materials.

The most frequent surface changes of the oral mucous membranes are ulcers and white patches. Increase in the layer of mucosal keratin produces white, thickened patches. Monilial infection will also produce white plaques. Ulceration is indicative of trauma or secondary lesions following initial vesicles of viral or other primary disease lesions.

Swelling of the oral mucosa and submucosa may be found on the basis of inflammation, reactive hyperplasia, cysts, congenital deformities, or neoplasm, in that order of frequency. Many submucosal swellings are detected only by careful palpation. These may stimulate differential diagnosis in the above categories.

Following palpation of the oral mucous membranes, the examiner should look for bleeding from any surface. Such a finding dictates great care in locating the source of bleeding and the tissue characteristics of the bleeding source (inflammatory or neoplastic).

Common Symptoms. Dry mouth (xerostomia) — a condition seen with atrophy of salivary glands with senility, disease states, radiation, and many drugs which decrease salivary function.

Excess saliva (ptyalism) — responses of salivation to any mucosal irritation, heavy-metal toxicity, or pilocarpinelike drug action.

Lips

The function of the lips in speech, oral intake, control of secretions, and contributing to facial expression is governed by the orbicularis oris group of muscles. Sensory nerve supply is abundant and there is a rich blood and lymphatic supply. Accessory salivary glands under the inner aspect of the lips provide lubrication. Because the lips closely cover the hard tooth structure, they are easily injured.

Common Symptoms. Ulcers — chiefly secondary to vesicular lesions of viral origin and to trauma.

Numbness — in lower lip, deficit due to anesthesia or damage to the inferior alveolar nerve in the mandible.

Drooling — motor loss due to facial nerve paralysis, either peripheral or central.

Swelling — rather pronounced response to any inflammatory process; it is sometimes subtle as in angioneurotic edema and other allergic phenomena.

The vermilion surface of the lips in the young shows slight vertical linear markings and a smooth, pliable surface. Atrophic changes of the vermilion with age erase the strial pattern and lose the sharp definition at the mucocutaneous junction. Surface keratosis, induration, and ulceration in older individuals should suggest the changes of squamous-cell carcinoma. Herpetic vesicles or ulcers of the lip are common. Fissures with inflammation at the angle of the mouth with loss of dental structures may be seen in the aged, and also may be a feature of nutritional deficiency. Superficial accessory salivary glands of the lip occasionally develop retention cysts (mucocele) following injury. Congenital anomalies include folds of the double lip and parasagittal scars in the upper lip from congenital cleft repair. The rich blood and lymphatic supply contributes to rapid edema with inflammation of the lips.

Oral Vestibule and Buccal Mucosa

Attachment of the upper and lower lip in the midline to underlying bone is demonstrated in normal frenula extending toward the attached gingiva. Posteriorly, similar frenula represent muscle attachments to the alveolar process of the jaws. Buccal mucous membrane inspection may reveal a horizontal white line extending from the commissure of the mouth to the retromolar pad, indicating the contact made by the occluding surfaces of the teeth. This may be a zone of hyperkeratotic reaction and shaggy superficial slough where cheek-biting habits are present. A generalized prominence of the posterior buccal mucosa with fairly large buccal fat pad structures may be seen. There often are small yellow macules or papules indicative of normal sebaceous gland deposition (Fordyce spots). Recent trauma of the buccal mucosa may produce small spots of submucosal hemorrhage, and similar lesions are produced readily in blood dyscrasias with bleeding tendency.

The parotid duct orifice is found in the posterior mucosal surface of the cheek opposite the maxillary second molar. The posterior lateral recess behind the tuberosity of the maxillary alveolar process requires mirror inspection for complete vision.

Palate

Morphologic and functional aspects of the hard and soft palates are quite different. The hard palatal vault is composed of underlying bony processes of the maxilla covered by dense fibrous tissue and mucosa. In the anterior one-third of the hard palate specialized ridges of normal palatal rugae are noted, with a midline anterior palatine papilla just behind the central incisor teeth. Cysts in this area are associated with the nasopalatine canal.

The most common variation in hard-palate structure is seen as a midline hard swelling or exostosis (torus palatinus). It occurs in 20 percent of the adult population. Such bony growths are benign and are significant only when the surface mucosa becomes ulcerated or dental prosthetic requirements necessitate their removal. The soft palate is muscular and has abundant submucosal accessory salivary glands. The normal central position of the soft palate is demonstrated by elevation and reflex.

Soft-palate function is coordinated with the pharynx in a constrictor mechanism functioning as the velopharyngeal valve. These actions are essential for normal swallowing and speech.

Abnormal. The palatal vault beneath a maxillary artificial denture may indicate changes of nodular papillomatosis from irritation and altered function of the region. Congenital clefts involve both the hard and the soft palates and may extend through the alveolar ridge between the canine and lateral incisor teeth. Degrees of original congenital deformity or scar tissue from surgical repair may be noted in these regions of potential cleft. A bifid uvula may be featured as part of a submucosal cleft palate, which usually includes hypernasal speech.

The chronic irritation of nicotine stomatitis may produce elevation of the accessory salivary glands with red dilated orifices, in contrast to general white mucosal hyperkeratosis. Palpation of the posterior palatal vault may reveal submucosal nodular swelling as the only sign of neoplasm in this region.

Tongue

Tongue mobility and function are essential to speech, mastication, taste, and swallowing. The specialized mucosa of the dorsum of the tongue presents papillations of filiform, fungiform, and circumvallate types, while at the posterior lateral borders of the tongue ridges of foliate papillae are noted. Many variations of the pattern of papillae are seen. Atrophy leaves a smooth-surfaced red appearance suggesting nutritional deficiency or pernicious anemia. Hypertrophy and hyperkeratosis of the filiform papillae may present a furred, hairy surface. Such a thick coat, which may be pigmented, is a condition associated with poor oral hygiene. A midline elevated area in the posterior dorsum of the tongue represents the congenital benign lesion of median rhomboid glossitis. A striking pattern of arcuate variations in papillary distribution is seen in transient forms in the benign condition known as "geographic tongue." The dorsum of the tongue is deeply furrowed in a congenital morphologic variation in some 5 percent of the population. Macroglossia may be indicative of hypothyroidism as well as a number of inflammatory, cystic, congenital, and neoplastic variations. Lesions that produce asymmetric tongue enlargement are hemangioma, lymphangioma, and neurofibroma.

Common Symptoms.   Coated tongue — thickening of mucosal keratin chiefly with filiform papillae hypertrophy is found in response to irritation and immobility and in association with poor oral hygiene.

Burning tongue — chiefly a psychogenic disorder.

Abnormal motility — neuromuscular disorders of stroke and myasthenia gravis induce changes and may cause speech disturbance from faulty tongue action. A fixed, firm tongue may result from infiltration with scar tissue or malignant neoplasm, usually squamous-cell carcinoma.

Ventrally the lingual frenum is noted at the midline attached to the gingiva at the symphysis of the mandible. The sublingual caruncles at the orifices of submaxillary ducts are noted and the flow of secretion is observed. The floor of the mouth may be the site of retention cysts of the sublingual glands, producing soft, translucent swellings. Occasionally a localized stone in the course of the submaxillary duct may be palpated and will produce obstructive symptoms. Exostosis of the mandible is seen as a hard mass projecting toward the floor of the mouth from the region of bone supporting the premolar teeth (torus mandibularis). Similar hard swelling may be noted in the midline of the mandible at the position of the genial tubercles, which are especially prominent when teeth are gone and the alveolar process has atrophied.

Gingivae

The gingival tissues covering the alveolar process normally have a pale coral-pink color and slightly stippled surface. Normal gingivae attach to the teeth, and gingival projections fill the interdental spaces as papillae. Gingivitis is a common inflammatory reaction which may be the result of local factors of irritation and infection. The most common irritant to this region is the deposition of dental calculus around the necks of teeth. This hard deposit is particularly abundant in the anterior mandibular teeth and the maxillary molar teeth near the orifices of the major salivary gland ducts. The epithelial attachment of the gingivae to the necks of the teeth is lost in the lesions of periodontal disease, with the production of pocket lesions adjacent to the teeth and loss of supporting soft tissue and bone (chronic periodontitis). The presence of pocket lesions and the level of gingival attachment are determined with a fine probe. A combination of painful gums with bleeding from the free gingival margins, pseudomembrane, and loss of interdental papillae is seen in ulcerative gingival stomatitis (Vincent's infection). Although the acute symptoms of this process may be attenuated by anti-infective medication, comprehensive dental treatment is required to eliminate the disease process. The most common localized gum inflammation is around the partially erupted third molar (pericoronitis). Local care and prompt removal of the third molar are usually required.

Common Symptoms. Gingival bleeding — local inflammation and infection or hemorrhagic disorders.

Gingival recession — gingivae recessed to a low position on the roots of the teeth with increased age, as a result of trauma from incorrect brushing, and from chronic periodontitis conditions.

Gingival swelling — a common sign of odontogenic infection, which also may produce sinus tracts draining dentoalveolar abscesses. Enlargement of the gingivae may be seen in generalized form in conditions of chronic inflammation, pregnancy, endocrine disturbance, Dilantin medication, blood dyscrasias, and as a familial tendency to gingival fibromatosis.

Teeth

The normal white enamel surface of the crowns of teeth becomes dark with surface stains and also with devitalization of the pulp of teeth through trauma or disease. The crowns of teeth may be irregular in form because of congenital hypoplasia, may become reduced in length by attrition, and may be broken down by destructive phases of dental caries. The classic hypoplasia of incisors in congenital syphilis presents the notched and barrel-shaped Hutchinson's incisor.

Normally the teeth are firmly anchored in the alveolar process by the periodontal membrane. Hypermobility of permanent teeth (adult) may be due to injury but is most frequently seen in advanced periodontal disease. Localized hypermobility of teeth should alert the examiner to consider alveolar bone destruction by neoplastic disease (primary or metastatic).

Jaws

The normal excursion of the jaws will admit the width of three contacting fingers of the patient's hand (3.5 to 4.5 cm.). The excursions of the mandible should be smooth and gliding in type. Restriction of the mandible may be caused by disturbances in the temporomandibular joint, extra-articular restriction by scar tissue, trismus from spasm of the elevating muscles of mastication from any inflammatory cause, and the specific contractions of hysteria or tetanus. Findings of crepitus or pain may be indicative of disturbances in the temporomandibular joint.

SPECIAL TECHNIQUES

Radiographic Examination. The examination of the mouth and jaw region is greatly assisted by radiographic examination. Detailed x-rays of teeth and alveolar bone structures are obtained by dental radiography. Numerous projections demonstrate the maxilla and mandible to advantage.

Pulp Vitality Tests. Stimulation of the teeth by thermal change and electric current devices is used by dentists to assess the viability of the dental pulp.

Percussion of Teeth. When tapped with a metal rod the normal tooth has a dull, limited sound. Teeth ankylosed to alveolar bone elicit a more resonant sound through bone amplification, especially in the maxilla. When dental disease has caused inflammation in the periapical tissues, percussion of the teeth is painful and serves to localize an offending tooth in a case in which symptoms are diffuse and may be poorly localized by the patient.

As in all diagnostic procedures, the combination of the facts presented by the patient's history, the clinical revelations of inspection and palpation, as well as radiographic survey of the region, should be correlated in obtaining preliminary impressions regarding the nature of apparent abnormality.

REFERENCES

1. Burkett, L. W. *Oral Medicine: Diagnosis and Treatment* (6th ed.). Philadelphia: Lippincott, 1971.
2. Colby, R. A., Kerr, D. A., and Robinson, H. B. G. *Color Atlas of Oral Pathology: Histology and Embryology, Developmental Disturbances, Diseases of the Teeth and Supporting Structures, Diseases of the Oral Mucosa and Jaws, Neoplasms* (3d ed.). Philadelphia: Lippincott, 1971.
3. Kerr, D. A., Ash, M. M., Jr., and Millard, H. D. *Oral Diagnosis* (4th ed.). St. Louis: Mosby, 1974.

4. Shafer, W. G., Hine, M. K., and Levy, B. M. *A Textbook of Oral Pathology* (3d Ed.). Philadelphia: Saunders, 1974.
5. Walsh, J. P. *A Manual of Stomatology.* Christchurch, New Zealand: N. M. Peryer Ltd., 1957.

EARS, NOSE, AND THROAT   8

William W. Montgomery

GLOSSARY

Aphonia Loss of the voice.

Deglutition Act of swallowing.

Dysphagia Difficulty in swallowing.

Dysphonia Difficulty or pain in speaking.

Epiphora Tearing.

Epistaxis Nasal hemorrhage.

Laryngopharynx Lower pharynx extending from the lingual surface of the epiglottis to the trachea and esophagus.

Mucocele Intrasinus cyst arising from mucosal lining.

Nares The openings into the nasal cavity.

Nasopharynx (epipharynx) Upper pharynx extending from the choanae to the inferior border of the soft palate.

Odynophagia Painful swallowing.

Oropharynx Portion of pharynx directly behind the oral cavity extending from the inferior border of soft palate to the lingual surface of the epiglottis.

Otalgia Pain in the ear.

Otitis Infection of the ear.

Otorrhea Discharge from the ear.

Pyocele Infected mucocele.

Rhinorrhea Nasal discharge.

Stridor Noisy respiration.

Vallecula Space between the base of the tongue and the lingual surface of the epiglottis.

Vestibule Of the ear, the oval cavity in the middle of the bony labyrinth; of the nose, the area just inside the nares.

TECHNIQUE OF EXAMINATION AND NORMAL FINDINGS

The patient should be seated for an ear, nose, and throat examination. A light source (60—100-watt bulb) and a head mirror are essential. The basic instruments for examination are shown in Figure 8-1.

Nose

Begin by an examination of the external nose. The examiner may stand or sit beside and face the patient. Observe and palpate for any loss of structure or support. A nasal speculum is necessary for adequate intranasal examination (Figure 8-2). Be sure not to overdilate the external nasal orifice or to touch the nasal septum with the tip of the speculum, for this will be quite painful. Observe

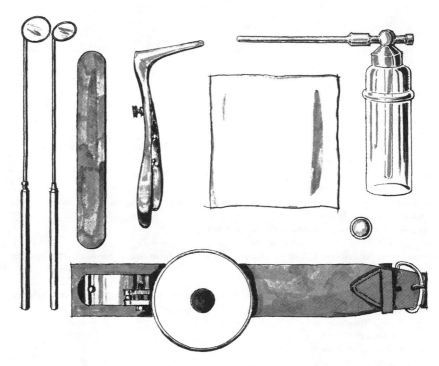

FIGURE 8-1. Basic instruments used for examinations of ear, nose, and throat. Clockwise from left these are: laryngeal mirror, nasopharyngeal mirror, throat stick, nasal speculum, gauze sponge, atomizer, finger cot, and head mirror.

the nasal vestibule; determine the adequacy of the airways. Observe carefully for a deviation of the nasal septum. Check the color of the nasal mucosa and determine whether the turbinates are normal, hypertrophic, or atrophic. Next, spray the nose with 0.25% Neo-Synephrine® or 1% ephedrine solution. Reexamine after a few minutes. The posterior aspect of the nasal cavities and the superior nasopharynx now can be visualized in most cases.

Examine the sinuses by palpating the roof of the orbit, the ascending processes of the maxillae, and the canine fossae. Tenderness may be elicited or masses palpated. Transillumination and x-ray examination of the sinuses will be discussed later.

The internal nose is the conditioner for inspired and expired air. There are two openings posteriorly, known as choanae, which lead into the nasopharynx and are sometimes referred to as the posterior nares. Usually the sinus orifices cannot be visualized during routine rhinoscopy, for they are located in the meati and obscured from vision by the turbinates (Figure 8-3).

A mucous blanket of viscid secretion covers the entire lining of the nasal cavities. This functions to collect debris and bacteria from the inspired air. The

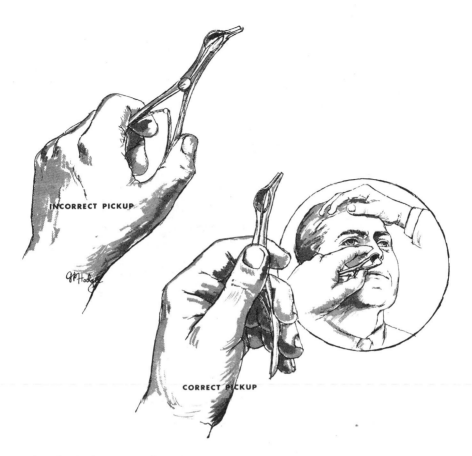

INCORRECT PICKUP

CORRECT PICKUP

FIGURE 8-2. Correct use of nasal speculum.

mucous secretion is continuously carried to the nasopharynx by ciliary action. When it reaches the pharynx, it is either swallowed or expectorated.

As the air enters the nasal cavities, it is warmed by heat from blood in the cavernous spaces in the turbinates. You will notice that one side of the nose remains more patent than the other at any given time. Blood entering and leaving the cavernous spaces is controlled by the autonomic nervous system. Air is also moistened as it enters the nasal cavities. The parasympathetic supply (vidian nerve) affects turbinate swelling and mucous secretion.

The olfactory organ is a small yellowish area on the roof of the nasal cavity. It is very difficult to visualize by ordinary rhinoscopy. There are two theories of olfactory function: the undulation theory, that energy waves, similar to light, impinge upon the olfactory nerve endings; and the chemical theory, that odorous substances initiate a chemical reaction in the olfactory epithelium.

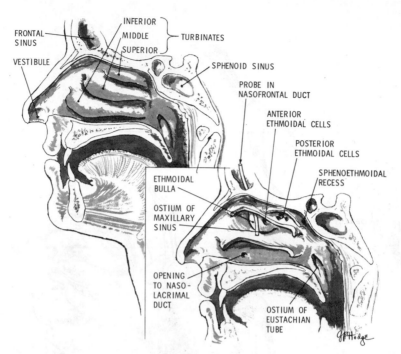

FIGURE 8-3. Sagittal view of nasal passages, frontal and ethmoid sinuses. Note that sinus orifices (see insert) are obscured from view by the turbinates.

Ear

Examine the lateral and medial surfaces of the auricle and the mastoid process. The ear canal and tympanic membrane are best examined with a head mirror and ear speculum. The battery otoscope is adequate, but somewhat cumbersome when instrumentation is necessary. The pneumatic otoscope serves to test the mobility of the tympanic membrane as well as to provide magnification. To obtain proper visualization of the canal and tympanic membrane, *the auricle must be pulled upward and backward*. In infants and small children, the auricle is pulled straight back. The reason for this is that the outer one-third of the canal is directed upward and backward, while the inner two-thirds is directed downward and forward. Tests for auditory and labyrinthine function will be discussed later in this chapter (pp. 102—103).

The auricle usually forms a 30-degree angle with the side of the head. Its lateral surface is irregularly concave. The main anatomic components of the auricle are the helix, antihelix, lobule, tragus, antitragus, and concha. The concha of the auricle receives its sensory innervation, as does the external auditory canal, from the auricular branch of the trigeminal nerve and the vagus (Arnold's nerve). In contrast to that of lower animals, the auricle of higher animals does not serve to directionalize and amplify sound. Its usefulness seems to be limited to "leading around little boys and hanging earrings."

The outer one-third of the canal contains hair follicles, sebaceous glands, and cerumen glands. At the junction of the middle and inner thirds of the canal is located a bony narrowing called the isthmus. Exostoses (benign bony projections usually associated with prolonged cold water swimming) are commonly found in the canal; however, these are rarely significant. The tympanic membrane is usually found on a slanted plane. The anterior inferior quadrant is the farthest away from the examiner. This accounts for the triangle of light being reflected anteroinferiorly from the umbo. Note the relation of the various landmarks in Figure 8-4.

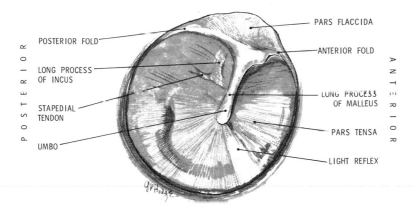

FIGURE 8-4. Right tympanic membrane showing important landmarks.

Nasopharynx

Although it is not part of a routine physical examination, the student should familiarize himself with the technique of looking into the nasopharynx. A size 0 through 3 mirror is used. This mirror is warmed by a flame, by being immersed in hot water, or by being held over an electric light bulb. If no heat is available, place a thick soapy solution on the mirror and wipe it off without rinsing. The patient sits directly in front of the examiner. The examiner's and the patient's heads should be at the same level. The patient is asked to sit erect and well back in the chair. His head is projected slightly forward.

The tongue is depressed into the floor of the mouth with the left hand, making sure not to extend the tip of the tongue blade posterior to the middle third of the tongue (Figure 8-5). Light is reflected into the pharynx with a head mirror. The mirror is grasped with the right hand as one would hold a pencil and is slipped behind and to one side of the uvula. The patient is encouraged to breathe naturally and not to hold his breath. With the mirror in the pharynx, asking the patient to breathe through his nose or to hum often relaxes a tense palate and opens the nasopharynx for examination. Be careful not to touch the base of the tongue. When necessary, 2% tetracaine (not to exceed 80 mg.) or 4% cocaine (not to exceed 20 mg.) solution may be sprayed into the pharynx to control the gag reflex.

The nasopharynx extends from the choanae to the inferior border of the soft palate. Looking anteriorly from the nasopharynx into the nose, we see the posterior border of the nasal septum dividing the two choanae. In each choana, the posterior tips of the middle and inferior turbinates can be visualized. Figure 8-6 shows the important landmarks.

Adenoid tissue is present on the posterior wall (usually absent by age 16 years). This mass of lymphoid tissue is also known as the pharyngeal tonsil. The adenoid is connected with the palatine and lingual tonsils by a band of lymphoid tissue extending down the lateral pharyngeal wall. This entire lymphoid complex is known as Waldeyer's ring.

The mucous blanket passes from the nose into the oropharynx by way of the nasopharynx. The nasal mucosa, under normal conditions, produces approximately a quart of seromucinous fluid a day. When this amount is decreased as a result of nasal or environmental factors, the mucous blanket becomes very thickened. It is then referred to as a postnasal drip, and the patient is quite conscious of this concentrated form of secretion. Smoking and air pollutants

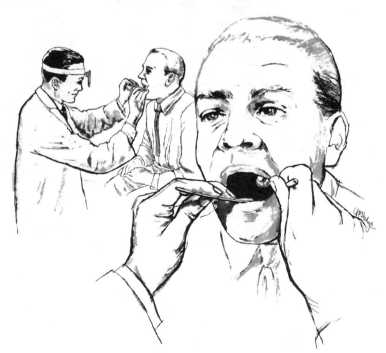

FIGURE 8-5. Technique of examination of nasopharynx.

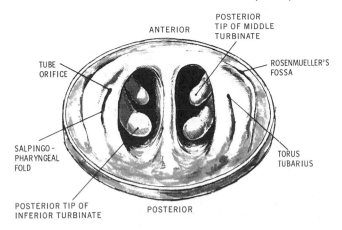

FIGURE 8-6. Mirror view of the nasopharynx.

tend to thicken the mucus and intensify the symptoms of postnasal drip, which at times is seen as white or yellow strands or webs in the nasopharynx.

Oropharynx

In examining the oropharynx, scrutinize the following structures in order: (1) the teeth, (2) the gums, (3) the salivary duct orifices, (4) the salivary glands, (5) the palate, (6) the floor of the mouth, (7) the tongue, and (8) the tonsils.

To examine the submaxillary duct (Wharton's) ask the patient to raise the tip of his tongue. Note the sublingual fold or frenulum. On either side of the frenulum is located an elevation known as the sublingual papilla. Saliva may be expressed from the duct orifice by massaging the submaxillary gland. At least ten sublingual ducts empty into the sublingual area on either side of the sublingual fold. The parotid duct (Stensen's) can be found entering the mouth through a small papilla located in the lateral buccal mucosa at the point opposite the lateral aspect of the second upper molar tooth.

The submaxillary gland can be palpated directly under the ramus of the mandible about halfway between the chin and the angle of the jaw. It has a firm, irregular consistency. This gland can be more accurately palpated bimanually. Place the index finger of one hand in the floor of the mouth, between the lateral aspect of the tongue and the teeth. The other hand palpates the gland externally. The submaxillary glands descend and become more prominent with advancing age; this is frequently misinterpreted as enlargement of the glands. The parotid gland is located anterior to and below the auricle. It normally extends from the sternomastoid muscle anteriorly to the masseter muscle. Unless abnormal, the numerous sublingual glands cannot be palpated with any degree of accuracy.

Note the shape and appearance of the hard palate. Check the function of the soft palate (e.g., for paralysis) by having the patient say "aah." Observe the size and length of the uvula. Is it in the midline?

Examine the floor of the mouth on each side by having the patient move his tongue to the opposite side. While examining the tongue, note if it protrudes in midline. The condition of the glossal mucosa should also be observed. Palpate the entire floor of the mouth, including the base of the tongue. Small lesions missed by visual examination are often found by palpation.

Estimate the size of the tonsils, if present. If they are cryptic, do the crypts contain debris (sebaceous material, pus, foreign bodies)? Is there inflammation of the anterior or posterior pillars? Note the status of the lingual tonsils by pulling the tongue forward or examining them with a laryngeal mirror.

There is a V-shaped row of circumvallate papillae at the junction of the anterior two-thirds and the posterior one-third of the tongue. These papillae receive taste fibers from the ninth cranial nerve (glossopharyngeal nerve) and have a more important taste function than the taste fibers from the lingual nerve (chorda tympani), which innervates the taste buds on the anterior two-thirds of the tongue. During ear operations, it is often necessary to sacrifice the chorda tympani nerve. The patient notes either no loss of taste or a transient loss. The lingual and glossopharyngeal nerves also distribute sensory fibers to the anterior two-thirds and posterior one-third of the tongue, respectively. *Since the glossopharyngeal nerve has to do with the gag reflex, it is best to keep well forward from the posterior one-third of the tongue with your tongue depressor.*

Laryngopharynx

Inspection of the laryngopharynx is not considered part of the routine physical examination. To examine the laryngopharynx, a size 4 through 6 mirror is used. The mirror is prepared as previously described. Instruct the patient to sit erect with his head projected slightly forward. Grasp the tongue with a piece of folded gauze. It is important that the thumb be on top of the tongue and the second finger underneath the tip of the tongue (Figure 8-7). The index finger elevates the upper lip. Insert the mirror, after testing the temperature on the back of the left hand. Place it in the oropharynx so that it elevates the uvula. *Touching the lateral walls, tonsils, or back of the tongue will cause gagging.* If the patient has a hypersensitive gag reflex, discontinue the examination and spray the pharynx with 2% tetracaine or 4% cocaine solution. Cetacaine spray is also useful and has a more rapid onset of anesthesia. Wait four or five minutes before resuming the indirect laryngoscopy.

At first ask the patient to breathe quietly and not to hold his breath. Especially reassure the patient that you will not obstruct his airway. Then ask him to say a-a-a-a-a and e-e-e-e-e. This will bring the larynx up and backward so that it may be more easily visualized. Observe the following: (1) the base of the tongue — lingual tonsils, (2) the lingual surface and margins of the epiglottis, (3) the arytenoids, (4) the aryepiglottic folds, (5) the false and true vocal cords, (6) the trachea, (7) the walls of the hypopharynx, (8) the pyriform sinuses, and (9)

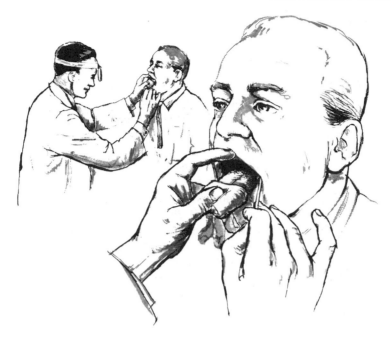

FIGURE 8-7. Technique of examination of laryngopharynx.

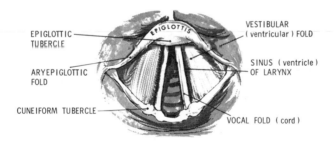

EPIGLOTTIC TUBERCLE

ARYEPIGLOTTIC FOLD

CUNEIFORM TUBERCLE

VESTIBULAR (ventricular) FOLD

SINUS (ventricle) OF LARYNX

VOCAL FOLD (cord)

FIGURE 8-8. View of larynx using laryngeal mirror.

the mouth of the esophagus. Figure 8-8 shows a mirror view of most of the important landmarks.

The intrinsic muscles of the larynx so act on its cartilaginous framework that they tense, relax, abduct, and adduct the vocal cords. This action can be observed during indirect laryngoscopy. The sphincteric action of the laryngopharynx can also be noted during swallowing. The trachea is guarded by three layers of sphincters: the epiglottis and aryepiglottic folds, the false cords, and the true cords.

ABNORMAL FINDINGS

Nose

Rhinitis, as in the common cold, is a frequent finding. Inflammatory disease of the nose is usually on an infectious, allergic, or vasomotor basis. Nasal allergy results in sneezing, watery rhinorrhea, stuffiness, and epiphora. The turbinates are pale and boggy. Vasomotor rhinitis is similar except that the turbinates tend to be more reddened. With infections, watery rhinorrhea suggests a viral etiology, while a thick, purulent discharge points to probable superimposed bacterial infection.

Sinusitis rarely if ever causes generalized headache. More frequently there is pain and tenderness over the involved sinus. There are numerous orbital complications of sinus disease which frequently are a first sign or symptom (for example, proptosis, pain, diplopia, epiphora, swelling of lid(s), and tumor mass). When the nose is examined the turbinates are found to be red and boggy, and a purulent nasal discharge is frequently present. The nasal mucosa is red and edematous. Fever and prostration may occur.

Nasal obstruction may be unilateral or bilateral. With unilateral obstruction one looks especially for a deviated septum, foreign bodies, or neoplasm. Bilateral obstruction is usually the result of rhinitis. In children adenoid hypertrophy is the common cause. A deviated septum which is S-shaped may actually obstruct both airways. Nasal polyps are another common cause of bilateral obstruction.

Perforation of the nasal septum may be on a traumatic or infectious basis. Anterior perforation occurs with tuberculosis, while posterior perforation is more common with syphilis. When epistaxis occurs it may arise posteriorly from a branch of the sphenopalatine artery or superiorly from an ethmoid vessel. However, by far the most common site of epistaxis is the anterior septum (Little's area) which is easily accessible to the examiner.

Ear

The auricle is commonly affected by frostbite, eczema, and sebaceous cysts. Tenderness in the external auditory canal usually indicates furunculosis or external otitis. The external canals may become obstructed by cerumen or foreign bodies, or by certain tumors, particularly exostoses, cysts, and malignant neoplasms.

The tympanic membrane mirrors past and present middle-ear disease. With bulging, the landmarks become obscure and there is usually some thickening and erythema, indicating acute otitis media. With retraction the landmarks are accentuated. This usually indicates obstruction of the eustachian tube or old scarring from past otitis. An amber-colored membrane indicates serous otitis media, and air bubbles or a fluid level can at times be seen. Perforations vary in size and usually point to old inflammatory disease. They may be central (usually benign), anterior, or marginal.

Any discharge should be classified as serous, mucoid, purulent, putrid, or sanguineous. A putrid, foul discharge usually indicates mastoid disease with bone destruction. A sanguineous discharge can occur with acute otitis, but the examiner should be careful to rule out a neoplasm or an injury. Pearly white cholesteatomas may occur in chronic otitis media.

There are three basic types of hearing losses:

1. Conductive hearing loss applies to any disturbance in the conduction of sound impulse as it passes through the ear canal, tympanic membrane, middle ear, and ossicular chain to the footplate of the stapes, which is situated in the oval window. As a general rule, a person with conductive hearing loss speaks softly, hears well on the telephone, and hears best in a noisy environment.

2. Sensorineural hearing loss applies to a disturbance anywhere from the cochlea, through the auditory nerve, and on to the hearing center in the cerebral cortex. A person with a perceptive hearing loss usually speaks loudly, hears better in a quieter environment, and does poorly in a crowd and on the telephone. He often states that he hears but does not understand (hears sounds but they are "garbled"), which is indicative of poor discrimination.

3. Mixed hearing loss is combined conductive and sensorineural.

Nasopharynx

The nasopharynx may be the site of inflammatory, neoplastic, or congenital disease. Polyps and cysts are not uncommon. When there is obstruction, there may be a change in the quality of the voice, since normal voice resonance is produced by the nasopharynx. The student should remember that inflammatory and neoplastic disease of the nasopharynx almost always produces obstruction of the eustachian tube orifice. This will result in hearing loss and otalgia. The resultant negative middle-ear pressure causes transudation of serum into the middle-ear space.

Oropharynx

Edema, ulceration, and purulent exudate are invariably signs of inflammatory disease, usually bacterial infection. Generalized redness of the oropharynx, however, does not necessarily indicate disease. There is also marked variation in the color of the normal oropharyngeal mucosa, especially in *smokers*.

Large tonsils are not necessarily diseased unless they interfere with respiration or swallowing. The typical appearance of acute tonsilitis is something that is easily recognized and will become familiar to you. It should be remembered that malignant and benign tumors may also arise from the tonsils.

Laryngopharynx

The most common indication for careful laryngeal examination is hoarseness of more than two weeks' duration. Hoarseness may occur as a result of acute inflammation (laryngitis). The laryngeal mucosa is red and edematous. The true vocal cords remain pale at first and later become fiery red. Secretions may be seen on the cords. Ecchymotic spots may develop due to coughing.

Hoarseness may also result from chronic laryngitis due to repeated infections, voice abuse, smoking, or poor nasal respiration. The laryngeal mucosa is dull red. The true vocal cords lose their pearly white appearance and are boggy. Engorged capillaries may be present on the cords, and thick stringy secretions are seen. With tuberculous laryngitis there may be, in addition, pallor and ulceration of the laryngeal mucosa.

Hoarseness may occasionally result from congenital abnormalities or benign tumors. It is an extremely important symptom of carcinoma of the larynx. Stridor, dysphagia, severe pain, halitosis, widening of the thyroid cartilages, hemoptysis, and cervical adenopathy are advanced signs. Hoarseness may be the only early symptom.

Paralysis of the vocal cords causes hoarseness when incomplete. Cord paralysis indicates interruption of the recurrent laryngeal nerve on the same side. This can result from a large number of traumatic, operative, inflammatory, neoplastic, or vascular abnormalities. It is occasionally a symptom of central nervous system disease as well.

SPECIAL TECHNIQUES

Transillumination of the Sinuses. Transillumination of the sinuses is used as a diagnostic tool for frontal and maxillary sinus disease. The light is placed under the medial aspect of the supraorbital rim for the frontal sinus and above the infraorbital rim for the maxillary sinus. The forehead is observed in the former and the hard palate in the latter. The test is not of true diagnostic value, for the frontal sinus is often underdeveloped. A sinus filled with clear fluid may transilluminate fairly well, but thickness of soft tissue and bone will interfere with transillumination. This test is used mostly to follow the patient's progress once a clinical and x-ray diagnosis has been made. Transillumination must be carried out in a dark room.

Sinus X-rays. Routine roentgenographic examination of the sinuses requires four views. These include an upright posterior-anterior view, a view with the tip of the nose and chin against the plate, a view with the forehead and nose touching the plate, and a lateral view. In general, look for fluid levels, filling defects, density of the sinuses, a thickness of the mucous membrane lining, fracture lines, bone destruction, and the status of the nasal and nasopharyngeal airways.

Whispered and Spoken Voice Test. The test is performed in a quiet room with the examiner facing the ear to be tested. The other ear is blocked with the examiner's hand. A rough hearing test is then performed one foot from the patient's ear. If a patient cannot hear a whispered voice at one foot, *he has at least a 30-decibel loss*. This loss is 60 decibels if he cannot hear a spoken voice at one foot.

The Weber Test. This test is accomplished by placing the vibrating tuning fork on the vertex, forehead, or front teeth. With a conductive loss, the sound lateralizes to the diseased ear. The reason is that the conductive loss is masking some of the environmental noise, and thus the cochlea is more efficient on the diseased side. The lateralization of the sound or vibrations to the better hearing ear signifies sensorineural hearing loss in the poorer hearing ear.

The Rinne Test. This test is a comparison of the duration of air conduction with that of bone conduction. The tuning fork is struck against a rubber object with maximum force so that the results will be consistent. It is first held against the mastoid bone until the patient no longer hears the sound. The fork is then held approximately an inch from the ear canal opening until the patient no longer hears the sound. *The patient should hear the fork twice as long by air as by bone* (air-time duration is, of course, measured from the striking of the fork). A positive but reduced Rinne test indicates sensorineural loss. A negative Rinne indicates a conductive hearing loss:

	Air	Bone
	(time in seconds)	
Normal hearing	100	50
Conductive loss	50	50
Sensorineural loss	60	25

A modified Rinne technique is to have the patient compare the loudness of the sound with the tuning fork one inch from the ear canal (air conduction) to the sound with the tuning fork pressed on the mastoid (bone conduction). Air conduction should definitely be louder than bone conduction (normal). Air conduction, equal or less than bone, indicates a conductive hearing loss.

Another good test is to strike the tuning fork lightly and compare the patient's air and bone conduction with your own. There are many other methods to test the hearing, such as electric audiometry, speech audiometry, and psychogalvanic skin resistance testing.

The Caloric Test. This is a measure of labyrinthine function in response to temperature. Water is placed in the external auditory canal and the eyes are observed for onset, direction, and duration of nystagmus. This technique requires 5 cc. of ice water. Make sure there is no cerumen. If there is a perforated tympanic membrane use iced 70% alcohol or a stream of cool air. Place the patient's head 60 degrees from the vertical axis (this is approximately achieved when the outer canthus of the eye and external canal are in a vertical plane). Douche the ear canal with ice water. Press stopwatch with onset of douching. Twenty-diopter spherical lenses prevent ocular fixation. Watch conjunctival blood vessels for direction of nystagmus.

The normal response is as follows:

1. Latent period of 15-30 seconds
2. Duration of nystagmus 30-75 seconds
3. Direction of nystagmus is toward the side opposite stimulation for cold fluid, toward the side of stimulation for warm fluid

Audiogram. The pure-tone audiogram is a more sophisticated method of testing for hearing loss. Basically, it is a test of air and bone conduction from 250 to 8000 cycles. One ear can be tested alone by masking the other ear with loud sound. Another hearing test, using words instead of pure tone, indicates how well the patient discriminates. If discrimination is very poor, a hearing aid will be of little use to the patient. More complex tests (Békésy, Sisi, and recruitment tests) help distinguish between cochlear and more central lesions.

REFERENCES

1. Boies, L. R., Hilger, J. A., and Priest, R. E. *Fundamentals of Otolaryngology: A Textbook of Ear, Nose and Throat Diseases* (4th ed.). Philadelphia: Saunders, 1964. Pp. 33—61; 196—207; 355—366.
2. DeWeese, D. D., and Saunders, W. H. *Textbook of Otolaryngology.* St. Louis: Mosby, 1960. Pp. 13—44.
3. Hollender, A. R. *Office Practice of Otolaryngology.* Philadelphia: F. A. Davis, 1965. Pp. 22—25.
4. Jackson, C., and Jackson, C. L. (Eds.). *Diseases of the Nose, Throat and Ear* (2nd ed.). Philadelphia: Saunders, 1959.
5. Mawson, S. R. *Diseases of the Ear* (2nd ed.). Baltimore: Williams & Wilkins, 1967. Pp. 78—114.
6. Shambaugh, G. E. *Surgery of the Ear.* Philadelphia: Saunders, 1959. Pp. 57—82.

RESPIRATORY SYSTEM 9

John G. Weg
Robert A. Green

GLOSSARY

Asthma A disease characterized by atopy with acute intermittent airways obstruction and bronchospasm, manifest clinically as wheezing, with a return to or toward normal between episodes; there is abnormal sensitivity to inhaled allergens and histamine.

Atelectasis Partial or complete airlessness with reduction in the volume of the affected segment, lobe, or entire lung; with or without locally obstructed bronchi; chronic or acute. Also refers to diffuse small areas with loss of volume, microatelectasis.

Breath sounds Sounds due to the movement of air through the lungs and air passages appreciated by auscultation; *asthmatic,* noisy musical respirations due to rhonchi of all types; *bronchial,* abnormal breath sounds, quite similar to tracheal sounds, heard over consolidated lung, also called *tubular; bronchovesicular*, sounds intermediate between vesicular and tracheal, heard over the major bronchi; *tracheal,* normal to-and-fro sounds heard over the trachea; *vesicular,* the normal breath sounds, predominantly inspiratory, heard over most of the lung.

Bronchiectasis A disease characterized by chronic dilatation of bronchial walls associated with chronic suppuration.

Bronchitis Acute or chronic bronchial inflammation.

Chronic bronchitis A clinical entity characterized by cough and sputum production on most days for at least three months in at least two successive years without a specific cause, usually associated with airways obstruction and a history of cigarette smoking.

Chronic obstructive pulmonary disease (COPD) A general term describing diseases characterized by diffuse airways obstruction; asthma, chronic bronchitis, and emphysema.

Clubbing Deforming enlargement of the terminal phalanges, usually acquired, associated with certain cardiac and pulmonary diseases; it is characterized by loss of the normal angle between the skin and nail base and sponginess of the nail base.

Collapse Reduction in lung volume, acute or chronic, due to bronchial or parenchymal disease.

Compression Mechanical reduction of lung volume by pressure.

Consolidation Diffuse replacement of a large zone of pulmonary parenchyma with liquid and solid products of inflammation, but only when the associated bronchus is patent. Preferably the term is used as a description of physical rather than pathologic or roentgenographic findings.

Cyanosis A blue color of the skin and mucous membranes due to severe arterial oxygen desaturation, e.g., a hemoglobin saturation of less than 75 to 85 percent, a PO_2 of less than 50 mm. Hg.

Dyspnea The subjective sensation of abnormal or inappropriate shortness of breath or difficulty in breathing.

*The writing of this chapter was supported in part by a Public Health Service Pulmonary Academic Award (1 K07 HL 70368—01) from the National Heart and Lung Institute.

Emphysema A disease characterized pathologically as destruction and dilatation of lung tissue distal to the terminal bronchioles and clinically by dyspnea, a minimally productive or nonproductive cough, airways obstruction, persistent hyperinflation of the lung, and decreased breath sounds; most commonly it is associated with chronic bronchitis; *Compensatory Emphysema,* a term referring to secondary dilatation of air spaces adjacent to areas of fibrosis; *Subcutaneous emphysema,* air or other gas in the subcutaneous tissues.

Empyema Pus in the pleural cavity; pyothorax.

Forced expiratory time (FET) The time required to empty the lungs completely with maximal effort from full inspiration to full expiration.

Forced vital capacity (FVC) The amount of air exhaled from the lungs from a maximum inspiration to a maximum expiration, with expiration as rapid and forceful as possible; the volume expired is measured against time.

Fremitus A palpable vibration or thrill; *rhonchal fremitus,* vibration due to rhonchi; *tactile* or *vocal fremitus,* voice sound vibration.

Friction rub Grating sensation, heard or palpated, which arises from inflamed serous surfaces.

Hemoptysis Expectoration of gross blood, by coughing, from the larynx or lower respiratory tract.

Kyphosis Increased convexity of the spine in the anteroposterior plane.

Müller's maneuver Production of increased negative intrapleural pressure by an attempt to inhale forcibly against a closed glottis.

Pleural effusion Fluid of any type in the pleural cavity; it includes empyema, hemothorax, and chylothorax. Thoracentesis is required for specific identification.

Pleurisy Any pleural inflammation; loosely, the pain associated with disease of the pleura.

Pneumoconiosis Condition of chronic pulmonary scarring resulting from inhalation of inorganic (mineral) or organic dusts; more broadly, parenchymal pulmonary disease of occupational cause, including reactions to fumes, gases, etc.

Pneumonia Clinical term which most commonly refers to pulmonary inflammation due to infection.

Pneumonitis Any pulmonary inflammation.

Pneumothorax Air or other gas within the pleural cavity.

Rale Abnormal sound which originates from the trachea, bronchi, or lungs; *crepitation, crepitant rale,* a rale of crackling quality; usually applied to moist rales; *consonating rale,* a loud, clear, ringing rale which sounds close to the ear; often associated with pulmonary consolidation; *fine, medium, coarse rale,* classificatory terms referring to loudness, duration, and quality of moist rales; *moist rale,* an interrupted sound arising from alveoli and smaller bronchi — the term *rale* is limited to this variety by some; *rhonchus,* a continuous, coarse sound arising from the trachea or bronchi — also termed *dry* or *musical rale; sibilant rale,* a high-pitched, very musical rhonchus; *sonorous rale,* a low-pitched, snoring rhonchus; *wheeze,* a nonspecific whistling sound associated with obstruction to air flow — applied to both rhonchi and coarse moist rales. (See discussion on pp. 131—132.)

Respiration, variations in: *apnea,* the absence of breathing; *bradypnea,* breathing at a slow rate; *Cheyne-Stokes,* cyclic breathing in which there is increased depth and rate of respiration between periods of apnea; *hyperpnea,* increased depth and usually rate of respiration; *orthopnea,* inability to breathe comfortably while supine, relieved by sitting up; *tachypnea,* increased rate of respiration.

Scoliosis Lateral deviation of the spine.

Stridor Difficult respiration due to upper airways obstruction, characterized by high-pitched crowing sounds in inspiration.

Valsalva maneuver Production of decreased negative intrapleural pressure by an attempt, after deep inspiration, to expire forcibly against a closed glottis.

Voice sounds The vibrations of the spoken voice transmitted through the lungs and appreciated by auscultation; also called *vocal resonance* and, rarely, *vocal fremitus; bronchophony,* increased intensity of voice sounds; *egophony,* a peculiar bleating nasal quality of voice sounds in which articulated long *e* (ee) simulates long *a* (ay); *pectoriloquy,* transmission of articulate speech; *whispering pectoriloquy,* distinct transmission of whispered sounds.

TECHNIQUE OF EXAMINATION AND
NORMAL FINDINGS

To master physical diagnosis of the chest, it is only necessary to have good eyes, good ears, one stethoscope, . . . a good roentgenographic unit, a ration of intelligence, a measure of determination and a mess of patients [5].

Examination of the respiratory system consists primarily of examination of the chest. Traditionally its four component parts are inspection, palpation, percussion, and auscultation. To these have been added a fifth: *study of the chest roentgenogram is indispensable to the physical examiner.* Certain conditions present with abnormalities either on physical or roentgen examination alone; the roentgenogram always helps in the evaluation of physical findings and cannot be properly interpreted itself without correlation with them. The patient is most often examined first and the roentgenogram seen subsequently. Some physicians prefer to see the roentgenogram first and then the patient; it can save valuable time to identify in advance the areas which require the most meticulous examination. The sequence is immaterial as long as the patient is reexamined following study of the roentgenogram, and the roentgenogram rechecked following careful physical examination. This procedure permits special attention to specific areas of the patient or film and search for unusual physical findings when indicated. To fail to elicit abnormal physical findings over an area shown subsequently by the roentgenogram to be significantly diseased is unfortunate but not rare; to fail then to reexamine carefully and determine what the findings really are is deplorable. The roentgenogram can be a major teacher of physical diagnosis. The chest examination is incomplete, in the healthy person undergoing a routine physical examination as well as in the symptomatic patient, without inspection of the chest roentgenogram. When a roentgenographic abnormality is found, every effort should be made to obtain previous roentgenograms for comparison.

Now a sixth component must be added: it is essential to evaluate *arterial blood gases* — the partial pressure of oxygen (Pa_{O_2}) and of carbon dioxide (Pa_{CO_2}), the pH, and usually the percent saturation of hemoglobin (Sa_{O_2}) — in patients with dyspnea, tachypnea (or other respiratory rate irregularities), or significant roentgenographic abnormality. Adequate oxygenation cannot be determined accurately clinically; cyanosis can be seen only when the hemoglobin saturation is reduced to 75 to 85 percent, (a P_{O_2} of 40 to 50 mm. Hg), e.g., equivalent to that in normal venous blood. The clinical estimation of hypo-ventilation, retention of carbon dioxide, hypercarbia, and acidosis is also un-satisfactory because the signs and symptoms are late and nonspecific.

The student must sharpen his perception to learn the normal, and appreciate the abnormal, findings. As discussed in Chapter 3, he must first learn what to look for; later, he must avoid the pitfall of prejudiced perception because of prior experience, specifically look for nothing yet see what is actually there. Develop a thorough systematic routine identical for each patient seen. Think of the pathologic changes, rather than specific diseases, which may account for the abnormalities elicited. Diagnosis awaits the correlation of clinical, roentgeno-graphic, and laboratory findings.

Knowledge of the underlying anatomy of the lungs is essential to a properly conducted examination, as each bronchopulmonary segment must be checked. The sketches and roentgenograms with bronchopulmonary anatomy super-imposed in Figures 9-1 A-D may be helpful. The major or long fissure delimits the lower lobe from the others. It extends from the fourth thoracic vertebra posteriorly obliquely downward, crossing the fifth rib in the midaxilla, to the sixth rib anteriorly. The right minor, horizontal, fissure separates the upper and middle lobes and extends from the major fissure in the midaxilla to the third intercostal space anteriorly. The planes between the segments vary considerably; the general zone of each segment can be identified. The inferior boundary of the lungs in the midinspiratory position is at the tenth vertebral body posteriorly, at the eighth rib in the axilla, and at the sixth chondrosternal junction anteriorly. The angle of Louis is a prominence on the sternum at the second chondrosternal junction; it is a helpful landmark for counting ribs and then identifying segments from which abnormal findings arise. Other helpful topographic aids are a series of imaginary lines, whose names are self-explanatory, projected onto the chest wall: midsternal line; midclavicular line; anterior, midaxillary, and posterior axillary line; midscapular line; and vertebral line. Record your findings by their relation to these lines, ribs, and interspaces. For example, rales may be heard 3 cm. to the right of the midsternal line at the level of the fourth anterior intercostal space. Record the finding as such; later you will interpret the rales as arising from the medial segment of the right middle lobe.

Routine physical examination allows excellent evaluation of the ventilatory function of the respiratory tract; it permits an estimate of the volume of exchanging gas and of the rate and distribution of air flow. Diffusion, perfusion, and the relation of ventilation to perfusion, the other functions of the lung, cannot be well evaluated, although the clinical findings in disease states in which these are involved are often characteristic.

A well-lighted, warm, quiet examining area is mandatory. Put the patient into the sitting position. A revolving stool is helpful. The patient should be relaxed, as cooperation is essential. The examiner must also be relaxed and not pressed for time. If the patient is too ill to sit upright by himself, have an aide support him in the sitting position, as examination with the patient on his back or side may introduce numerous undesirable variables.

The patient is stripped to the waist; a drape is utilized for female patients. Sit facing the patient, but also inspect from behind, from the side, and while standing behind the patient and looking down over his shoulders at the anterior chest.

Inspection

Inspection has its major value in observation of abnormalities of respiration and of symmetry, both of ventilation and structure, of the two sides of the thorax.

RIGHT LUNG

Upper Lobe

 1. Apical segment
 2. Posterior segment
*2'. Axillary subsegment
 3. Anterior segment
*3'. Axillary subsegment

Middle Lobe

 4. Lateral segment
 5. Medial segment

Lower Lobe

 6. Superior segment
†7. Medial basal segment
 8. Anterior basal segment
 9. Lateral basal segment
10. Posterior basal segment

LEFT LUNG

Upper Lobe

Superior division
††1—2. Apical-posterior segment
 3. Anterior segment

Inferior (lingular) division
 4. Superior lingular segment
 5. Inferior lingular segment

Lower Lobe

 6. Superior segment
††7—8. Anteromedial basal segment
 9. Lateral basal segment
10. Posterior basal segment

*These axillary subsections are listed as such because of the frequency with which they are diseased together without the remainder of their respective segments.
†The medial basal segment does not present on the surface of the lung under the chest wall.
††The apical-posterior and anteromedial basal segments on the left are considered segments with subsegmental subdivisions because they are each supplied by a single segmental bronchus. In some classifications the posterior segment is referred to as 3 and the anterior segment as 2.

First and of signal importance is the respiratory rate. Is it increased or decreased? Is it irregular? What is the length of inspiration? Of expiration? Is the patient in pain? Is he in respiratory distress, or is respiration noisy? Observe the chest during quiet respiration. Note the status of the skin and breasts, the muscular development, the state of nutrition, localized areas of bulging or retraction, the presence of thoracic deformities, especially if unilateral, the position of the apex impulse, scars or sinus tract openings, evidence of vascular pulsation or dilatation. What are the size and the shape of the chest? Pay special attention to scoliosis, as its presence may influence other physical findings and interpretation of the chest roentgenograms. Note the angle of the costal margins at the xiphoid, the angle that the ribs make posteriorly with the vertebrae, and the slope of the ribs. Note whether the interspaces retract or bulge generally or in limited areas during ventilation. Does one area lag or flare more than another? Is respiration predominantly costal or abdominal?

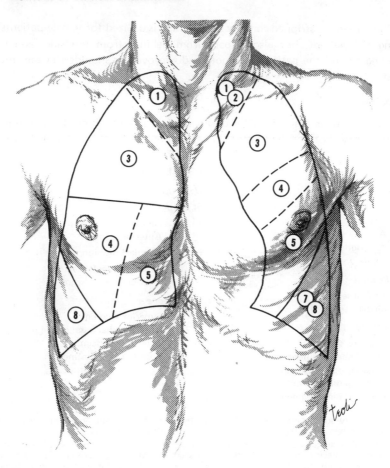

FIGURE 9-1A. Anterior.

FIGURES 9-1, A—D. Segmental pulmonary anatomy. The bronchopulmonary segments are outlined on the roentgenograms as they project as the surface of the lung and thus on the chest wall where their physical findings would present separately. The lines do *not* necessarily represent the roentgenographic projection of each entire segment. All projections on the external chest are drawn to correspond to the position in which the associated roentgenogram is taken. The continuous lines represent interlobar fissures; the interrupted lines represent segmental planes. The bronchopulmonary segmental nomenclature is dependent upon the anatomy and nomenclature of the supplying bronchi.

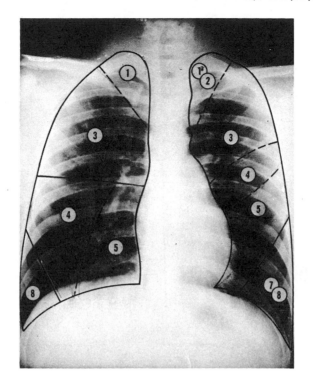

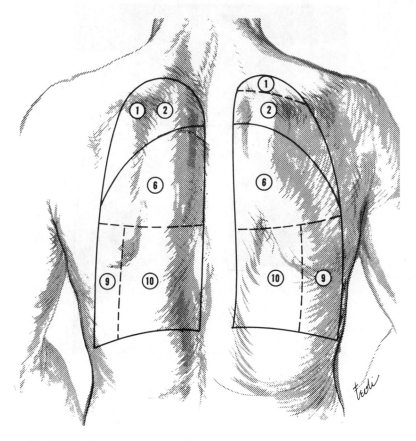

FIGURE 9-1B. Posterior.

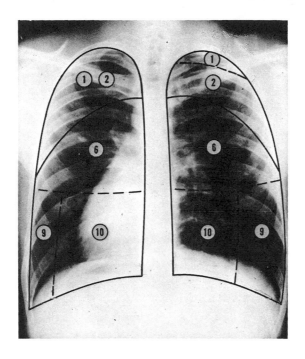

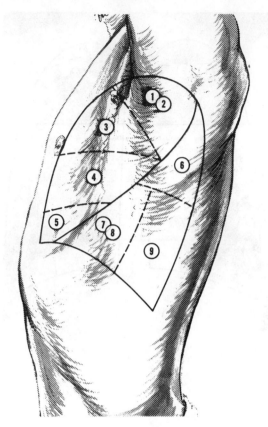

FIGURE 9-1C.   Left.

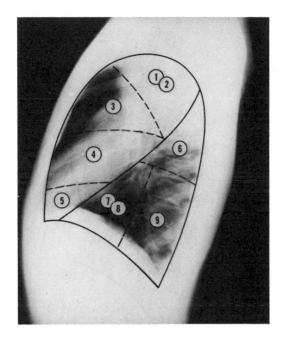

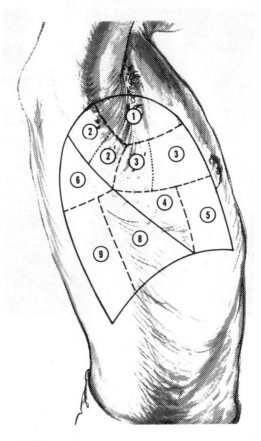

FIGURE 9-1D. Right.

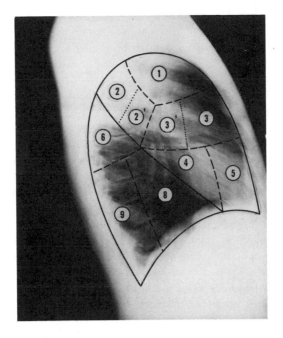

The respiratory rate in the resting patient varies between 12 and 16 breaths per minute; respiration is regular and quiet; the respiratory phase lasts half again as long as the expiratory.

After inspecting the chest with quiet breathing ask the patient to take in a maximum breath and force it out with his mouth wide open until his lungs are completely empty, i.e., a *forced vital capacity (FVC) maneuver.* This should be preceded by a demonstration, and the patient should perform the maneuver two or three times to assure maximal effort and complete emptying. The examiner should measure and record, in seconds, the time to complete this maneuver, the *forced expiratory time (FET),* by listening near the mouth or with the stethoscope placed over the cervical trachea. A normal person can empty his lungs in 3.5 seconds or less. A prolonged FET, 4 seconds or greater, provides a most sensitive and quantitative clinical measure of diffuse airways obstruction (see Figure 9-2) (1,4).

The hallmark of normal inspection is symmetry — of structure and of movement. However, due to the frequency of *scoliosis* (curvature of the spine) in the normal population, symmetry is rarely perfect. The thorax is broader from side to side than from front to back, and its shape varies with the build of the individual, being short and broad in the stocky, and long, flat, and narrow in the asthenic. Minor variations from the normal in the structure of the thorax include a funnel-shaped depression of the lower portion of the sternum, referred to as *pectus excavatum,* and the reverse condition, in which the sternum projects

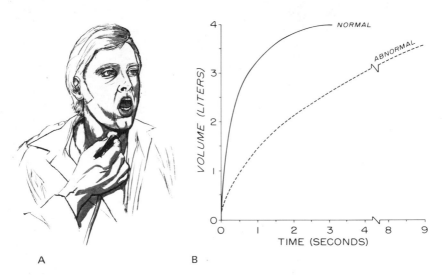

A   B

FIGURE 9-2. Forced expiratory time. A. Patient forcibly exhaling from full inspiration with mouth wide open; examiner times expiration with stethoscope over the trachea. B. Spirometric recording of forced vital capacity maneuver; volume is measured in liters on the ordinate and time in seconds on the abscissa. A normal person can exhale approximately 80% of his vital capacity in one second, the forced expiratory volume at 1 second divided by the forced vital capacity is recorded as $FEV_1\%$.

beyond the frontal plane of the abdomen, referred to as *pectus carinatum* or *pigeon breast.* Slight thoracic *kyphosis* resulting from poor posture is not uncommon and is often associated with scoliosis.

Normally the interspaces between the ribs do not particularly retract or bulge during inspiration. The subcostal angle is less than 90 degrees and widens with inspiration. The ribs themselves make an angle of approximately 45 degrees with the vertebral column. Most women tend to breathe largely with their costal cage, and thoracic expansion is therefore easily noted. In most males respiration is largely diaphragmatic, and bulging of the abdomen with inspiration is seen. The entire rib cage moves laterally and upward with inspiration, gently rising with each breath. Ventilation is symmetric in onset and depth.

Palpation

Palpation excels in the evaluation of the degree and symmetry of expansion with respiration, as well as in appreciation of the transmitted vibrations of the spoken voice. It is complementary to inspection in evaluation of respiratory excursion. Use both hands simultaneously, and palpate symmetric areas of the thorax. Use the fingertips, the palm, or the ulnar edge of the hand. Palpate from above down during normal and deep respiration and while the patient phonates. The examiner's hands must be warm. Standing behind the patient, place the hands on the lower chest with the thumbs adjacent near the spine. Tell the patient to inhale deeply, and compare the symmetry of onset and depth of inspiration. Place the hands similarly over the lower lateral chest, and standing behind the patient place them over the shoulders onto the anterior chest below the clavicles. This procedure also can be performed over the low anterior chest, but is of less value there.

Palpate the chest wall, and note any masses. Note the general turgor of the skin, its temperature, moistness, or the presence of edema. Note the character of the musculature. Palpate each rib, noting whether tenderness is elicited. Note the presence of any spontaneous vibrations that occur with respiration or the cardiac beat. Note the position of the trachea in the episternal notch relative to the midline, and its distance from the posterior surface of the sternum. This is generally ascertained most accurately by standing behind the patient and letting each index finger slide off the head of the clavicle deep into the sternal notch on either side of the intrathoracic trachea (Figure 9-3). At this time the supraclavicular areas should be palpated for the presence of lymph nodes. Check your impressions on inspection regarding scoliosis and deformities.

With the palmar aspect of the fingers or the ulnar aspect of the hand appreciate tactile fremitus by asking the patient to phonate. It is standard practice to have the patient repeat "1, 2, 3" or "99" to elicit fremitus; "blue moon" is also a helpful phrase. As with other chest examinations evaluate vocal fremitus over all lung segments, and note its comparative increase or decrease or its absence. It will be elicited again during auscultation, and the examiner may wish to check his tactile findings with the auscultatory findings at this point.

FIGURE 9-3. Palpating the intrathoracic trachea.

On palpation no abnormalities of the skin, chest wall, or ribs are noted. The sound vibrations are best transmitted to the examiner's hand in patients with thin chest walls and deep voices. *Fremitus is most prominent over areas where the bronchi are relatively close to the chest wall.* It increases as the intensity of the voice increases and as its pitch drops. It is normally symmetrical except for a slightly greater intensity over the right upper lobe than over the left. The trachea is in the midline, admitting almost an index finger on either side at the episternal notch when palpated from behind the patient; the space between it and the sternum does not allow more than one finger at the episternal notch. The lower palpable trachea may deviate slightly to the right in older patients.

Percussion

Percussion has its greatest value in determining the relative amount of air or solid material in the underlying lung and in delimiting the boundaries of organs or portions of the lung which differ in structural density. Direct percussion involves tapping the chest with the middle or ring finger. It is especially valuable in percussing over the clavicle. Mediate percussion is more popular and is best learned by direct demonstration by an instructor. Figure 9-4 shows the basic method for mediate, or indirect, percussion.

Place the distal two phalanges of the middle finger of the left hand (the finger is then pleximeter) firmly against the chest wall in the intercostal spaces parallel to the ribs. Strike the distal interphalangeal joint with a quick, sharp stroke with

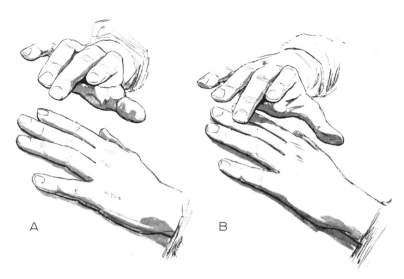

FIGURE 9-4. Basic method for mediate percussion of the chest.

the tip of the middle finger (which becomes the plexor) of the right hand, one or two rapid staccato blows in succession. Hold the forearm stationary, and make the striking motion with the wrist. Note both the sound elicited and the sense of resistance and vibration underneath the finger. Percuss from side to side, comparing symmetrical areas of the chest. Percuss gently though firmly, and make a special effort to apply equal force at all points. It will be necessary to percuss more firmly in individuals with thick chest walls, more lightly in the thin. Skillful percussion requires much practice, until the technique becomes automatic and the examiner can concentrate completely on the sounds and sensations he elicits.

Percuss the lower margins of the lungs and the width of the heart and upper mediastinum. The extent and equality of diaphragmatic excursion are well evaluated in most patients by percussion. Fluoroscopy is more reliable though less available. Tell the patient to inhale deeply and hold his breath. Note the line of change in the percussion note between the resonant lung and the abdominal viscera posteriorly. Then ask the patient to exhale completely and hold his breath. The lower level of pulmonary resonance has now moved upward (Figure 9-5). The distance between these two points represents the extent of diaphragmatic excursion. Complementary to this is the actual measurement of costal expansion, which requires a flexible tape measure placed around the chest at the nipple level.

The normal percussion note varies with the thickness of the chest wall and the force applied by the examiner. The clear, long, low-pitched sound elicited over the normal lung is termed *resonance.*

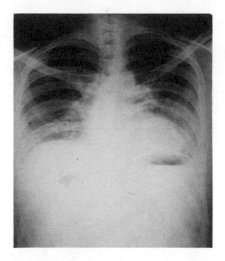 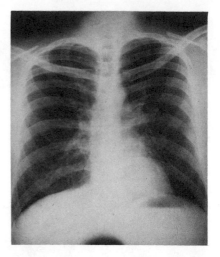

FIGURE 9-5. Roentgenogram of a normal patient showing the striking alterations in the intrathoracic viscera produced by respiration. At left, full inspiration; at right, full expiration.

Dullness occurs when the air content of the underlying tissue is decreased and its solidity increased. The sound is short, high-pitched, soft, and thudding, and lacks the vibratory quality of a resonant sound. It is heard normally over the heart and is accompanied by an increased sense of resistance in the pleximeter finger (Figure 9-6). Variations of the note intermediate between dullness and normal resonance are sometimes termed "slight dullness" or "impaired resonance" and are elicited normally over the scapulae, over thick musculature, and occasionally at the right apex. Asymmetry of apical resonance, however, is rarely elicited during the routine examination, and slight differences often require perusal of the roentgenogram to determine which side, if either, is normal or abnormal. When patients are examined while lying on the side, the dependent lung tends to have less resonance.

Flatness is absolute dullness. When no air is present in the underlying tissue the sound is very short, feeble, and high-pitched; flatness is found over the muscle of the arm or thigh.

Hyperresonance refers to a more vibrant, lower-pitched, louder, and longer sound heard normally over the lungs during full inspiration.

Tympany is difficult to describe but implies that the sound is moderately loud and fairly well sustained, with a musical quality in which a specific pitch is often noted. It is normally heard in the left upper quadrant of the abdomen over the air-filled stomach or over any hollow viscus. The pitch of tympany is variable, but it is usually high-pitched, clear, hollow, and drumlike. No dullness is usually elicited under the upper third of the sternum. (The borders of cardiac dullness will be described in Chapter 10.) The lower lung margin may vary, but

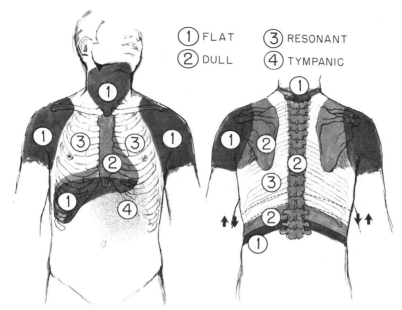

FIGURE 9-6. Quality of percussion note over the normal chest. Note diaphragmatic excursion in posterior view.

in midinspiration it is usually at the tenth vertebra posteriorly, at the eighth rib laterally, and at the sixth cartilage anteriorly. Figure 9-6 shows the normal diaphragmatic excursion, which averages about 3 cm. in females and 5 or 6 cm. in males. Costal expansion of 4 to 6 cm. is considered normal.

Auscultation

Auscultation has no peer in a comparison of the state of bronchial patency of various lung divisions and in appreciation of the normal breath sounds as well as abnormal respiratory murmurs arising from diseased areas. A quiet room is essential. Direct auscultation by application of the ear to the chest wall is unnecessary. Use a stethoscope, preferably one fitted both with a bell and a diaphragm; some examiners prefer one or the other — the bell is particularly helpful in very thin patients with deep intercostal spaces which do not permit the larger diaphragm to make full contact. Apply the stethoscope firmly to the chest wall and make certain that factitious sounds due to abnormal muscular movement, hair on the chest wall, or rubbing against the stethoscope do not occur. Cover systematically all portions of both lung fields posteriorly, laterally, and anteriorly, from above down, comparing areas side to side. *Be certain that each bronchopulmonary segment is auscultated.*

Instruct the patient to breathe, with his mouth open, a little deeper and faster than normal; demonstrate it to him yourself. Notice especially the character of the breath sounds and the presence of abnormal sounds. Notice any changes that occur during more rapid and deep respiration; in a normal person the breath sounds should increase in intensity (*recruitment*). Demonstrate the method to the patient. Compare each area examined with the symmetric area of the opposite thorax and with other adjacent pulmonary zones. Direct the patient with instructions such as "a little deeper, please," "a little faster, please," "not quite so hard," until you are convinced that the abnormalities heard are not factitious. Ask him to phonate and to whisper a phrase such as "1,2,3" or "99" and note the character of the vocal resonance perceived.

The importance of the character of normal breath sounds is often inadequately emphasized. *Vesicular* breath sounds are breezy or swishy in character. The inspiratory phase predominates, and the pitch is high. Expiration is heard only as a short, fainter, lower-pitched puff less than one-quarter as long as inspiration. Vesicular breathing is normal over most of the lungs. *Bronchial* breathing is for practical purposes synonymous with tracheal breathing and is heard normally over the trachea and the main bronchi. In bronchial breathing, inspiration is louder and higher pitched, but the great change is in expiration, which is increased in duration so much that it actually is longer than inspiration; its pitch is high and its intensity greater. Its quality is hollow or tubular and rather harsh. *Bronchovesicular* breath sounds represent an intermediate stage; they are heard normally in the second interspaces anteriorly, in the interscapular area posteriorly, and often at the medial right apex. Inspiration is unchanged from vesicular breathing, but expiration is as loud, equal in length, and similar in pitch. *Exaggerated vesicular* breathing is heard normally at times in thin people and children, during exercise, and in unnecessarily loud and rapid respiration. Here the expiratory phase is more prominent than in vesicular breathing but is not of bronchovesicular character. The normal relationship of inspiration to expiration must be learned with respect to each area of the lung examined.

You must expect differences in vocal resonance due to thickness of the chest wall as well as the area to which the stethoscope is applied. Voice sounds are heard best near the trachea and major bronchi. Exaggerated voice sounds — *bronchophony* — are a normal finding over the trachea and right upper lobe posteriorly. Speech is heard only as indistinct noise.

Rales are not heard in the normal chest. A description of the normal chest roentgenogram is beyond the scope of this text.

Following completion of these physical maneuvers, examine the roentgenogram carefully. Be systematic, checking the (1) soft tissues, (2) bones of the shoulder, (3) neck, (4) spine and ribs, (5) visible abdomen, (6) mediastinum, (7) heart, (8) diaphragm, and (9) lungs. Give special attention to areas over which physical abnormalities were elicited. If the film reveals a surprising unsuspected abnormality, return to the patient and repeat the physical examination. Perform additional studies, such as changing the patient's position, shaking him, or checking for a tracheal tug, if the film suggests that these may be helpful [2]. Go back and forth, if necessary, until careful correlation leads to accurate diagnosis.

CARDINAL SYMPTOMS AND
ABNORMAL FINDINGS

Cough is the commonest symptom of pulmonary disorders; a cough is abnormal. It is caused by stimulation of afferent vagal endings, and helps clear the airways of extraneous material. It occurs with focal anatomic pulmonary lesions anywhere in the respiratory system, and is also frequently due to diffuse airways or parenchymal abnormality. Coughs should be characterized as productive or nonproductive of sputum, paroxysmal, brassy, loud and high-pitched, or whooplike in character.

Sputum production often accompanies cough when irritation or inflammation of any portion of the pulmonary system leads to transudation or exudation of fluids; it is abnormal. The nature and quantity of sputum should be recorded (see Special Techniques, pp. 137–138).

Hemoptysis is frequently due to a single anatomic lesion, proximal or distal, which inflames or destroys the lung or bronchus involved. It can occur with diffuse increase in pulmonary capillary pressure, as in cardiac disease. Blood loss into the lung can be great despite minimal external evidence of bleeding. Blood streaking of sputum commonly accompanies diffuse or localized inflammatory disorders and can be associated with paroxysmal coughing. Careful questioning and examination are required to separate hemoptysis from hematemesis and oral or nasopharyngeal bleeding.

Chest pain due to pulmonary disease most often results from involvement of nerve endings in the parietal pleura, as the pulmonary parenchyma itself is insensitive. Pain can arise from major bronchial and peribronchial disease, in which case it tends to be constant, deep, and aching. Pleural pain, which varies with respiration, is sharp and intermittent. Chest pain due to pulmonary hypertension may simulate angina pectoris. Pain may also arise from the chest wall; it is usually associated with localized tenderness.

Dyspnea, unlike the other symptoms described above, is more often correlated with physiologic abnormalities than with anatomic; that is, it is usually due to diffuse and extensive rather than focal pulmonary disease. Ventilation, gas exchange, and relationships between air flow and blood flow may all be awry and lead to dyspnea. Disease of any anatomic portion of the lung, if extensive enough, may be responsible. Dyspnea is the patient's perception of shortness of breath or difficulty in breathing in an inappropriate setting, e.g., shortness of breath after running a mile is appropriate and thus not dyspnea, while shortness of breath in a young person after walking up a flight of stairs is dyspnea. Although all symptoms require quantitation, this is essential in describing dyspnea — to be meaningful each occurrence of dyspnea should be recorded in relation to everyday activity with precise details of respiratory rate and length of activity, time to recovery, and progression over time. Since dyspnea is subjective, special care is necessary in understanding the patient's description; he may only complain of fatigue or of a tightness or heaviness in the chest. The respiratory rate is almost always increased with dyspnea of an organic nature, while a normal rate associated with sighing usually indicates an anxiety state.

Other symptoms of particular importance in chest disease are: *hoarseness*, which may indicate damage to a laryngeal nerve by tumor or inflammation or be secondary to vocal cord trauma with severe coughing; *fever*, indicative of inflammatory or mitotic disease; *chills* or a *rigor* (a shaking chill); weight loss; and edema.

A *past history* of pneumonia in childhood, recurrent pneumonia, or whooping cough may indicate a predisposition to respiratory tract infection, bronchiectasis, or congenital anomalies; the use of oily nose drops, poor oral hygiene, a recent dental extraction, alcoholism, or unconsciousness may predispose to aspiration pneumonia and/or lung abscess. A history of close exposure to infectious tuberculosis, occupational exposure to certain mineral or organic dusts, travel to areas with endemic fungal infections, or inhalant allergy, e.g., hay fever, provides potentially important diagnostic information.

A single abnormal finding is rarely diagnostic; correlation with other physical abnormalities and with the chest roentgenogram is necessary before the status of the underlying lung can be deduced with confidence. Dullness and absent breath sounds, for example, occur with both pleural effusion and an obstructed lobe, often requiring the roentgenogram for differentiation, whereas a small infiltrated lobe on the roentgenogram may result from neoplasm or inflammatory fibrosis (Figure 9-7). Differentiation is made by the physical findings of dullness and diminished breath sounds in the former and dullness and bronchial breath sounds in the latter.

Inspection for respiratory disease requires inspection of more than the thorax. Clubbing, cyanosis, use of accessory respiratory muscles, respiratory distress, and marked sweating are important extrathoracic signs of intrathoracic disease; poor veins with scarring may indicate drug addiction; peripheral thrombophlebitis may correlate with chest pain and dyspnea; a rash following an intercostal nerve root distribution suggests herpes infection; a prominent bulging of the dorsal spine in a kyphotic individual, or a gibbous deformity, suggests tuberculosis; the presence of Horner's Syndrome (see Chapter 20, Nervous System, p. 341) with pain in the shoulder radiating into the arm with muscle atrophy suggests malignancy in the lung apex; a prominent venous pattern of the chest wall may indicate vena caval obstruction; and so on.

Clubbing of the fingers is especially important. The ends of the fingers and toes are swollen, at times grotesquely so. The nails are curved convexly from

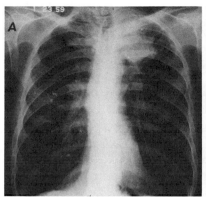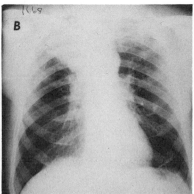

FIGURE 9-7. The necessity for correlating roentgenograms with physical findings is demonstrated in these illustrations. A. Elderly man with infiltration of left upper lobe with extensive calcification. Sputum had tubercle bacilli. Diagnosis: Active tuberculosis. Physical findings, however, showed marked dullness and *absent breath sounds* over the left upper thorax (especially anteriorly), which suggested obstructive pneumonia secondary to bronchial obstruction; neoplasm suspected. Bronchoscopy established diagnosis of carcinoma in addition to tuberculosis. B. Elderly man with dense right upper lobe infiltration. Carcinoma suspected, yet onset of illness acute. *Breath sounds are bronchial over right upper lobe with moist rales.* Diagnosis: Pneumonic consolidation without obstructing neoplasm. Abnormality cleared completely with treatment.

base to tip and from side to side. The tissue at the base of the nail is spongy and the nail itself loosely attached; its free edge may be palpable. Helpful in diagnosis is a change in the normally obtuse angle the skin makes with the base of the nail: in clubbing the nail is raised and this angle lost and flattened. If the syndrome of pulmonary osteoarthropathy accompanies the clubbing, the periosteum over the ends of the long bones of the forearm and leg may be palpably tender. This is almost always a manifestation of carcinoma of the lung. See Figure 9-8 and Chapter 10, Tables 10-2A and 10-2B, for further classification of clubbing.

Asymmetry due to localized prominence of the chest wall will occur occasionally with large tumors or large pleural effusions; in chronic conditions in which considerable scarring of the lung and pleura has occurred, localized contraction is not uncommon. Scoliosis must be eliminated as the cause of such asymmetry. A barrel-shaped appearance of the chest is common in older patients and those with emphysema. The ribs are more horizontal, and the subcostal angle is greater than 90 degrees without variation during respiration. Thus this is neither a specific nor a sensitive finding. Localized areas of diminished ventilation may reflect either local disease or the effects of severe pain with splinting on thoracic motion. Local inspiratory retraction of the intercostal spaces means local bronchial obstruction; generalized retractions are common in symptomatic chronic obstructive pulmonary disease, asthma, chronic bronchitis,

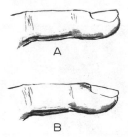

FIGURE 9-8. Clubbing. A. Normal finger. The angle between the nail and a straight line through the long axis of the digit is 20° or more. The nail base is firmly sealed. B. Clubbing. The angle between the nail and finger is less than 20°; it may be less than 0°. The base of the nail feels spongy when pressure is applied.

and emphysema. Dilated veins may indicate compression of the venous flow within the mediastinum, and subcutaneous swellings may represent metastatic tumor nodules, abscesses, or the pointing of an empyema. Sinus tract openings or scars similarly suggest underlying infection.

Note *abnormalities of respiratory rate and rhythm.* Extremely slow respiration usually indicates central nervous system respiratory depression due to disease, drugs, or carbon dioxide retention. Periodic or Cheyne-Stokes respiration occurs with serious cardiopulmonary or cerebral disorders. It is characterized by periodic regular increasing depth of respiration between shorter periods of apnea; the arterial PCO_2 is higher during breathing and lower during apnea; it is characteristically in the normal or hypocarbic range. In contrast Biot's respiration is irregularly irregular, almost spasmodic, with longer periods of apnea than of breathing; it is almost always associated with hypoventilation, increased PCO_2 (Figure 9-9). Extreme tachypnea, present during most acute pulmonary conditions, may be due to chronic pulmonary or cardiac disease or systemic disorders such as shock, severe pain, and acidosis; while it may represent undue excitement or nervousness, especially with sighing, an organic cause should be excluded. The patient's preferred position is important. Can he lie flat comfortably? Cardiacs prefer the sitting position, as do patients with

FIGURE 9-9. Altered Respiratory Rhythms. A. Periodic or Cheyne-Stokes Respiration. This is characteristically very regular and associated with hypocarbia. B. Biot's Respiration. This is characterized by its irregularity and is associated with hypercarbia.

COPD (chronic obstructive pulmonary disease), during acute attacks. Patients with pericardial disease often sit and lean forward. While cardiac patients often awake dyspneic several hours after reclining due to redistribution of blood volume, people with large amounts of sputum frequently awake shortly after reclining to clear retained secretions. Preference for one side or the other may explain the localization of pathologic processes; secretions or foreign bodies may be aspirated into a specific area because of position.

Palpate the chest wall and ribs for areas of tenderness, which commonly result from local trauma, tumor, or underlying pleural inflammation, as with pneumonia and pulmonary infarction. A costal cartilage may be exquisitely tender when inflamed (Tietze's syndrome) as is the site of a rib fracture; firm anteroposterior compression of the chest (sternum to vertebrae) will often cause exquisite tenderness at the site of a rib fracture. Fractures may occur with heavy coughing.

Notice the position of the trachea and the cardiac apex beat. Conditions such as large bulky tumors or pleural effusion and pneumothorax which increase the volume of one hemithorax *usually cause a shift of the entire mediastinum to the opposite side,* as evidenced by shift of the trachea and the apex beat. Conversely, conditions in which lung volume on one side is reduced, such as obstructing tumors and chronic inflammatory parenchymal and pleural disease, *draw the mediastinum toward the same side.* Estimate the distance from the posterior sternum to the trachea; it is increased in older patients with emphysema and with anterior mediastinal tumors. Notice if subcutaneous emphysema is present: the fingers palpate a peculiar crackling sensation of bubbles of air underneath the skin, implying a leak somewhere from the lung or other air-containing viscus. It is usually felt earliest in the supraclavicular area but may spread into the neck and face, over the trunk, and into the scrotum; sometimes it can be heard more easily with the stethoscope than felt. This finding may occur spontaneously with trauma; it is usually associated with mediastinal emphysema or a pneumothorax. The supraclavicular area should be palpated for lymph nodes such as do occur in sarcoidosis or with tumor metastases. Nodes are sometimes brought up to the fingertips with the Valsalva maneuver.

Notice if coarse vibrations associated with noisy respiration are palpable. This finding, termed *rhonchal fremitus,* implies the presence of exudate in the trachea or larger bronchial tubes which will also produce rhonchi on auscultation. Auscultation and palpation are both necessary to differentiate rhonchal fremitus from a *pleural friction rub.* The pleural rub is more grating in quality and is commonly associated with pain. Palpable rhonchi are predominantly inspiratory, whereas rubs involve both phases of the respiratory cycle. Rhonchi may clear with cough; rubs are unaffected. Rubs are usually more localized, are heard or felt unilaterally, sound close to the ear, and are accentuated by pressure of the hand or stethoscope, features which are all absent with rhonchi.

The evaluation of *tactile* or *vocal fremitus* is at times unreliable. Take special care to compare symmetric areas of the chest, as symmetric decreases or increases in fremitus are rarely significant. Localized diminution of fremitus occurs when the transmission of the voice sound from the trachea through the vibrating lung to the examining hand or finger is interfered with by any cause. It is therefore absent in patients with an obstructed bronchus, when air or fluid in the pleural space is interposed, or when the chest wall is considerably thickened or edematous. Generalized diminution in fremitus occurs with diffuse bronchial obstruction. Localized areas of increased fremitus occur with consolidation — denser lung with a patent bronchus — and occasionally when the lung is compressed above a pleural effusion. Increased fremitus over consolidated lungs is appreciated only if the increased density of the lung extends to the pleural surface.

Remember that the intensity of fremitus and the breath sounds are influenced in the same direction by the same physical factors. They should therefore correlate at all times: diminished fremitus with diminished breath sounds, increased fremitus with loud or bronchial breathing. Minor differences in correlation, if present, are often due to technical factors of a stronger stimulus to production of breath sound or fremitus, or easier appreciation by hand or ear. The auscultatory findings are usually more reliable.

The normal percussion note is resonant over all of the lungs, except at the right apex, where occasionally slight dullness is detected. Generalized hyper-resonance not due to full-held inspiration may be found when the lung contains more air than normally (hyperinflation); however, unless the hyperresonance is pronounced, the considerable variations between observers, and between one observer's impressions at different times, impairs the usefulness of this finding. Localized areas of hyperresonance are noted when pneumothorax is present, and occasionally over solitary bullae. Impaired resonance is difficult to evaluate but may be noted adjacent to areas of greater pathology and over lungs partially consolidated, as in diffuse bronchopneumonia. Dullness is elicited with pulmonary infiltration of almost any cause, regardless of the state of patency of the supplying bronchus, whenever the air content of the lung is partially or completely replaced by fluids or solids. When correlated with the character of the breath sounds on auscultation and fremitus during palpation, the status of the underlying parenchyma in a dull area can be reliably predicted. Dullness also is noted when the pleurae and pleural cavity are thickened or filled with anything except air. (Rarely, air under great tension will give a dull rather than a hyperresonant note.) Extreme dullness, or flatness, is noted when no air at all underlies the pleximeter, as occurs, for practical purposes, only with pleural effusion. Tympany is rare over the lungs themselves. It occasionally occurs over a large pneumothorax.

The margins of the lungs can be determined by percussion. When they are lower than normal or when the mediastinum is narrower than normal, hyperinflation such as occurs with emphysema is suggested; when higher than normal, fibrosis or increased abdominal contents are suggested. Measured diaphragmatic excursion, when symmetrically reduced, may indicate generalized bronchial obstruction or muscular weakness. Unilateral limitation of motion implies paralysis or splinting. Limited costal expansion indicates diffuse obstruction or fibrosis, muscle weakness, or ankylosing spinal disease.

Considerable auscultatory experience with the wide range of normal breath sounds is necessary before you can distinguish abnormal sounds with confidence. When breath sounds differ at symmetric sites of the thorax, it may be impossible to state which side is pathologic without the aid of a roentgenogram. Symmetric findings may be misleading; chronic bilateral apical tuberculosis may be present with symmetric bronchial or bronchovesicular breathing misinterpreted as normal because of the symmetry. Nevertheless, comparison of one side with the other and of the upper with the ipsilateral lower lobe is an extremely helpful maneuver during auscultation. Factitious sounds are confusing; every student should apply the stethoscope to the hairy chest and hear the varied sounds made by hair rubbing against the stethoscope. These can be minimized by wetting the skin. In addition, apply the stethoscope to the biceps muscle while it contracts, and note the striking similarity to rales. It is for this reason that the room and the stethoscope must be warm and the patient relaxed so that involuntary muscular contractions of the chest wall will not occur and masquerade as pulmonary abnormalities.

Abnormalities of the breath sounds have great significance. What is normal in one area may be pathologic elsewhere. Bronchial and bronchovesicular breath sounds are due to increased transmission of sound through partially solidified lung, of any cause, provided the bronchus is patent. They are therefore heard with consolidation, compression, and fibrosis. Breath sounds are diminished with local or diffuse bronchial obstruction and with pleural disease. Amphoric and cavernous breathing are terms, of little practical value, that refer to exaggerated forms of bronchial breathing heard over large cavities.

When bronchial breathing is heard over the upper anterior chest, the finding may be due to increased transmission through consolidated lung or merely to normal tracheal breathing due to shift in tracheal position. This distinction (which may have great importance in differential diagnosis, as the upper lobe bronchus would then be interpreted as obstructed or patent) is unfortunately often difficult to make even after correlation with the roentgenogram. Transmitted tracheal sounds rather than bronchial breath sounds are probably present if the sounds are not heard in the axilla, if they are quite harsh, without a pause between inspiratory and expiratory phases, or if an area of vesicular breathing seems to be interposed between the trachea and the abnormal breath sounds. The presence of rales, of course, strongly favors bronchial breathing.

Rales, which are abnormal additional or adventitious sounds, are always pathologic. They are subdivided into continuous coarse sounds termed *rhonchi* and interrupted crackling sounds termed *moist rales* or simply *rales,* the same as the parent term. The inescapable confusion thus caused is difficult to clarify [3].

Although rhonchi and some loud, interrupted crackling rales are sometimes confusingly called "dry," all of these sounds indicate the presence of fluid somewhere within the respiratory tract. The exudate may result from infection, inflammation, aspiration, edema, or retained secretions. Spasm of the bronchial musculature and edema of the bronchial wall may exaggerate rhonchal sounds caused by exudate within the lumen.

Rhonchi imply disease of the trachea or larger bronchi, whereas moist medium and fine rales imply bronchiolar and alveolar disease. Rhonchi are usually heard earlier in inspiration than are rales, as sound is produced when the column of inspired air meets the exudate at its anatomic location. Sonorous rhonchi are low-pitched, as are coarse rales which are bubbling in quality. Both frequently clear with cough, while fine and medium rales due to inflammatory conditions are usually increased by cough.

Fine rales are short, high-pitched sounds simulated by rubbing the hair between the fingers near the ear; medium rales are louder and lower pitched. Any of the fine moist rales may have a loud, clear consonating quality in which they sound close to the ear; these tend to occur only in inflammatory conditions. The fine rales of pulmonary congestion or impaired pulmonary ventilation may disappear following a few deep breaths or coughs, but congestive rales reappear shortly thereafter. Congestive rales will move with change in position while inflammatory rales will not. Fine moist rales are also heard in interstitial fibrosing conditions, as the alveoli are usually involved, too; the rales are usually generalized and often sound very close to the ear. Inflammatory rales may not be heard during quiet respiration but can be brought out by a special maneuver. Ask the patient to inhale, then to exhale almost completely, and then to give a short barking cough. The rales are then heard in the subsequent inspiration and are called *posttussive rales.* The maneuver must often be demonstrated as most individuals cough normally following inspiration. Similarly, some wheezes are elicited only by asking the patient to exhale forcibly until he empties his lungs of air.

Generalized rales and rhonchi imply conditions which affect the entire lung: bronchitis, asthma and emphysema with rhonchi, pulmonary congestion, diffuse infection, or fibrosis with moist rales. Localized moist rales occur with localized inflammatory conditions, such as tuberculosis, pneumonia, and bronchiectasis; a localized wheeze implies localized extrinsic or intrinsic bronchial obstruction which may occur with a tumor or foreign body. Be sure that the localized or unilateral wheeze is not merely a transient localized component of generalized wheezing. A rhonchus rarely may be palpable but inaudible.

The *pleural friction rub* is a characteristic low-pitched, coarse, grating, loud sound heard close to the ear usually during both phases of respiration. It disappears when the patient holds his breath. It is pathognomonic of inflammation of the pleura. Differentiation from rhonchi is occasionally necessary.

Rarely, an extremely loud, knocking, crunching sound unlike a rub is heard over the mediastinum synchronous with cardiac action. This is Hamman's sign, and it occurs with mediastinal emphysema and *left pneumothorax.* The diagnosis is supported if subcutaneous emphysema is palpated in the neck.

Abnormal vocal resonance occurs for the same reasons as its tactile counterpart, but is frequently more discriminating and reliable. The intensity of the voice sounds is decreased with pleural disease and when pulmonary ventilation is either diffusely or locally obstructed. Greater intensity and slightly increased clarity of the transmitted voice (*bronchophony*) is noted over areas of consolidation. Often a more sensitive way to define small early partial consolidation is to have the patient whisper while you listen with the stethoscope for distinctly heard articulated syllables (*whispering pectoriloquy*). *Egophony* refers to a change in the quality of the spoken voice, which now sounds bleating or nasal. In addition to the usual phrases, elicit it by asking the patient to say "ee," which will then sound like "ay." Egophony is rarely found except over an area of compressed lung above a pleural effusion.

In this section the abnormal physical findings of each type of examination have been presented separately with their possible causes. In practice, however, it is only the combination of abnormalities plus the roentgenogram that leads to understanding of the state of the underlying lung, and the importance of one or another single finding may outweigh or be strengthened by others in correlative diagnosis. The following discussion describes together the abnormalities that occur in various major pathologic states. Specific diseases are not emphasized.

Diffuse Acute Bronchial Obstruction . The patient is in extreme respiratory distress. With laryngeal obstruction crowing sounds are heard; with bronchial obstruction wheezing is appreciated. Respiration is rapid. The intercostal spaces retract on inspiration. Accessory respiratory muscles may be used. Fremitus is decreased, but rhonchal fremitus may be present. The percussion note is normal or hyperresonant throughout. The breath sounds are diminished in intensity, although the chest may be extremely noisy because of rhonchi and coarse rales. Acute exacerbations in patients with COPD (asthma, chronic bronchitis, emphysema), bronchiolitis, and foreign bodies produce this condition.

Diffuse Chronic Bronchial Obstruction. The patient may be in no respiratory distress, although the respiratory rate may be increased. The chest may be barrel-shaped; the interspaces are wide and retract with inspiration. Fremitus is decreased, the percussion note hyperresonant, and the breath sounds diffusely diminished or almost inaudible. Rhonchi or rales may not be heard with quiet breathing except for a few scattered rales at the lung bases. Rhonchi may be elicited by deep breathing or the forced vital capacity maneuver, but the breath sounds do not increase. Prolongation of the FET may be the only evidence of airways obstruction. The diaphragms do not move well. While these findings may be elicited from time to time in any patient with COPD because of air trapping and hyperinflation, persistently decreased breath sounds in a large chest with appropriate roentgenographic findings correlate well with the presence of emphysema (see Figure 9-10).

Local Bronchial Obstruction. With obstruction of a large bronchus a lag may be present, the thorax contracted, and the trachea and mediastinal contents shifted toward the side of the obstruction. The shift may increase further with inspiration and decrease with expiration. Fremitus is decreased, percussion note dull, and breath sounds diminished or absent. No rales are heard. *Bronchogenic carcinoma* and, occasionally, other *tumors, foreign bodies, or inflammatory stenosing lesions* of the bronchi are responsible.

Local Partial Bronchial Obstruction. It is remarkable how rarely partial bronchial obstruction manifests itself clinically. In spite of the presence of air within the parenchyma of an obstructed lobe on the roentgenograms, the physical findings are often those of complete bronchial obstruction. At times, however, rales may be present and the breath sounds still heard although diminished. A localized high-pitched inspiratory or expiratory wheeze may be present; the wheeze may persist after the patient abruptly stops a forced expiration or inspiration, i.e., when breath sounds have ceased elsewhere in the lung.

Compression. An otherwise normal compressed lung above a pleural effusion may be dull or tympanic to percussion. Fremitus is increased, and breath sounds

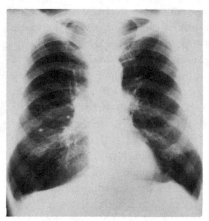
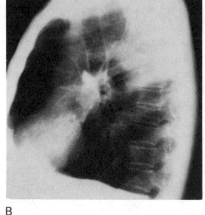

A   B

FIGURE 9-10. Roentgenogram of a man with extensive emphysema. A. Posteroanterior view. The diaphragms are flattened and the vascular markings are not uniformly distributed. B. Lateral view. The diaphragms are flattened as determined by an angle of greater than 90° between the sternum and diaphragm; and the anterior clear space — the distance between the sternum and the aorta — is greater than 3 cm. (on original roentgenogram).

are bronchovesicular or bronchial. Egophony is elicited. The findings of pleural effusion often merge gradually with those of a normal lung without the intervening abnormality of compressed lung.

Consolidation. The patient is usually ill, depending upon the underlying condition. Ventilation is usually deep and rapid. Fremitus is increased; the percussion note is dull; the breath sounds are loud and bronchovesicular or bronchial. Fine and medium moist rales are heard which are consonating in quality. Bronchophony and pectoriloquy may be present. Consolidation occurs mainly in *pulmonary infections* and with large areas of *pulmonary infarction.*

Mediastinal Mass. The patient may be either asymptomatic or quite ill; he may be unable to lie on his back. The veins may be distended if the great vessels within the mediastinum are compressed. The distance from the sternum back toward the trachea is increased if the mass is in the anterior mediastinum. The lungs may be entirely normal; however, percussion may reveal increased width of dullness anteriorly extending laterally across the midline.

Pleural Effusion. The patient may or may not appear ill, depending upon the cause of the effusion and the rapidity with which it has developed. The affected side of the hemithorax may bulge; ventilation may lag and be diminished. The trachea and mediastinum may be shifted to the opposite side. The percussion note is flat, fremitus is markedly decreased or absent, and breath sounds are absent. No rales are heard. With a moderate effusion breath sounds may be heard underlying the effusion; physical findings of compression may be noted above it.

Pleural Inflammation without Effusion. Primary forms of this condition are rare, and physical findings of associated pathology are noted. Slight dullness and a pleural friction rub may be the only abnormalities elicited if the responsible adjacent inflammation is minimal.

Pneumothorax. Whether or not the patient is symptomatic depends on the extent, acuteness, and rapidity of progression of the pneumothorax and on the adequacy of the remainder of the lung. Fremitus is absent, but the percussion note is hyperresonant. Breath sounds are also usually absent. When hydro-pneumothorax is present, a splash may sometimes be elicited by abruptly shaking the patient from side to side (a *succussion splash*); the dullness or flatness of the effusion shifts with change in position to a greater extent, and more rapidly, than in pleural effusion alone. Table 9-1 summarizes physical findings due to localized disease.

Pulmonary Congestion. If the congestion is acute and severe, the patient is in extreme distress; if partial or chronic, minimal dyspnea may be present. The respiratory rate and use of accessory musculature reflect this difference.

Table 9-1. Correlation of Physical Findings to Localized Disease (The presence of findings will vary with the extent of involvement)

Disease	Tracheal Deviation*	Percussion	Vibration†	Rales
Pneumothorax	Away or N	Hyperresonant	↓ or 0	0
Obstructed bronchus				
With atelectasis	Toward or N	dull	↓ or 0	0
With consolidation	N	dull	↓ or 0	0
Open bronchus				
With atelectasis	Toward or N	dull	↑	0
With consolidation	N	dull	↑	+
Pleural effusion	Away or N	flat	↓ or 0	0

*Toward or away from side of lesion.
†Vibration includes both fremitus and breath sounds which vary in the same way.
N = Normal; + = Present; 0 = Absent; ↑ = Increased; ↓ = Decreased.

Fremitus may be normal or slightly decreased. Associated signs of pleural effusion may be present at the base. Breath sounds are of fair quality, and resonance is normal or only slightly impaired. Fine and medium rales that do not sound close to the ear and that shift after prolonged change of position are characteristic. With marked pulmonary edema, bubbling coarse rales are also heard. In patients with emphysema, breath sounds may increase in intensity toward normal.

Compare the abnormal findings on physical and roentgen examination and correlate them. Recheck the physical examination carefully if any discrepancy between physical and roentgen findings is noted. If the roentgenogram alone was abnormal, additional physical abnormalities may be elicited; for example, the posttussive inspiration may be rewarding in the demonstration of rales. If only abnormal physical findings were present, and they persist on repeat examination, recheck the roentgenogram more carefully.

A change in the patient's position is a helpful special maneuver. It is at times difficult to distinguish the rales of inflammatory and fibrosing conditions from those of chronic pulmonary congestion. Note carefully the distribution of the rales. If they are posterior and basal, place the patient flat on his stomach and instruct him to remain stationary for at least one-half hour. Recheck the examination. Congestive rales will often disappear posteriorly and appear anteriorly, whereas inflammatory rales will not be affected. Certain wheezes due to extrabronchial compression or intrabronchial polypoid lesions appear only with change in position. Dullness and absent breath sounds at one lung base may mean either fluid or obstruction; if the patient lies on the same side for a half-hour, dullness will now extend high into the axilla if free fluid is present; obtain a roentgenogram in this lateral decubitus position to confirm the finding. The physical findings of pleural effusion change slowly with positioning; rapid change suggests combined pneumothorax and effusion. Place the patient on the contralateral side. Dullness should now be rapidly replaced with hyperresonance or tympany. A shifting linear fluid level on the roentgenogram confirms these findings. Shaking such a patient vigorously may demonstrate a succussion splash.

If a mediastinal mass is noted in the film, check for a tracheal tug. Place your fingers on the thyroid cartilage and attempt to raise it cephalad. A pulsatile resistance suggests *aortic aneurysm.*

If a nodular shadow connected to the hilum with broad vascular bands is seen and an *arteriovenous malformation* suspected, listen over the proper anatomic area for a bruit with the patient holding his breath. Note if it increases with inspiration; it may disappear completely with a held forced expiration (Valsalva maneuver).

If the differential diagnosis between pneumothorax and a large bulla is unclear from the roentgen and physical examination, perform the coin test. Have an aide place one coin flat against the anterior chest wall and tap it with another while you listen with the stethoscope posteriorly, or vice versa. Normally, and with bulla, you will hear only a dull thudding sound, but with pneumothorax the note will have a distinct metallic ringing quality.

SPECIAL TECHNIQUES

Chest Fluoroscopy. Chest fluoroscopy is far less important than roentgeno-graphy. It should never be used by untrained observers. Although image-intensification fluoroscopy eliminates the need for dark adaptation, both patient and physician are subject to an increased degree of radiation exposure (about 5 to 10 times that of a set of chest roentgenograms for 3 to 5 minutes of fluoroscopy using good technique). It should never be used to "screen" the chest for the identification of pulmonary lesions. It does have certain major values, however, and these should not be neglected.

It is the most accurate way to demonstrate complete or partial paralysis of one diaphragmatic leaflet. Have the patient inspire rapidly through the nose a few times (the "sniff" test). The paralyzed leaflet will not move. With normal or deep breathing it may be entrained by movement of the other lung and diaphragm. Localized areas of "trapping" of air and differences in ventilation and the direction of mediastinal shift with ventilatory movement may be seen. Lesions that appear to be solid on the roentgenogram may enlarge during Müller's maneuver and diminish in size during the Valsalva maneuver if they are vascular or cystic. Note abnormal vascular pulsations. Anatomic localization of lesions is occasionally facilitated by rotating the patient into various oblique projections. Perhaps the most important use of image intensification is the visualization of a localized lesion while obtaining a biopsy or cytologic specimen.

Sputum Examination. Gross examination of the sputum is an extremely helpful, often neglected aid to physical diagnosis. Note the volume, color, odor, turbidity, and consistency. Some "sputum" is colorless, clear, and watery. Mucoid sputum is gray-white, translucent, and slimy. Globs of thicker white mucus or pus may be intermixed, mucopurulent sputum. Purulent sputum is thick and opaque; it may be green, yellow, or brown; mixed varieties are common. Sputum may be so viscous that it sticks to the inverted specimen container.

For adequate gross sputum examination the patient must save his secretions, since very poor estimates of both the quantity and the quality are common. Expectoration should be directly into clear plastic or glass-capped containers. The patient's course should be followed by daily measurements of volume or weight as well as the changes in the sputum character. A gram stain of sputum should be part of the initial evaluation.

Sputum may vary in amount from a few teaspoons daily, raised predominantly in the morning, to a pint or more. Often patients are unaware of even copious sputum production if it has increased very gradually; they should be asked about it in several ways: smoker's cough, clearing throat in morning, and so forth. Morning expectoration usually implies accumulation of secretions during the night and is common to many chronic bronchopulmonary suppurative disorders. Copious sputum may be mucoid, as occurs commonly in bronchitis, or may be largely purulent with superimposed infection. A large volume of purulent sputum expectorated into a glass jar ordinarily settles into four distinct layers. At the bottom is amorphous debris; a translucent layer of thin pus, mucus, and saliva is next; above this pus globules float freely and hang suspended from the surface which is covered with a layer of froth. Sputum of this type usually results from pulmonary destruction and suppuration, and is common in lung abscess and bronchiectasis. It is an interesting example of the influence of prior knowledge upon perception that this sputum is commonly seen as three-layered, in spite of the four clear-cut zones.

Note specifically whether blood is present. Copious bleeding results in bright-red, frothy sputum; blood may present as small streaks on the surface or as tiny globules of darker blood intermixed with the sputum. The brighter the blood, the fresher the bleeding. Note whether the specimen is pure blood or blood mixed with pus or mucus. Note the presence of black anthracotic particles, concretions (in answer to direct questioning the patient may describe the expectoration of sand or small stones — broncholiths), casts, and foreign material.

Sniff the opened sputum jar. Most sputum is neither foul nor offensive to the patient or to the examiner. However, with certain necrotizing infections an extremely fetid odor is present, often at a considerable distance from the patient or his sputum jar.

Bacteriologic, mycobacteriologic, and mycologic examination, by smear and culture, is obviously basic to specific diagnosis of infectious diseases. In recent years cytologic examination of the sputum for malignant cells has developed into a reliable and noninvasive method of screening patients with symptoms or abnormal roentgenograms for cancer of the lung. Two cytologic specimens with malignant cells are equivalent to a tissue diagnosis.

Bronchoscopy and Bronchography. Bronchoscopy with the rigid bronchoscope is invaluable for the inspection and biopsy of all varieties of lesions of the proximal tracheobronchial tree. Employ it freely when localized bronchial obstruction is suggested by the physical and roentgenographic examination. The recent addition of the flexible fiberoptic bronchoscope permits visualization to the third order (subsegmental) bronchi; lesions more distal to this can be brushed for cytologic examination or biopsed with the aid of image-intensification fluoroscopy.

Bronchography permits the visualization of the distal bronchial tree by instillation of a radio-opaque dye. It may be helpful in demonstrating bronchial obstruction, patency, or dilatation, local or diffuse. It is usually performed following other more basic procedures.

Pulmonary Arteriography. Arteriography is useful in defining the blood flow to different areas of the lung; it is an accurate and sensitive means of detecting pulmonary emboli.

Ventilation and Perfusion Lung Scans. These lung scans provide a noninvasive, graphic, geographic, semiquantitative estimate of the interrelationship between two important lung functions; they are an accurate and sensitive means of detecting pulmonary emboli; they can be most helpful in anticipating the loss of function that will result from resectional lung surgery.

Percutaneous Needle Lung Biopsy. In many diffuse and localized lesions a definite diagnosis can be made only with tissue or, in the case of tumor, a cytologic specimen. This procedure should follow adequate noninvasive diagnostic efforts.

Pulmonary Function Studies. A variety of tests, simple to sophisticated, are available to measure the functions of the lung: ventilation, diffusion, perfusion, and the relationship of ventilation to perfusion. They provide useful quantitative information of functional impairment and response to therapy. Since no single study measures all of the functions, a group of studies is generally selected.

The most readily available tool is the *spirometer.* It is used to measure volume change related to time such as the FVC and FEV_1 and average flow rates (Figure 9-2).

Utilizing different instrumentation the rate of flow related to volume can be measured when the patient performs an FVC maneuver; this generates a maximal expiratory flow volume curve, which is a more sensitive measure of expiratory airways obstruction. Instrumentation is available to measure lung volumes, diffusing capacity, airways resistance, and lung compliance.

Arterial Blood Gas Studies. Arterial blood gas measurements are essential in evaluating the effectiveness of the cardiopulmonary apparatus. Four values are generally obtained from a specimen of arterial blood: (1) the partial pressure of oxygen (Pa_{O_2}) in mm. Hg; (2) the partial pressure of carbon dioxide (P_{CO_2}) in mm. Hg; (3) the pH in pH units or nanomoles; and (4) the percent saturation of hemoglobin. As mentioned earlier there are no reliable early signs of respiratory failure — hypoxemia or hypercarbia (alveolar hypoventilation), or the acid-base status of the individual. The presence of respiratory distress, dyspnea, tachypnea, or bradypnea is a good indication for obtaining arterial blood gases; waiting for cyanosis or major cerebral dysfunction to herald respiratory failure exposes the patient to unwarranted risk of damage to vital organs and even death.

REFERENCES

1. Lal, S., Ferguson, A. D., and Campbell, E. J. M. Forced expiratory time: A simple test for airways obstruction. *Br. Med. J.* 1:814, 1964.
2. Rabin, C. B. New or neglected physical signs in diagnosis of chest diseases. *J.A.M.A.* 194:546, 1965.
3. Robertson, A. J., and Coope, R. Rales, rhonchi, and Laennec. *Lancet* 2:417, 1957.
4. Rosenblatt, G., and Stein, M. Clinical value of the forced expiratory time measured during auscultation. *N. Engl. J. Med.* 267:432, 1962.
5. Waring, J. J. Physical examination, helps and hindrances. *Ann. Intern. Med.* 28:15, 1948.

CIRCULATORY SYSTEM 10

Richard D. Judge
Ron J. Vanden Belt

GLOSSARY

Aneurysm Saccular dilatation of an artery or cardiac chamber.

Angina pectoris Literally, strangulation of the chest; a paroxysmal, constricting substernal pain of brief duration which frequently accompanies myocardial ischemia.

Apex (cardiac) Pointed, most lateral portion of the heart, usually located near the left fifth intercostal space.

Arrhythmia Any variation from the normal, regular rhythm of the heart.

Atrial fibrillation Grossly irregular ventricular rhythm associated with rapid, uncoordinated movements of the atria.

Atrial flutter Cardiac arrhythmia associated with rapid (about 300 per minute), regular, uniform atrial contractions and a ventricular rate and rhythm that vary with the grade of A-V block.

Atrial tachycardia An arrhythmia arising in the atria, usually characterized by rapid, extremely regular beating of the entire heart.

A-V block Slowing or interruption of impulse conduction from atria to ventricles.

Base (cardiac) Region of the aortic and pulmonic outflow tracts; the second and third intercostal spaces parasternally.

Bradycardia Slow heart beat (less than 60 per minute).

Bruit Extracardiac blowing sound heard at times over peripheral vessels, usually arterial.

Cor pulmonale Heart disease secondary to pulmonary disease.

Coronary heart disease Heart disease resulting from narrowing or occlusion of one or more coronary arteries.

Cyanosis Bluish discoloration of the skin produced by inadequate oxygenation of the blood.

Diastole Dilatation; period of "relaxation" during which ventricles fill with blood; technically ends with the onset of the first heart sound.

Dyspnea Difficult or labored breathing.

Edema Presence of abnormal amounts of interstitial fluid in soft tissue or lungs.

Embolism Sudden occlusion of a vessel by clot or other obstruction carried to its place by the current of blood.

Friction Characteristic grating adventitious sound, usually pericardial or pleural, simulating noise made by friction between two rough surfaces.

Gallop rhythm Characteristic cadence produced by three heart sounds (first and second heart sound with one or more extra sounds) in conjuction with tachycardia.

Hypertension Persistent elevation of arterial blood pressure.

Infarction Ischemic necrosis of tissue resulting from interference with its circulation.

Murmur Adventitious sound resulting from turbulent blood flow within the heart or great vessels.

Orthopnea Inability to breathe comfortably when supine.

Palpitation Awareness of the heart beat.

Paroxysmal Sudden, unexpected.

Presystolic Immediately preceding the first sound; occurring in the latter one-third of diastole.

Protodiastolic Immediately following the second sound; occurring in the initial one-third of diastole.

Pulse Expansile wave felt over an artery. It is propagated at a speed approximately ten times that of the actual flow of blood in the system.

Raynaud's phenomenon Paroxysmal pallor or cyanosis of a distal extremity, induced by chilling or emotion.

Shock Acute circulatory collapse, with pallor, hypotension, and coldness of the skin (also, a palpable heart sound).

Syncope Temporary unconsciousness due to cerebral ischemia.

Systole Contraction; period of contraction during which the atria or ventricles eject blood; *ventricular systole* includes the first and second sounds and the period between them; when the term *systole* or *diastole* is used alone, it is assumed to refer to the ventricles.

Tachycardia Rapid regular heart beat (over 100 per minute in adults).

Thrill Palpable vibrations (palpable murmur).

Varicose Dilated.

Ventricular tachycardia Arrhythmia originating in the ventricles, characterized by rapid, relatively regular heart beat.

TECHNIQUE OF EXAMINATION

Clinical examination of the circulatory system has been very significantly modified by information made available through the advent of cardiac catheterization. A sound, modern interpretation of both normal and pathologic cardiovascular phenomena requires a clear understanding of the functional anatomy and physiology which underlie their generation. The unique aspect of clinical cardiovascular observation has to do with its transient nature. Phenomena pass by relatively rapidly (though recurrently) on a time axis which is ever moving. It is understandable, then, that cardiovascular examination presents a challenge to our perceptive ability unrivaled by other systems. The examiner must frequently use two senses simultaneously: one for identification of a movement; the other for timing purposes. This correlation of vision, touch, and sound requires infinite patience and practice. It is, in a sense, the essence of bedside observation.

Complete evaluation of the circulatory system extends beyond the simple examination of the heart itself. It must include a careful analysis of the peripheral arterial and venous circulations. Occlusion of a coronary artery may produce alarming signs and symptoms easily recognized by the physician; occlusion of a deep femoral vein, on the other hand, frequently results in more subtle clinical findings which may go unnoticed but are nevertheless equally threatening to the life of the patient. In this discussion the circulation will be considered comprehensively.

Nowhere in medicine is the "diagnostic negative" of greater significance than in the assessment of the circulatory system. Subtle, transient cardiovascular phenomena too often escape the casual, open-minded observer. Your strategy must therefore be directed toward the conscious exclusion of certain possibilities as you progress through a systematic evaluation. Is the arterial pulse increased; is there an A wave; is there a presystolic extra sound? A "no" answer to this type of self-inquiry is justified only if you have actively ruled out the possibility. To begin with, then, you must guard against undue preoccupation with listening to the heart and particularly with listening only for murmurs. This common trap inevitably awaits the inexperienced observer. Rushing for the stethoscope prematurely will certainly result in an inadequate examination. A detailed history coupled with careful preliminary observations (particularly inspection and palpation) often permits the examiner to predict the auscultatory findings with remarkable accuracy. More important, when he places his stethoscope on the chest, he knows what he is looking for and why.

Begin with the patient sitting. Make your initial observations, including examination of the fundi and lungs; then have the patient recline with arms at sides. Station yourself on the right and take the blood pressure. Then examine the extremities, neck, and precordium, in that order. The male patient will be stripped to the waist. A towel is very satisfactory for draping the female chest; it can be manipulated to allow all necessary observations without causing undue embarrassment. Have adequate light and a quiet room, and eliminate all distractions as far as possible.

Blood Pressure

This is a logical first step and will be considered separately. Routine measurement should be made with the patient *recumbent*. If you find an abnormality, compare the determination in both arms with the patient supine, sitting, and standing. Any reliable sphygmomanometer may be employed.

Fit the cuff evenly around the upper arm, with the lower edge one inch above the elbow. The *palpatory method* is first employed to determine the systolic pressure. Rapidly inflate the cuff until the brachial pulse disappears and then deflate it slowly. The level at which the brachial pulse first reappears is the systolic pressure. A gross estimate of the diastolic pressure is sometimes possible by palpation. With further deflation of the cuff, the brachial pulse assumes a bounding quality and then abruptly becomes normal. This point of change, when evident, roughly approximates the diastolic pressure. The *auscultatory method* is then employed to estimate both the systolic and diastolic pressures. Place the stethoscope lightly over the brachial artery and inflate to 20 or 30 mm. above the palpatory systolic pressure. The highest level at which sounds are heard (phase I of Korotkoff) is the systolic pressure. With further lowering by decrements of 2 or 3 mm., the sounds are replaced by a bruit (phase II) and then by loud, sharp sounds (phase III). Finally, the sounds suddenly become damped (phase IV), and a few millimeters below this they disappear. The point of complete disappearance of sound is considered the best index of diastolic pressure. Under hemodynamic conditions in which sound does not cease, the point of muffling should be taken as the diastolic pressure, with the point of total disappearance recorded as well, e.g., 140/60—0.

When it is advisable to measure the blood pressure in the leg, the palpatory method should be employed with the patient prone. Apply the cuff to the calf and estimate the systolic level by palpating the posterior tibial artery. (Since it is never critical to measure the diastolic pressure in the leg, this practice is sufficient and accurate.) Applying the standard cuff to the thigh and listening for sounds over the popliteal artery in the average adult is not only mechanically difficult, but also may result in falsely elevated readings.

Inspection and Palpation

Preliminary inspection begins immediately during history taking. Body build, character of respiration, general color, obvious pulsations, and signs of emotional tension may all be casually observed at this time. Inspection and palpation are best conducted together. Organize your observations according to three major areas: (1) the extremities, (2) the neck, and (3) the precordium. The logic is simple and clear. Begin your observations peripherally and gradually move toward the heart itself.

Extremities. First examine the arms and legs, paying particular attention to the hands and feet. Make your observations in the following order: (a) fingernails and toenails, (b) skin color, (c) hair distribution, (d) venous pattern, (e) presence of swelling or atrophy, and (f) obvious pulsations. Develop the habit of comparing sides. Estimate skin temperature by light, quick palpation with the tips of the fingers or with the back of the middle phalanges of the clenched fist. Compare identical sites on both sides and the gradation of temperature along the limb. Figure 10-1 illustrates several maneuvers for examining the peripheral arterial pulses. Use the three middle fingers (not simply a single finger) and vary the pressure. Occlude the vessel completely at first and release gradually. The brachial, radial, femoral, posterior tibial, and dorsalis pedis should be examined routinely, the ulnar and popliteal pulses under special conditions. Note each of the following in order: (1) pulse rate, (2) rhythm, (3) amplitude, (4) any special quality, and (5) elasticity of the vessel wall. Amplitude may be classified as increased, normal, diminished, or absent. Examination of the peripheral arterial system should include auscultation over the femoral arteries, abdominal aorta, and carotid and subclavian arteries for bruits which might indicate occlusive disease.

Neck. Make three separate observations in the neck: (1) carotid pulse, (2) jugular venous pulse (wave form), and (3) jugular venous pressure. With the patient supine, palpate each carotid medial to the sternocleidomastoid muscle just below the angle of the jaw. The three-finger method may be used, although some examiners prefer to use the thumb (Figures 10-2A and B). Next observe the base of the neck for venous pulsations. Their distinguishing characteristics will be considered later in greater detail; however, these four simple maneuvers

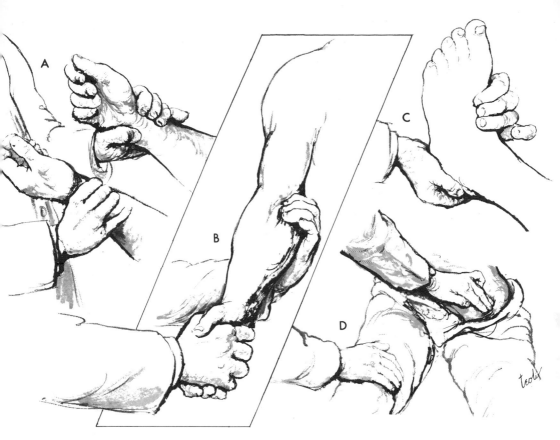

FIGURE 10-1. Method of palpating the arterial pulses. A. Radial. B. Brachial. C. Dorsalis pedis. D. Femoral.

should be followed in order: (1) note the general character of the pulse, which is normally diffuse and undulant; (2) observe any variation produced by respiration; (3) apply light pressure to the root of the neck, observing the effect on the pulsations; and (4) gradually elevate the head of the bed or the examining table in increments to the 45-degree level and note the overall effect. Last, the jugular venous pressure (JVP) is grossly estimated by thinking of the neck veins as manometer tubes directly attached to the heart (right atrium). The internal jugular vein is more reliable than the external for this purpose. Both will collapse, however, at the point where the intraluminal pressure falls below the level of atmospheric pressure. With practice, this point can be identified and its level above an arbitrary reference point such as the sternal angle (angle of Louis) can be estimated as the head of the bed is adjusted gradually upward. The sternal angle has been universally chosen for this purpose since its distance above the midpoint of the right atrium remains relatively constant (about 5 cm.) in all positions (see Figure 10-11). More accurate estimate of the right atrial pressure can be achieved by use of a central venous pressure (CVP) catheter (see page 198).

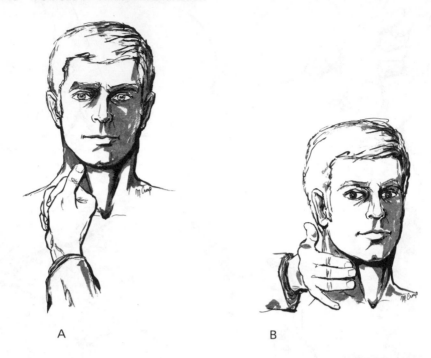

A   B

FIGURE 10-2. Carotid palpation. A. Some examiners prefer to stand in front of the patient and use the thumb on the carotid artery for analysis of the arterial pulse contour. This also allows one to examine the neck veins easily without changing position while using the carotid pulse for timing purposes. B. Others prefer to use three fingers from behind or at the side of the patient.

Precordium.   Move toward the foot of the bed and observe the precordium from the level of the anterior chest. The character and location of any visible cardiac impulses should be noted. Minor precordial movements can be amplified by observing during expiratory apnea. The commands, "Now take in a breath, let it all out, hold it," will provide the optimal setting for noticing these pulsations. At times it may be helpful to place a tongue blade or stiff card over the precordium, observing the movement of the free edge. This simply acts as a mechanical amplifier. Now palpate the precordium carefully with the hand lightly applied, using primarily the middle and ring fingers. Four major areas are explored: apex, left sternal border, pulmonic area, and aortic area (Figure 10-3). The apex impulse is always more forceful with the patient on his left side, which usually displaces it 2 to 3 cm. to the left and brings it closer to the chest wall. Figure 10-3 also shows a simple method of amplifying this movement by holding the tip of a tongue blade over the point of thrust. It is impossible to overemphasize the importance of careful palpation. Its value lies not only in the estimation of heart size, but also in the identification of vibrations originating in the heart which are of low frequency and therefore inaudible to auscultation.

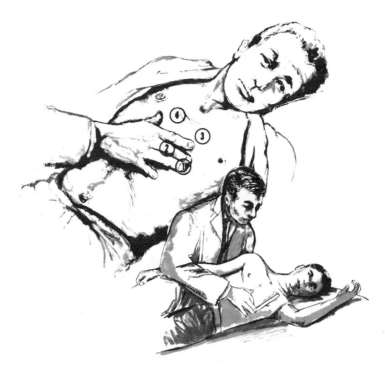

FIGURE 10-3. Palpation of the precordium. Upper: (1) Apex, (2) tricuspid area, (3) pulmonic area, (4) aortic area. Lower: Method of amplifying apical impulse using tongue blade.

Percussion

The methods of direct and indirect percussion described in Chapter 9 are occasionally applicable to the estimation of heart size. However, as more detailed knowledge of normal and abnormal precordial movements has been accumulated, palpation has largely replaced percussion in cardiac examination. This, coupled with the superiority of the roentgenogram for evaluation of overall heart size and chamber enlargement, has relegated percussion of the heart largely to a place of historical interest.

Auscultation

Cardiac auscultation must be learned at the bedside. Diagrams, tape recordings, and phonocardiograms are of great help to the learner but are no substitute for personal experience. Proficiency requires years of studied practice coupled with a clear understanding of the origin of normal and abnormal cardiac sounds. Certain frustrations are inevitable at first, but do not be disheartened.

Stethoscope. There is a frequently quoted saying that what you put in your ears is of less importance than what lies between them. While this is true to a great extent, a well-designed, efficient stethoscope is indispensable. Several crucial factors must be considered in its selection; therefore, it is wise to try a variety of instruments before selecting your own. It will be a very close friend for a long time, so choose it as you would a mate!

Earpieces must fit snugly. An earpiece which is too small or too tight may partially or completely occlude itself against the anterior wall of your external auditory canal. For this reason, it is valuable to try a variety before making your final choice and, if necessary, tailor the earpieces so that they feel comfortable and parallel to the long axis of your external auditory canal (Figure 10-4A).

Tubing of less than 1/8-inch inside diameter will attenuate high frequencies. Thick-walled tubing, particularly plastic, will reject outside noise better than a thin, flexible rubber variety. Double tubing extending all the way to the end

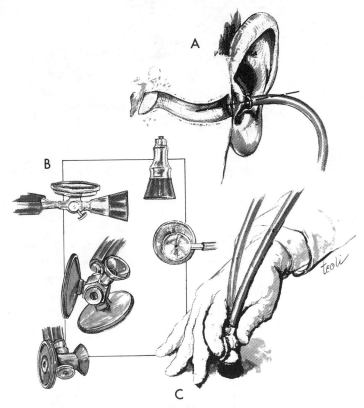

FIGURE 10-4. The stethoscope. A. Correct fit in external auditory meatus. B. Common types of stethoscope endpieces. C. Correct method for holding stethoscope endpiece, permitting variable pressure against chest wall.

piece is slightly superior to the Y configuration. Tubing length should not exceed 12 to 15 inches so that the overall distance from ear to chest piece is no greater than 21 inches. All of these are critical considerations in assessing this portion of a stethoscope.

Endpieces are of two standard types, the bell and the diaphragm (Bowles). They are available in many sizes; the 1-inch bell and the 1-1/2-inch diaphragm are most generally selected for examining adults (Figure 10-4B). The rigid diaphragm has a natural frequency of around 300 cps. It therefore acts as a band-pass filter, eliminating low-pitched sounds. High-pitched sounds, such as the second heart sound, and high-pitched murmurs are best heard with the diaphragm. The vaulted "trumpet" bell has been shown to have a slight advantage over the more shallow varieties. With the bell, the skin becomes the diaphragm and the natural frequency varies depending on the amount of pressure exerted, much as the timpani player varies the pitch of his instrument. It probably ranges from 40 cps with light pressure to 150 to 200 cps with firm pressure. When you try to detect low-pitched sounds and murmurs, therefore, *the bell should be applied as lightly as possible* (Figure 10-4C). A rubber nipple cut to size and fitted to the end of the bell may promote a better seal with lighter pressure.

Cardiac Sound. Sound intensity is objective; loudness is subjective. Loudness therefore depends on both the intensity of the sound at its origin and the sensitivity of the ear. Needless to say, we vary greatly in our auditory acuity, particularly at the upper and lower limits of audibility. Although the human ear may at times perceive frequencies extending from 20 to 16,000 cps and higher under special conditions, its maximal sensitivity occurs at 1000 to 2000 cps. Figure 10-5 compares the average hearing threshold with the average intensity of cardiovascular sounds. Notice that only vibrations ranging from 25 to 600 cps achieve intensity levels high enough to become audible by the average listener, although the heart produces vibrations ranging from 1 to 1000 cps.

In simple terms, this means that: (1) Vibrations originating in the heart with frequencies below 25 cps (and there are many) are entirely inaudible. This reinforces the importance of palpation since with experience much information from the inaudible range can be ascertained by this method. (2) A sound of 500 cps will be louder to the ear than one of 60 cps, even when they each have the same intensity at the stethoscope earpiece. This means that great care must be taken to train the ear to perceive low-pitched sounds, as they are easily overlooked.

The ear adjusts itself to the intensity of sound. A loud sound causes it to protect itself by cutting down on its receptive ability. This produces the phenomenon called "masking," for if a faint sound follows a loud one, it may actually be entirely inaudible. High levels of extraneous room noise may mask faint sounds and murmurs for the same reason.

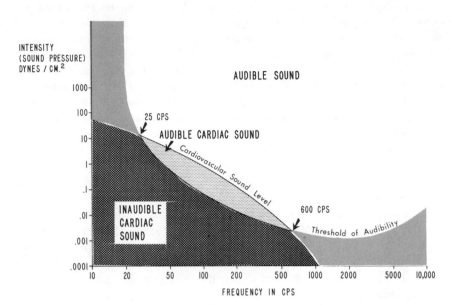

FIGURE 10-5. Comparison of auditory threshold with average intensity of cardiac sound at various frequencies. Note that only sounds of 25 to 600 cps are in the audible range.

Technique of Auscultation. After eliminating extraneous noise as far as possible, begin listening, with the patient supine. The four reference points used for localization of sounds on the surface of the chest are shown in Figure 10-6. Because the pulmonic and tricuspid valves are located near the chest wall, their sounds are transmitted to auscultatory areas close by. The aortic and mitral valves, however, are situated deep in the chest, and their sounds are transmitted in the direction of blood flow to points closer to the chest wall. The mitral sounds are referred to the apex; the aortic sounds follow the ascending aorta as it curves forward and are well heard in the second right interspace, where the aorta most closely approximates the anterior chest wall. However, sounds generated at the mitral and aortic valves are usually heard over much of the precordium.

Begin by listening at the apex, but do not simply skip from one major valve area to another. This is a common mistake, which results in missing much important information. Many intermediate points and satellite areas must be scrutinized as the stethoscope is moved slowly and systematically from the apex to the tricuspid area, up along the left sternal border to the pulmonic area, and thence to the aortic area. Maneuvers that bring the heart closer to the chest wall will increase the loudness of certain sounds. Two standard accessory positions should *always* be used: the left lateral position, which usually makes louder sounds at the apex; and the sitting position, which may bring out otherwise inaudible murmurs at the base and along the left sternal border. In emphy-

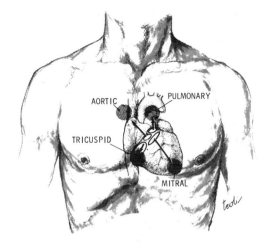

FIGURE 10-6. Reference points for localizing heart sounds. Diagram shows localization of the valves and the points to which their sounds are usually referred.

sematous patients, heart sounds usually faint over the precordium may be well heard over the epigastrium. Always notice the effects of position changes and of respiration on the quality of the sounds. Exercising the patient may be of great help, but this is not necessarily routine. Listening to the heart with the patient squatting may also be a useful procedure in special instances.

The key to successful auscultation lies in *listening to one thing at a time*. It is frequently worthwhile to close your eyes and take a relaxed, comfortable position while focusing all your attention on listening. The following routine will eliminate many errors of omission:

1. Observe and time the rate and rhythm. Is the rate unusually fast or slow? Are there any irregularities of rhythm?
2. Identify the first and second sounds and listen to them separately. Is the first sound normal, accentuated, diminished? Next, listen solely to the second sound and establish its characteristics. Compare the first and second sounds in the various valve areas.
3. Now that the sounds are familiar, focus first on systole and then on diastole. Listen specifically for *extra sounds* only (not murmurs).
4. Listen now for the presence of murmurs — first in systole, then in diastole — scanning all major areas. Notice the point of maximum intensity and the transmission of any murmurs heard.

At times it is difficult even for the experienced clinician to identify systole and diastole. This is particularly true with rapid heart rates where the systolic and diastolic intervals tend to become equal in duration. Two techniques may be of value. One is to correlate the heart sounds with the apex impulse or the carotid pulse, both of which are systolic in timing. The first sound should just precede these phenomena. Because of the time required for the mechanical transmission of the pulse wave, peripheral arteries such as the radial or femoral are *not* suitable for timing heart sounds. The second technique is to familiarize yourself with the second sound at the base, which is invariably the loudest, and to inch the stethoscope downward toward the apex, keeping this sound clearly in your ear.

NORMAL FINDINGS

To recognize and to understand the signs of vascular disease, you must first develop an appreciation of the enormous variability of the so-called normal range. The circulatory system, perhaps more than any other, is constantly adapting to internal and external factors with changes in cardiac rate, intravascular pressure, and stroke volume. Keep in mind that while the cardiac output may triple with heavy exercise, it may also double with excitement. If the patient is angry or fearful, if he has just smoked a cigarette or finished a large meal, if he has had to hurry to his appointment, if he is chilled or overheated, his vascular system will respond accordingly. Even the simple act of standing or of lying down on the examining table calls into play several major hemodynamic mechanisms. Thus, it is difficult for the physician to be certain of a basal or steady state at the time of examination. He must learn to recognize the physiologic variables, so as not to confuse them with signs of disease. The young, conditioned athlete, the overweight businessman, and the slightly anemic young housewife must be expected to present far different patterns of blood pressure, pulse rate, or heart sounds.

A comprehensive review of cardiac physiology would probably be in order at this point, but it is beyond the scope of this work. Table 10-1 lists sample normal values for pressure and oxygen saturation from several key points in the circulatory system. Keep in mind that the systemic and pulmonary circulations are basically quite different; the former is a high-resistance, high-pressure circuit

Table 10-1. Representative Normal Adult Measurements

Site	Pressure (mm. Hg)	Oxygen Saturation (%)
Brachial vein	8 (mean)	78
Right atrium	3 (mean)	72
Right ventricle	25/3	72
Pulmonary artery	25/15	72
Pulmonary veins	9 (mean)	95
Left atrium	7 (mean)	95
Left ventricle	115/7	95
Aorta	115/80	95
Brachial artery	120/75	95
Basal cardiac output 5 liters/minute		
Stroke volume 65 ml.		

requiring relatively high stroke work, while the latter is a more distensible, low-pressure, low-resistance system. Despite these differences, however, the basic configurations of the pressure pulses are qualitatively similar for both the right and the left heart; and they occur in three basic patterns: atrial, ventricular, and arterial. *As you proceed through this section you should practice drawing diagrams of these basic patterns, relating them to the heart sounds and electrocardiogram.* This simple exercise will help you to understand the important temporal relationships. The skilled examiner develops a mental image of the wave form of movements he observes. This should be your goal as well.

Blood Pressure

The normal adult blood pressure varies over a wide range. The normal systolic range varies from 95 to 140 mm. Hg, generally increasing with age. The normal diastolic range is from 60 to 90 mm. Hg. Pulse pressure is the difference between the systolic and diastolic pressures. Mean pressure can be approximated by dividing the pulse pressure by three and adding this value to the diastolic pressure.

Typical normal values would be: systolic, 120 mm. Hg; diastolic, 80 mm. Hg; pulse pressure, 120 − 80 = 40 mm. Hg; mean pressure, 80 + 40/3 = 93 mm. Hg.

A difference of 5 to 10 mm. Hg between arms is common. The systolic pressure in the lower extremities is usually about 10 mm. Hg above that in the upper extremities. The patient's assuming an erect posture may not necessarily affect the arm pressure taken at heart level, but it is not uncommon for the systolic pressure to fall by 10 to 15 mm. Hg on standing; and about half of the time the diastolic pressure will rise slightly (by 5 mm. Hg).

These common sources of error should be recognized and avoided:

1. Discrepancies between the relative cuff size and limb size. The relatively obese arm (or thigh) yields falsely elevated values. Check the systolic level with the cuff on the forearm, palpating the radial artery, in obese patients. The standard cuff with an emaciated arm or in a child may give falsely depressed values.

2. Applying the cuff too loosely will give falsely elevated values.

3. The anxious patient may have an elevated level. If it is high, always leave the cuff in place and recheck the blood pressure several times. Never rely on a single determination.

4. It is possible to fail to recognize an "auscultatory gap." Sounds may disappear between the systolic and diastolic pressures and then reappear. If the cuff pressure is raised only to the range of the gap, the systolic reading will be falsely low. This error is eliminated by first determining the systolic level by the palpatory method.

5. Feeble Korotkoff's sounds may make your determination unreliable. In this event deflate the cuff and have the patient elevate his arm, reinflating the

cuff in this position. Then lower the arm and repeat. The sounds may be louder now, because of diminished venous pressure. If they are not, you may have to settle for a palpatory systolic pressure only.

Inspection and Palpation

The level of vascular tone in the extremities is, of course, under sympathetic control. With vasodilatation there is rubor, warmth, throbbing of the distal digits, and capillary pulsations of the nailbeds, with distention of the superficial veins. Digital throbbing is best appreciated by interlocking your fingertips with the patient's (right hand to right hand). Capillary pulsation is best seen by transilluminating the tip of the thumb. Vasoconstriction results in pallor, coldness, and collapsed superficial veins. Smoking cigarettes, chilling, or apprehension may cause vasoconstriction. In the dependent position the superficial veins become distended, and the venous valves may be identified as nodular bulges; with elevation of the limb they collapse. These changes in caliber are at times a cause of undue concern to patients.

Arterial Pulse. The normal resting pulse rate ranges from 60 to 100. It may be in the 50's in a conditioned athlete or 100 or over in an excited patient. The lability of the pulse rate is well known to all. Usually the rhythm is relatively regular. Occasional premature beats are so common that they are not necessarily considered abnormal. They are perceived as transient skips or breaks in rhythm. Sinus arrhythmia can be identified in most patients under the age of 40. It refers to a transient increase in pulse rate with inspiration followed by slowing with expiration. This phenomenon can be rather marked in some normal patients.

In addition to rate and regularity, the arterial pulse should be examined for its amplitude and contour. The amplitude of the pulse is largely a function of the pulse pressure, which is related to stroke volume, elasticity of the arterial circulation, and peak velocity of the ejection of blood from the left ventricle. If the stroke volume increases, as with excitement, heat, alcohol, exercise, or slowing of the heart rate, the pulse pressure widens, giving a throbbing quality on palpation.

The amplitude of the pulse obviously contributes to its contour. However, important information about the characteristics of left ventricular ejection can be learned from assessing the rate of rise and the shape of the arterial pulse wave. Because of the distortion which occurs when the pulse wave is transmitted distally, the carotid arteries must be used for accurate evaluation of pulse contour. An excellent way of analyzing pulse contour is to correlate it with a graphic record. Figure 10-7 shows such a tracing recorded from a normal carotid artery. The initial percussion wave is separated from the dicrotic wave by a dicrotic notch caused by aortic valve closure. The record is qualitative, but it helps the beginner to understand better what he feels. The upstroke normally takes no more than 0.10 second and the rounded crest takes another 0.08 to 0.12 second. The brachial and femoral pulses are usually synchronous.

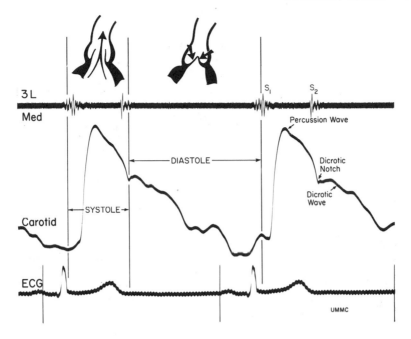

FIGURE 10-7. Characteristic arterial pulse wave with simultaneously recorded phonocardiogram and electrocardiogram.

Valsalva Maneuver. The Valsalva maneuver is a test that you can perform on yourself right at this moment. It will challenge your ability to observe and to interpret a rather complex series of physiologic variations. Close your glottis and strain down hard while palpating your carotid or radial pulse. What happens?

The normal response to the Valsalva maneuver is shown in Figure 10-8. The explanation is as follows: (a) The high intrathoracic pressure decreases venous return, causing an abrupt drop in cardiac output after several seconds. (b) The normal response to the decreased output is twofold: peripheral vasoconstriction and tachycardia. (c) Despite this compensatory mechanism, the pulse pressure dwindles, and this causes the pulse to become feeble. (d) After release the cardiac output is suddenly restored, causing an abrupt increase in blood pressure due to the temporary increase in peripheral vascular resistance. (e) This in turn triggers a reflex bradycardia (slowing), which is transient and somewhat delayed. Finally, the normal situation is restored.

Responses (d) and (e) disappear with many forms of heart disease.

Neck Veins. The bedside examination of the neck veins is a noninvasive method of assessing right atrial hemodynamics. The internal jugular veins connect without valves to the right atrium and accurately display right atrial wave form and pressure to the careful observer. Two major waves occur in the normal right atrium during each cardiac cycle. (See Figure 10-9A.) The larger of these is the A

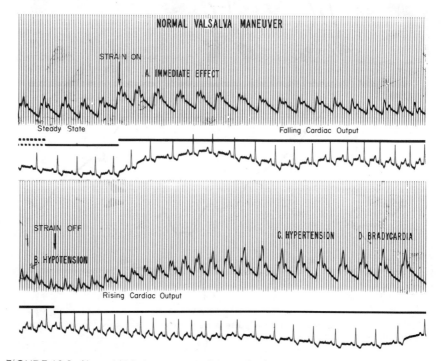

FIGURE 10-8. Normal Valsalva maneuver (see text).

wave, which is generated as the atrium contracts to fill the ventricle just prior to systole. As the atrium relaxes and its pressure falls, the x descent is seen while ventricular systole begins and the tricuspid valve is closed. The right atrial pressure rises as the atrium is filled from the periphery, and the V wave is generated, which is smaller than the A wave. As the atrium empties during early diastole, a fall of the V wave or y descent is seen.

Normally, mean right atrial pressure is quite low. As the right heart fails, atrial pressure increases to maintain adequate ventricular filling.

Examination of the neck veins. Satisfactory examination of the neck veins requires a bed or table which will allow the patient's head and trunk to be elevated to various heights. Clothing should be removed from the neck and upper thorax and good illumination should be available. If pillows are removed and the head turned slightly away from the examiner (without putting a marked stretch on the neck), the neck veins can be visualized in the majority of patients, the head of the bed being raised and lowered to bring the patient to the position where venous pulsations are most easily seen (Figure 10-10).

Differentiation of venous and arterial waves. The venous pulse is diffuse and undulant and usually disappears or markedly decreases in the sitting position. Venous waves can usually be obliterated by moderate pressure at the base of the

A. VENOUS PULSE

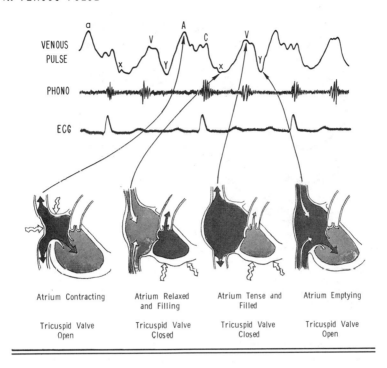

B. NORMAL APEX IMPULSE (supine)

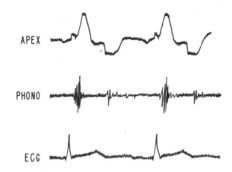

FIGURE 10-9. A. Relationship of jugular venous pulse curve to atrial hemodynamics.
B. Normal apex impulse as recorded by kinetocardiography. Notice minor movement with
atrial contraction before first sound.

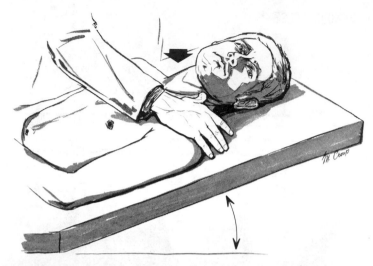

FIGURE 10-10. Examination of the neck veins. Maximum information can be gained only with proper technique. Note the use of an examining table with an adjustable head, the removal of clothing from the upper torso, and the absence of a pillow under the patient's head. The patient's head is turned *slightly* away from the examiner and the thumb placed on the opposite carotid artery for timing purposes. Good illumination is essential. Occasionally, tangential lighting with a pocket flashlight is helpful.

neck. Venous pressure decreases with inspiration, although the venous waves may become more prominent during inspiration. Arterial pulsations in the neck are localized and brisk and are usually seen best high and medial to the sternocleidomastoid muscle, while venous pulsations are seen lower and more laterally, either under or just behind the sternocleidomastoid muscle. Arterial pulsations are unaffected by position and do not vary with respiration, nor can they be obliterated by pressure at the base of the neck as the venous pulse can be.

To time the jugular pulse you *must* use two senses in order to answer the simple question, "Is this wave occurring before or after the first sound?" Perhaps the simplest method is to use vision and touch, observing the venous pattern while palpating the opposite carotid pulse. If the wave precedes the arterial pulsation, it must be an A wave; if it is synchronous or a little delayed, it is a V wave. A second method is to use vision and hearing. Listen to the heart while making similar observations. Exceptions to this rule are caused by the occasional rhythm disturbance which superimposes atrial and ventricular contraction. Such exceptions are rare, however, and they will be considered later.

Jugular venous pressure (JVP). It is less confusing to think of the jugular venous pressure as a separate phenomenon. The neck veins are frequently distended with the patient supine, tending to collapse with inspiration and to refill during expiration. Light pressure over the jugular bulb causes them to distend further, and release is followed by collapse to the previous level. Because of the effects of gravity, distention usually disappears when the head of the bed is elevated to 45 degrees. This position is satisfactory for determining jugular venous pressure in the majority of instances. With severe elevation of jugular venous pressure greater degrees of elevation of the bed may be needed.

Since the sternal angle bears a constant relationship to the right atrium in most positions, it is taken as the reference point for measuring the jugular venous pressure (Figure 10-11). While a reasonable estimate of the jugular venous pressure in centimeters above the sternal angle can frequently be achieved, a simpler scale of *normal, mildly,* and *markedly increased* will suffice. Pregnancy invariably increases jugular pressure. Exertion, anxiety, premenstrual increases in blood volume, abdominal pressure (corsets and binders), and even obesity may produce mild elevations.

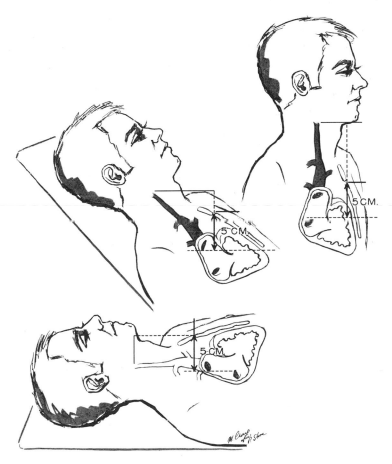

FIGURE 10-11. Determination of jugular venous pressure. Note that in all positions the vertical height of the sternal angle remains 5 cm. above mid-right-atrial level, making this easily identifiable bony landmark a constant reference point for estimating the jugular venous pressure. With increasing degrees of venous pressure the patient's head must be elevated to ascertain the mean venous pressure.

Precordium. What is seen and felt over the precordium varies markedly with position, phase of respiration, amount and distribution of muscle and fat, and thoracic cage configuration. In normal people the cardiac apex is usually the most lateral impulse of cardiac origin that can be felt on the chest wall. It is often referred to as the *point of maximum impulse (PMI).* In some patients with cardiac disease, however, the cardiac apex may *not* be the point of *maximum* impulse.

The normal apical impulse is in the fourth or fifth interspace, should not be felt in more than one interspace, usually occupies less than the first one-half of systole, and should not be felt further to the left than halfway between the midsternal line and the lateral thoracic border. Placing the patient in the left lateral decubitus position (see Figure 10-3) brings the cardiac apex closer to the chest wall. For purposes of auscultation and analysis of the palpatory configuration of the apical impulse this is a useful maneuver, but assessment as to location and duration of the apical impulse should be made with the patient supine. A graphic recording of the normal apex cardiogram is shown in Figure 10-9B.

An outward systolic movement along the left sternal border is usually thought to be indicative of right ventricular pressure or volume overload (see Figure 10-15); however, young normal subjects frequently have a palpable impulse along the left sternal border, and this sign must be carefully interpreted in light of the rest of the findings. Likewise, pulmonary closure may be palpable in this group of patients without reflecting pulmonary hypertension.

Auscultation

You should clearly understand the physiologic origin of normal cardiovascular sound. Individual differences in loudness, quality, and pitch occur in patients of varying age and build. Sounds may be remarkably loud and clear in young, thin-chested patients or with tachycardia due to exercise or excitement. They may be quite muffled and practically inaudible in the obese, heavy-chested individual. The heart may sound very different as the stethoscope position is changed from point to point in the same patient. This is partially due to the relative proximity of the various valves. The first sound as heard at the apex may not only become softer at the base but may also seem shorter and have a different quality due to the damping effect of the interposed soft tissues. Similarly, the second sound loses intensity and changes quality as the stethoscope is moved downward toward the apex.

The cardiac cycle is schematically represented in Figure 10-12I. The sounds of greatest importance are the first and second heart sounds which divide the cardiac cycle into systole and diastole (technically, ventricular systole and diastole). Although there is controversy concerning the origin of the first sound, it may be considered here to be due to closure of the mitral and tricuspid valves and the related phenomena of tensing of the myocardium, A-V valve

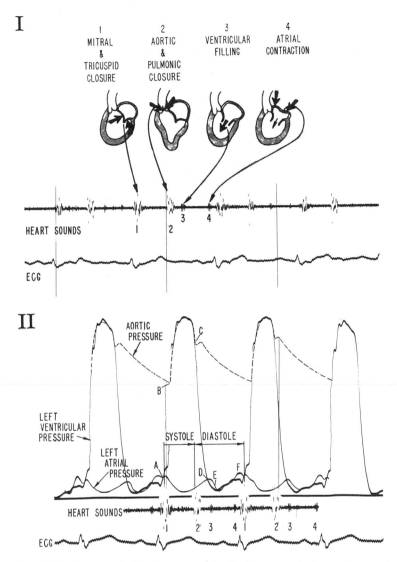

FIGURE 10-12. Origin of the heart sounds. I: Diagram of the cardiac cycle. II: Relationship of heart sounds to intracardiac pressure. A. The first sound occurs when the left ventricular pressure exceeds the left atrial pressure. At this point the mitral valve closes. B. Opening of the normal aortic valve is inaudible. C. Aortic closure is synchronous with the second sound. D. Mitral opening is normally inaudible. E. The physiologic third heart sound occurs during the period of early rapid filling of the left ventricle. F. Left atrial contraction produces the physiologic fourth sound. Similar relationships are true of the right heart. Together they produce the normal sounds.

(tricuspid, mitral) supporting structures (papillary muscle, chordal apparatus) as well as to changes in blood velocity. The second sound results from aortic and pulmonary valve closure. A physiologic third sound is at times audible during the period of rapid filling of the ventricles. A physiologic fourth sound in late diastole coincides with atrial contraction; *it is normally inaudible* but can sometimes be recorded by phonocardiography. The normal first and second heart sounds range in frequency from 60 to 200 cps. The physiologic third sound is low-pitched and usually around 40 cps or less. In Figure 10-12 II these events are correlated with their corresponding systemic or left-sided pressure relationships.

The normal range of cardiac sounds includes those shown on the following pages.

The *first sound* is louder, longer, and lower pitched than the second sound at the apex.

APEX		LUB dup
(TRICUSPID AREA)		T - LUB dup

The mitral valve closes from 0.02 to 0.03 second before the tricuspid. Splitting of the first sound is therefore common normally, particularly in the tricuspid area. Since tricuspid closure is usually not well heard at the apex, a single-component first sound related to the mitral valve is usually heard there.

BASE   lub DUP

The *second sound* results from combined aortic and pulmonary valve closure. It is almost always louder than the first sound at the base. The aortic component is widely transmitted to the neck and over the precordium. It is as a rule entirely responsible for the second sound at the apex. The pulmonary component is softer and is normally heard only along the high left sternal border. Although the second left intercostal space is called the pulmonic area, analysis of the second sound and its splitting is frequently best accomplished in the third or fourth interspace.

PHYSIOLOGIC SPLITTING
(Pulmonic Area)

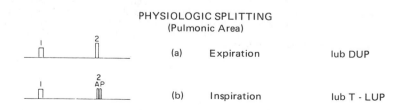

(a)   Expiration   lub DUP

(b)   Inspiration   lub T - LUP

Physiologic splitting of the second sound may be demonstrated in most normal people during inspiration. Closure of the aortic and pulmonary valves during expiration is synchronous or nearly so because right and left ventricular systole are approximately equal in duration. With inspiration, venous blood rushes into the thorax from the large systemic venous reservoirs. This increases venous return and prolongs right ventricular systole by temporarily increasing right ventricular stroke volume. This, of course, delays pulmonary valve closure. At the same time venous return to the left heart diminishes due to the increased pulmonary capacity during inspiration. This shortens left ventricular systole and permits earlier aortic closure. The two factors combine to produce transient "physiologic splitting" of the pulmonic second sound.

PHYSIOLOGIC THIRD SOUND   LUB duppa

The *physiologic third sound* is a faint, inconstant, low-pitched sound audible at or near the apex. It is best heard with the patient in the left decubitus position, and it frequently disappears when he sits erect. It varies in intensity with respiration, usually becoming louder with expiration. It is extremely common in children and young adults, but is seldom heard in persons over 30. On the other hand, fourth heart sounds are so frequent over the age of 50 that their presence may not necessarily imply cardiac disease.

FUNCTIONAL SYSTOLIC MURMUR

The *functional or innocent systolic murmur* is commonly heard in children and young adults, especially women. It is characteristically soft, early, short, and variable. It is usually heard best at the pulmonic area or along the left sternal border but at times is audible only at or medial to the apex. Functional murmurs vary with position and respiration and frequently have a peculiar vibratory quality. Their origin will be discussed later (see p. 186). They are not associated with any structural abnormality or recognizable heart disease.

Venous hum refers to a phasic roaring heard in about 25 percent of normal adults at the lower border of the sternocleidomastoid muscle in the sitting position. It is accentuated in diastole and may be transmitted to the high precordium. Hums are interrupted by compression over the jugular vein, the Valsalva maneuver, and recumbency. They are an innocent finding of no consequence.

Bruits are common in normal young people. They may be heard in the supraclavicular area or abdomen. *Abdominal murmur* refers to a soft systolic bruit often heard on careful auscultation using the diaphragm endpiece, between the xyphoid and umbilicus in the midline. It is usually very soft and of medium or low pitch. Its interpretation is difficult since at times it may also be an important sign of underlying vascular disease.

CARDINAL SYMPTOMS

Dyspnea. Dyspnea of cardiac origin is characteristically related to effort until the advanced stage of heart disease, when it may become present even at rest. The labored respirations result from the increased work of breathing caused by decreased compliance of the lungs due to an abnormal increase in pulmonary venous pressure. Since we all experience breathlessness with heavy exertion, mild dyspnea early in the course of cardiac disability requires careful, detailed assessment. Any unexplained reduction in exercise tolerance should arouse suspicion. The inability to keep up with associates of the same age suggests significance. The level of activity capable of bringing on the symptom must be explicitly determined and quantitated; for example, the number of flights of stairs resulting in distress. Is the complaint abnormal for the patient's age, build, and apparent physicial condition. Has there been a slow progression, rapid progression, or sudden onset of the present state? How long does it take to restore the natural breathing with rest? To what extent does it interfere with daily activity? Are there associated symptoms — pain, palpitation, or cough?

Paroxysmal dyspnea describes the abrupt onset of breathlessness for no apparent cause. When it occurs at night (*paroxysmal nocturnal dyspnea*), it usually awakens the patient with an alarming smothering sensation which invariably prompts him to assume the sitting position. There may be associated wheezing and cough. It is sometimes precipitated by a disturbance in cardiac rhythm with an excessively rapid or slow heart rate.

Orthopnea. Dyspnea precipitated by assuming the recumbent position is referred to as orthopnea. It is a relatively late symptom which is frequently relieved or improved when the thorax is again elevated. In clinical practice, this is referred to as two-, three-, or four-pillow (and so on) orthopnea. It may disappear with improvement which follows therapy. It is a useful index of the patient's status.

Pain. Pain is an important symptom of circulatory disease. Most commonly it is a consequence of ischemia, but it may arise from the pericardium or aorta or from elevation of the pulmonary arterial pressure.

Angina pectoris is a common manifestation of coronary artery disease and is the symptom of inadequate oxygen delivery to the myocardium. It is usually not over the precordium but beneath the sternum, and may radiate into the neck or into either or both arms, usually the left. It is characteristically constrictive or oppressive in character (from the Latin *angere*, to strangle), although patients will describe anginal discomfort quite variably (e.g., burning, aching, squeezing) and some will refuse to call the sensation *pain*. Factors which increase cardiac work, such as exercise, excitement, cold weather, and meals, tend to provoke angina. The essence of the diagnosis of angina pectoris centers about distress which occurs in the substernal area, is related to effort, and is relieved within *minutes* by rest. Fleeting stabs or jolts of pain or pain lasting for hours to days is not angina pectoris. Severe persistent pain of this same anginal quality may develop with myocardial infarction, at which time it is often accompanied by vomiting, weakness, diaphoresis, and other symptoms.

Claudication refers to ischemic pain caused by an inadequate arterial circulation to a muscle group, usually the legs (from the Latin *claudicare*, to limp). Its character is similar to angina in that it is precipitated by exercise and relieved by rest.

Pericarditis may cause retrosternal pain which mimics that of myocardial ischemia in most respects. It is persistent and, unlike angina, it is often intensified by inspiration (pleuritic) and relieved by changes in body position such as sitting up.

Palpitation. This is a general term describing an awareness of the heart beat which may occur with increased stroke volume, irregularity of rhythm, or tachycardia or bradycardia. The precise identification of the underlying disturbance requires electrocardiographic confirmation. Nonetheless, a close clinical estimate may at times be possible by tapping out various cadences on the back of the hand and asking the patient to select the one which most closely simulates what he felt. Palpitation often is of no consequence.

Cough. Cough, especially nocturnal, is a common complaint with pulmonary congestion. It may be a dry hack or may produce clear, thin sputum. *Hemoptysis* may accompany cough when pulmonary venous pressures are greatly elevated, as in mitral stenosis or severe left ventricular failure. There is usually associated dyspnea with cough of cardiac origin.

Other important cardiac symptoms include fatigue, edema, anorexia, weight loss, vomiting, abdominal pain, and syncope. These will not be considered in detail.

ABNORMAL FINDINGS

Disease modifies circulatory function in many ways and in many places. The signs we look for are those directly or indirectly due to this disordered function. The objective is simple enough — to identify significant dysfunction as early as possible in order to correct it or at least delay its progression — but cardiovascular disease is a clever adversary which will elude even the wisest physician should he fail to apply basic physiologic principles.

Blood Pressure

Hypertension is a term used to describe *persistent* elevation of the systemic blood pressure. When the pressure in the arms consistently exceeds 140/90, it is usually considered to be elevated, although this is a gross oversimplification. Labile elevations are to be expected and do not constitute true hypertension.

Widened pulse pressure is common to all conditions producing an *increased stroke volume.* Simple bradycardia widens the pulse pressure, for example, 150/70, as do fever, anemia, and hypermetabolic states. These disorders all have in common an increased stroke volume. Incompetence of the aortic valve widens the pulse pressure due to lowering of the diastolic pressure, for example, 150/30. With aging, the elasticity of the great arteries diminishes, producing an increase in the systolic pressure at times referred to as "systolic hypertension," for example, 165/80.

Hypotension is present if the pressure falls below 90/60. Some normal adults and many children will have systolic blood pressure in the vicinity of 90 mm. of mercury. Shock cannot be said to be present unless there is evidence of decreased regional blood flow, e.g., syncope, sweating, oliguria, and obtundation. Systemic hypotension or shock may result from inadequate cardiac output, decreased peripheral resistance, or an inadequate blood volume. This is usually associated with dizziness, visual blurring, sweating, and at times syncope.

Inspection and Palpation

Extremities. Clubbing of the digits has been previously mentioned in Chapter 9. (See Figure 9-11.) Three common cardiovascular causes of clubbing are cyanotic congenital heart disease, infective endocarditis, and advanced cor pulmonale. The best early sign of clubbing is loss of the normal angle at the base of the nail. Later there is an increase in the longitudinal curvature of the nail. These changes are better pointed out at the bedside than described with words or pictures. The etiology of clubbing is still uncertain, although it is probably related to increased blood flow through multiple arteriovenous shunts in the distal phalanges. A useful classification of clubbing is given in Tables 10-2A and 10-2B.

Cyanosis is a bluish skin color caused by a relative decrease in oxygen saturation of the cutaneous capillary blood. Normal arterial blood is about 95 percent saturated; venous blood is about 70 percent saturated. When the arterial saturation falls below 85 percent, cyanosis usually becomes manifest, provided that the patient is not severely anemic. Cyanosis may be central or peripheral. *Central cyanosis* has three major causes: congenital heart diseases with right-to-left shunts; pulmonary arteriovenous fistulae; and advanced pulmonary disease with cor pulmonale and incomplete oxygenation of the blood in the lungs. Central cyanosis is generalized and associated with arterial oxygen desaturation. Polycythemia and clubbing are frequently present. *Peripheral cyanosis* (sometimes called acrocyanosis) is limited to the hands, feet, tip of the

Table 10-2A. Causes of Symmetric Clubbing

Cardiovascular disease
  Congenital, cyanotic
  Subacute bacterial endocarditis
  Advanced cor pulmonale
  Pulmonary arteriovenous fistula
Pulmonary disease
  Inflammatory
    Bronchiectasis
    Abscess
    Empyema
  Pneumoconiosis
  Neoplasm
    Carcinoma, primary
    Pleural mesothelioma
  Interstitial fibrosis
Extrathoracic disease
  Gastrointestinal
    Sprue
    Ulcerative colitis
    Regional enteritis
    Dysentery
  Hepatic
    Biliary cirrhosis
    Liver abscess
    Amyloidosis
  Toxic
    Arsenic
    Phosphorus
    Alcohol
    Silica or beryllium
  Familial
  Miscellaneous
    Chronic pyelonephritis
    Syringomyelia
    Chronic granulocytic leukemia
    Hyperparathyroidism

Table 10-2B. Causes of Asymmetric Clubbing

Unidigital — median nerve injury
Unilateral
  Aneurysm of the innominate artery
  Recurrent subluxation of shoulder
Unequal
  Anomalous aortic arch
  Reversed patent ductus

nose, ear lobes, and lips. It results from a critical reduction in systemic blood flow, usually due to diminished cardiac output (heart failure) or obstructive peripheral arterial disease. The extremities are usually cold and mottled. Clubbing is not a consequence. Light pressure will produce a sustained white print in the bluish background which fades slowly.

Edema has many causes. Cardiac edema is usually dependent (feet and ankles). It is graded 1+ through 4+ on the basis of pitting produced by sustained, light pressure with the thumb over the medial malleolus or pretibial area. It is a common sign of congestive heart failure. Edema also occurs unilaterally following the occlusion of a major vein. Peripheral arterial occlusion may cause mild "brawny" (nonpitting) edema.

Venous Circulation. *Varicose veins* is a term usually restricted to dilatation of the superficial leg veins. *Varicose* means "dilated, swollen." The rate of blood flow through these vessels is diminished and the intraluminal pressure is increased. There are two types of varicose veins: primary, due to an inherent weakness of the vessel wall and venous valves; and secondary, due to proximal obstruction in the vena cava, pelvic veins, or iliofemoral veins. Both the greater and lesser saphenous systems may be involved. The greater saphenous vein lies superficially on the anteromedial aspect of the thigh and lower leg. It drains into the common femoral vein in the groin. The lesser saphenous vein lies superfically on the posterolateral aspect of the calf from the ankle to the popliteal space. Both saphenous veins communicate with the deep femoral venous system by means of multiple communicating or perforating veins which pierce the fascia. When the valves in the perforating veins are incompetent, the superficial saphenous varicosities may fill from the deep venous system.

Diagnosis, usually simple, is made by inspection of the dependent limb. In severe cases, pigmentation, edema, and even ulceration of the skin in the region of the medial malleolus point to significant venostasis. It is important to determine two additional facts in patients with varicose veins. Are the valves incompetent in the communicating veins between the superficial and deep systems? Are the deep veins patent?

The presence of incompetent communicating or perforating veins can be demonstrated simply. A tourniquet is applied around the thigh with the patient in the recumbent position. When the patient assumes an erect position, the incompetent valves in the communicating veins permit the varicosities to fill rapidly from above (Brodie-Trendelenburg test).

Patency of the deep veins may be established by the use of the Perthes test. A tourniquet is used to occlude the subcutaneous veins at knee or thigh level. This tourniquet must be at or below the lowest significant incompetent perforator. Filling of the superficial varicosities from above is prevented by this tourniquet. As the patient walks, the muscles exert a pumping action on the deep veins and drain the dilated superficial varicosities. Failure of the varicosities to empty means either that the tourniquet is placed too high and a large incompetent communicating vein permits the varices to fill, or that the deep veins have been damaged by an inflammatory process, disturbing their normal function.

Venous thrombosis may be acute (thrombophlebitis) or silent (formerly termed phlebothrombosis), deep or superficial. Superficial thrombophlebitis produces redness, induration, and tenderness adjacent to the involved venous segment, which is thickened and cordlike. Diagnosis is easily made, but it should be remembered that there may be associated deep thrombosis.

Acute inflammatory thrombosis of a major vein (deep thrombophlebitis) results in rather striking pain, tenderness, warmth, and swelling of the involved limb. Sensation is preserved and superficial veins may be distended. There may be considerable reflex arteriospasm, which at times may cause the extremity to become pale and the peripheral pulses to be reduced or even absent (phlegmasia cerulea dolens). This must be differentiated from acute arterial occlusion, which usually occurs without swelling of the extremity.

Deep venous thrombosis involving the deep femoral and pelvic veins may be entirely asymptomatic, and fatal pulmonary embolism may occur without warning. It is worthwhile for the physician to become "thrombosis conscious" and constantly to watch for minor suggestive signs in his bedfast patients. These include (a) tenderness along the iliac vessels and below the inguinal ligament, along the femoral canal, in the popliteal space, over the deep calf veins, and over the plantar veins; (b) minimal swelling detectable only by measuring and comparing the circumference of both calves and both thighs at several levels; (c) unexplained low-grade fever and tachycardia; (d) a trace of ankle edema; and (e) calf pain on sharp dorsiflexion of the foot with the knee slightly flexed (Homans' sign). With the latter maneuver, it is important to distinguish between calf pain and Achilles tendon pain. The latter is common in women who wear high heels and in patients wearing flat slippers.

Arterial Circulation. Arterial occlusion may be complete or partial; it may occur acutely or gradually. Chronic arterial insufficiency results from gradual reduction in vessel caliber. It may be due to degenerative or inflammatory processes of the vascular wall. Examination of the affected extremity may show the following:

1. Diminished or absent pulses. Audible systolic bruits extending into early diastole heard over major arteries (femoral or subclavian).
2. Reduced or absent hair peripherally (over the digits and dorsum of the hands or feet).
3. Atrophy of muscles and soft tissues.
4. Thin, shiny, taut skin.
5. Thickened nails with rough transverse ridges and longitudinal curving.
6. Mild brawny edema.
7. Coolness on palpation.
8. Intense grayish pallor on elevation of the extremity. Dependency after a minute or two of elevation produces a dusky, plum-colored rubor which develops very gradually (30 seconds to one minute).
9. Flat, collapsed superficial veins.

10. Delayed venous filling time. Empty the superficial veins by elevating the extremity. Prompt filling (less than 10 seconds) occurs normally with dependent lowering.

In examining an upper or lower extremity for suspected underlying arterial insufficiency, it is particularly important to *assess the effects of exercise* in order to identify early disease. Three important changes may be elicited in this way: (1) Pallor of the skin may occur over the distal limb. (2) Arterial pulses may disappear. (3) Systolic bruits not present at rest may become apparent over the major arteries. These important diagnostic signs may become apparent only in the postexercise state.

Advanced arterial insufficiency shows all the above features, and in addition there may be:

1. A bluish-gray mottling of the skin unchanged by position.
2. Early ulceration between or on the tips of the digits.
3. Tenderness to pressure.
4. Stocking anesthesia.

These signs indicate that gangrene is imminent.

Chronic occlusion of the aortic bifurcation is associated with (1) absent femoral pulses, (2) intermittent claudication extending into the buttocks, and (3) sexual impotence (Leriche syndrome).

Acute arterial occlusion usually begins with agonizing pain in the affected extremity, which below the occlusion site is pale, cyanotic, and pulseless. It may also be tender and exhibit stocking anesthesia. This is a serious emergency.

Abdominal bruits, formerly considered only curiosities, have recently received due emphasis. Because of their demonstrated occurrence in young healthy individuals, it is important to be cautious about overinterpretation of such a bruit as an isolated finding. In the presence of hypertension, however, a bruit heard in the epigastrium or anterior lumbar region may be an important sign of renal artery stenosis. Similar systolic bruits have been pointed out in patients with mesenteric arterial disease (abdominal angina) as well as over greatly enlarged spleens. Venous hums are at times identified over the cirrhotic liver due to torrential flow through venous collaterals. In addition, neoplasms of the pancreas, stomach, and liver have been demonstrated to produce abdominal systolic bruits due to local arterial involvement.

Aneurysm of an artery produces a pulsatile swelling along the course of the vessel. The aorta and popliteal artery are the vessels most often involved. A systolic thrill may be felt over the tumor. The aorta should always be carefully palpated for the presence of aneurysm. This is usually felt as an expansile mass in the midabdomen. A lateral film of the abdomen may show calcification in the wall of the aneurysm, and the lumbar spine may be eroded. Rupture of an aortic aneurysm is usually indicated by severe, constant back pain, and is often associated with pain in one or both groins and a mass in the flank. If the thrill noted over such a mass is continuous, one should suspect an *A-V fistula.*

The quality of the arterial pulse may provide clues to many circulatory disorders other than peripheral vascular disease.

Water-hammer pulse (Figure 10-13) is characterized by a wide pulse pressure, a low diastolic pressure, and a dicrotic notch which is absent or displaced downward on the descending limb of the tidal wave. The pulse has a collapsing, shocking quality which is reinforced by elevating the arm above the head. It is a classic sign of aortic insufficiency, but it is seen in other conditions where there is a low-resistance runoff due to a "leak" in the arterial system — patent ductus

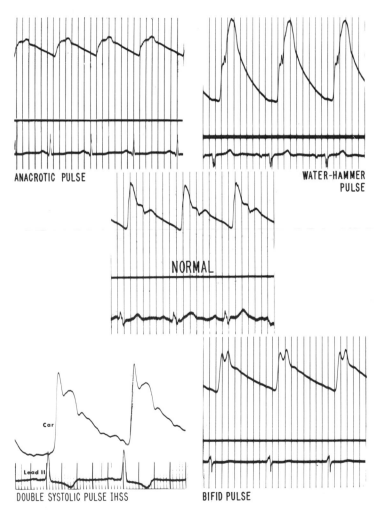

FIGURE 10-13. Representative arterial pulse tracings comparing the normal with the anacrotic pulse of aortic stenosis; the water-hammer pulse of aortic insufficiency; the double systolic pulse of idiopathic hypertrophic subaortic stenosis; and bifid pulse.

arteriosus and peripheral A-V fistula. It is the basis for other peripheral signs of aortic insufficiency, including "hopping carotids," "pistol-shot sounds," Duroziez's sign, de Musset's sign, and others. Although interesting they add little of diagnostic importance.

Bounding pulse is characterized by a wide pulse pressure with a normal or slightly lower diastolic pressure. Fever, anemia, hepatic failure, thyrotoxicosis, and complete heart block are all capable of producing a bounding pulse. They have in common an increased stroke volume and a diminished peripheral resistance. An extremely bounding pulse becomes, of course, a water-hammer pulse.

Weak pulse (pulsus parvus) has a normal contour but a low amplitude. It feels weak and "thready." The pulse pressure is narrowed by a low stroke volume and associated peripheral vasoconstriction. It is present with low-output failures of all types. Common causes include mitral stenosis, acute myocardial infarction, shock, and constrictive pericarditis.

Anacrotic pulse (Figure 10-13) is associated with valvular aortic stenosis. The ascending limb is delayed and the summit is broad. The pulse pressure may be narrowed. The slow rise and delayed peak can often be appreciated with careful practice.

Bifid pulse (pulsus bisferiens) (Figure 10-13) is characterized by double systolic peaks which can usually be felt with the palpating finger. Bifid pulse is found with aortic regurgitation which may be isolated or, more frequently, may be associated with some degree of aortic stenosis. Twin peaks felt at the top of a slowly rising arterial pulse may be the first clue that a murmur or aortic regurgitation will be heard in addition to that of aortic stenosis.

Another type of pulse with two systolic waves is the *spike-and-dome pulse* of idiopathic hypertrophic subaortic stenosis (IHSS) (Figure 10-13). This pulse is characterized by the extremely brisk initial wave (spike) followed by the rounded dome which occurs after the muscular obstruction to left ventricular outflow develops. The rounded dome is not as easily felt as the second peak of the bifid pulse of aortic regurgitation, and consequently the two waves of this pulse may be difficult to appreciate.

The *dicrotic pulse* has two waves, but only one is systolic and the second is a very prominent dicrotic wave occurring in diastole. This pulse is best appreciated in a more peripheral artery (brachial or femoral) and is usually seen in younger patients with myocardial disease.

Pulsus alternans means alternating large- and small-amplitude beats with a *regular rhythm*. It can be detected by palpation or by using a blood pressure cuff. Alternate systolic pressures may vary by as much as 25 mm. Hg. It is an important sign of left ventricular failure.

Pulsus bigeminus (bigeminy) is a coupling of two beats separated by a pause. It results most often from alternating normal and premature beats. The second beat is weak due to reduced diastolic filling time.

Pulsus paradoxus is an important sign of cardiac tamponade. It is found with tense pericardial effusions and with chronic constrictive pericarditis. The term refers to a weakening of the pulse during normal inspiration. It is really a misnomer, however, since what happens is an exaggeration of the normal inspiratory decline in systolic blood pressure. Multiple mechanisms for pulsus paradoxus have been postulated in the past. The observed fall in systolic blood pressure with inspiration is most likely caused by normal inspiratory augmentation of venous return to the right heart chambers in a heart with restricted capacity for diastolic filling. As a result, there is a reciprocal decrease in left ventricular filling and a fall in systolic blood pressure. The systolic pressure may fall by 10 mm. or more with inspiration. The estimation of blood pressure is more reliable than palpation for distinguishing pulsus paradoxus. As the pressure is reduced, the first Korotkoff's sounds appear only during expiration (upper systolic level). Further reduction produces a point at which all beats are heard (lower systolic level). A difference of greater than 10 mm. suggests cardiac tamponade. A common noncardiac cause of pulsus paradoxus is the labored respiration of the patient with obstructive pulmonary disease (asthma, emphysema).

Neck. Carotid pulsations may be striking in hyperkinetic states and particularly with aortic insufficiency (Corrigan's sign). A *carotid thrill* is commonly felt with aortic stenosis. It also occurs with partial occlusion of the orifice of the common carotid artery. A bruit may be heard over a partially occluded carotid artery.

Diminished or absent carotid and brachial pulsations may result from a diffuse process involving the aortic arch and its major branches. This is called the *aortic arch syndrome.* It may be due to arteriosclerosis, aortitis, aortic aneurysm, or congenital anomalies. Isolated occlusion of the common carotid artery may cause neurologic symptoms and signs. Internal carotid pulsation may be estimated only by donning a sterile surgical glove and palpating the vessel in the tonsillar fossa.

Carotid sinus pressure (CSP) increases vagal tone and results in slowing of the heart rate. On occasion, this response may be exaggerated and CSP will result in extreme bradycardia (even cardiac standstill) or hypotension. This is called the *carotid sinus syndrome.* Valsalva's maneuver combined with CSP results in more pronounced vagal tone. CSP is frequently used in the diagnosis and treatment of certain arrhythmias. It is also helpful in transiently slowing the heart rate to aid in auscultation. CSP should be applied cautiously if at all in the elderly, who are prone to have carotid artery atherosclerosis.

Jugular venous distention is an important sign of congestive heart failure, although elevations of venous pressure may result from mechanical obstruction to venous inflow by intrathoracic neoplastic, inflammatory, or vascular masses. The presence of jugular venous hypertension may escape notice in obese or "bull-necked" individuals. Also, extreme venous hypertension may go unnoticed because the veins may be distended all the way to the top of the head and pulsations are not seen. Qualitative grading of venous distention as *mild, moderate,* or *severe* is ordinarily sufficient. The following diagrams show pulmonary hypertension, A-V dissociation with cannon waves, V-waves of tricuspid insufficiency, and constrictive pericarditis.

Giant A waves result from very forceful right atrial contraction. They are seen with tricuspid stenosis. More commonly they result from the loss of diastolic compliance which accompanies right ventricular hypertrophy, as in pulmonary stenosis and pulmonary hypertension. The diagram shows the venous waves seen with marked pulmonary hypertension. The right atrium contracts vigorously to fill the poorly compliant right ventricle and a large A wave is seen in the neck veins. With the onset of atrial fibrillation and the loss of the *effective* atrial contraction, giant A waves as well as normal A waves are no longer seen in the jugular venous pulse.

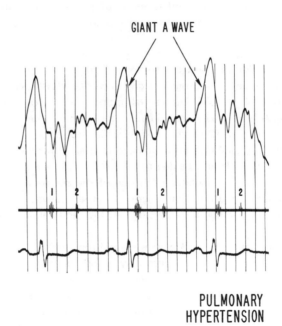

GIANT A WAVE

PULMONARY
HYPERTENSION

Cannon waves occur with certain arrhythmias characterized by atrioventricular dissociation (complete A-V block, ventricular and nodal tachycardia and premature ventricular beats). When atrial contraction occurs during ventricular systole, cannon (C) waves result. The origin is the same in all these, namely, contraction of the right atrium on a closed tricuspid valve due to synchronous atrial and ventricular systole. Since the blood cannot move forward, there is striking backward regurgitation into the jugular system. Note the variation in intensity of S_1 as the P waves vary their relationship to the QRS.

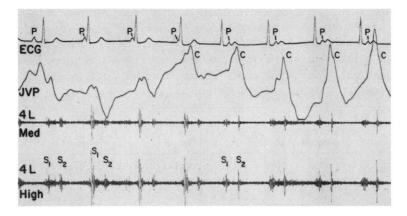

A-V DISSOCIATION WITH CANNON WAVES

Large V waves are transmitted to the neck with tricuspid regurgitation, which reflects abnormal right atrial filling through an incompetent tricuspid valve during systole. Jugular venous pulse (JVP) tracing shows V wave from patient with rheumatic tricuspid insufficiency. Atrial fibrillation is often present with this condition, in the presence of which no A wave is seen. The large positive V wave occurs during systole and can be timed by observation of the neck veins with simultaneous palpation of the carotid pulse of auscultation of the heart.

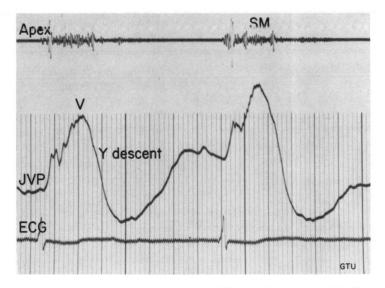

V-WAVES OF TRICUSPID INSUFFICIENCY*

*Courtesy Division of Cardiology, Georgetown University Hospital.

Sharp y descent associated with an elevated mean pressure is seen with chronic pericardial constriction and *acute pericardial effusion with tamponade.* It may be seen in severe right heart failure of any cause. In these conditions the pressure drops only briefly following tricuspid opening, remaining elevated during the rest of the cardiac cycle. The sharp descents results in a characteristic "M" shape in the jugular venous wave form.

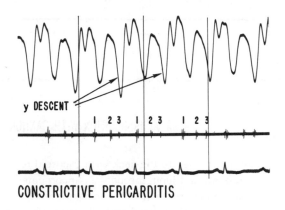

y DESCENT

CONSTRICTIVE PERICARDITIS

Precordium.   Abnormal precordial movements are of prime diagnostic importance. They may be systolic or diastolic, outward or inward. Sustained low-frequency movements are usually visible and palpable; high-frequency vibrations are palpable only. Sustained high-frequency vibrations are called "thrills," which are simply palpable murmurs. Short, high-frequency vibrations are "shocks," which are palpable heart sounds.

Left ventricular hypertrophy produces a *sustained,* systolic apex impulse which may be displaced laterally and downward. The graphic record of this movement is shown in Figure 10-14. Forceful atrial contraction necessitated by the decreased compliance of the thick-walled left ventricle may result in a palpable presystolic component to the apical impulse of left ventricular hypertrophy. This presystolic distention can usually be felt with the palpating hand, especially with the patient in the left lateral decubitus position. A tongue blade held lightly over the apex will amplify this double impulse, making it easily demonstrable at the bedside without special equipment. The excursions may be timed by simultaneously listening to the heart, which usually discloses the presence of a presystolic sound in these patients. Hypertension, aortic stenosis, and coronary disease are common causes.

Left ventricular dilatation due to heart failure also produces an abnormal, sustained apex impulse. This movement may be double for yet another reason — forceful passive filling in early diastole. A protodiastolic extra sound (S_3) often accompanies this abnormality. Advanced coronary disease, cardiomyopathies, and certain valvular diseases, especially severe mitral regurgitation, are major causes of this abnormality.

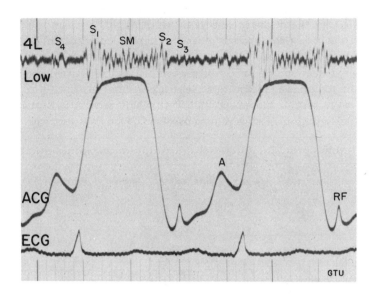

FIGURE 10-14. Top. Apex cardiogram (ACG) from 44-year-old man with severe mitral regurgitation demonstrating both presystolic (A) and rapid filling (RF) waves as well as a sustained systolic impulse. The corresponding filling sounds (S_4 and S_3) can be seen in the phonocardiographic tracing. This abnormal contour of the apical impulse usually can be felt with the examining hand. (Courtesy Division of Cardiology, Georgetown University Hospital.) Bottom. Double apex impulse amplified by "tongue blade" method.

Right ventricular hypertrophy and dilatation produce similar movements which are usually more diffuse and can be felt along the left sternal edge (Figure 10-15). The impulse of the volume- or pressure-loaded right ventricle frequently can be felt with a finger placed under the xyphoid process while the patient gently inhales. Right ventricular impulses are often single but may exhibit presystolic distention or a rapid filling wave just as the left ventricle does. The impulse of the pressure-loaded right ventricle is more sustained than the rapid outward early systolic movement found in patients with atrial septal defect (a lesion which results in selective volume overload of the right ventricle).

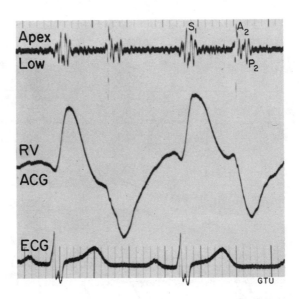

FIGURE 10-15.   Right ventricular lift recorded along the left sternal border from 7-year-old boy with an atrial septal defect. The outward movement which occupies only the early part of systole is typical of this right ventricular volume overload. (Courtesy Division of Cardiology, Georgetown University Hospital.)

Pulmonary artery dilatation produces a localized systolic lift in the second or third left intercostal space.

Constrictive pericarditis may produce a typical early diastolic outthrust along the sternal border or over the midprecordium. This is sometimes associated with a short inward movement during systole. It is often misinterpreted as a systolic event, but when unquestionably present it is pathognomonic of this one condition. Palpate and listen together!

Maneuvers that bring the heart closer to the chest wall will accentuate most of the precordial movements. Having the patient hold his breath in full expiration and/or rolling him a quarter of a turn to the left may be very helpful. Long-standing cardiac enlargement, when present during childhood, may cause a visible precordial bulge.

Percussion

When the left border of cardiac dullness falls outside the midclavicular line, the heart is usually enlarged. A greatly increased area of cardiac dullness, extending to the anterior axillary line, suggests marked ventricular dilatation or pericardial effusion. Percussion is not as reliable as palpation for estimating selective ventricular enlargement. The chest x-ray is undoubtedly superior to percussion in determining cardiac contour. It too has shortcomings, and perhaps the only completely reliable method of determining specific chamber enlargement is angiocardiography.

Auscultation

Abnormal Heart Sounds. Abnormal heart sounds are illustrated below.

 ACCENTUATED FIRST SOUND   LUB dup

The *first heart sound* is accentuated in many conditions, some of which are not abnormal. Sinus tachycardia in a healthy young person may result in a prominent first sound. Anemia, hyperthyroidism, and other disorders resulting in a hyperkinetic circulation cause a similar change. The increased left atrial pressure of mitral stenosis or a short P-R interval which causes the mitral and tricuspid valves to close from a relatively open position will also increase the intensity of the first sound. Low output states or delayed A-V conduction (long P-R interval) may result in a quiet first sound. A-V dissociation (complete heart block) with ventricular systole being variably related to atrial contraction is associated with marked changes in intensity of the first sound.

The aortic or pulmonic components of the *second sound* may be accentuated with elevation of the pressure in either of the respective circuits, i.e., systemic or pulmonary artery hypertension. With aortic or pulmonic stenosis they may become diminished or inaudible due to decreased mobility of the diseased cusps. Delayed pulmonic valve closure results in persistent splitting of the second sound which may not vary at all with respiration (fixed splitting) or may vary with it in the usual manner, e.g., not fusing in expiration but widening more than usual during inspiration. Atrial septal defect with volume overload causing prolonged right ventricular ejection is the most frequent cause of fixed splitting of the second sound. Pulmonic stenosis and ventricular septal defects often cause wide splitting of the second sound with retention of normal respiratory variation. Delayed electrical activation of the right ventricle, i.e., right bundle branch block, causes persistent splitting of the second sound with normal respiratory movement.

REVERSED SPLITTING
(Pulmonic Area)

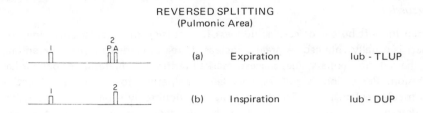

(a)   Expiration   lub - TLUP

(b)   Inspiration   lub - DUP

Reversed or paradoxic splitting of the second sound results from delayed left ventricular activation (left bundle branch block or an artificial pacemaker in the right ventricle) or prolonged emptying time of the left ventricle for mechanical reasons (aortic stenosis, marked hypertension, or poor contractility as seen in cardiomyopathy or myocardial ischemia). In paradoxic splitting, the second sound is split in expiration and single during inspiration. The basic cause of reversed splitting of the second sound is late aortic valve closure resulting from prolonged or delayed left ventricular ejection. Thus, in expiration the normal right ventricular ejection is shortest and the pulmonic valve closure precedes the aortic. When normal prolongation of right ventricular ejection occurs with inspiration, the two closures sound superimposed. Examination for paradoxic splitting is best accomplished during normal quiet respiration or *minimally* exaggerated breathing with the mouth open.

PRESYSTOLIC EXTRA   da - LUB dup
SOUND (Apex)

Presystolic extra sounds (S_4, atrial gallop) are dull and low-pitched. They are best heard at the apex with the patient rolled onto his left side and the bell of the stethoscope lightly applied precisely over the apex impulse. At times they are more easily felt than heard. The fourth heart sound is caused by vigorous atrial contraction in patients with decreased left ventricular compliance, which may be caused by left ventricular hypertrophy or myocardial ischemia. With pulmonary hypertension or pulmonic stenosis, a fourth heart sound originating in the right ventricle may be heard along the left sternal border. As with most right-sided auscultatory events, a right ventricular fourth sound increases with inspiration.

PROTODIASTOLIC EXTRA   LUB duppa
SOUND

Early diastolic extra sounds (S_3, ventricular gallop, or protodiastolic extra sound) occur during the period of rapid ventricular filling. They are commonly heard in mitral regurgitation, constrictive pericarditis, and ventricular failure of varying etiology. The S_3 is even lower pitched than the S_4 and is often more difficult to detect. Careful location of the cardiac apex in the left lateral decubitus position with light application of the bell of the stethoscope is imperative for detection of this sound. As with the S_4, an S_3 may originate in the right ventricle and exhibit the same respiratory variation. Right ventricular S_3's occur with tricuspid regurgitation or right ventricular failure.

When an S_3 and/or an S_4 is associated with a rapid heart rate, the heart sounds assume the cadence of a galloping horse. *Gallop rhythm* is regarded as a sign of congestive heart failure. With more rapid rates, the third and fourth sounds are superimposed and result in a single loud prolonged "summation" gallop.

The physiologic *third heart sound* is frequently heard in healthy young people. The occurrence of a third heart sound at an older age and/or in association with cardiac disease is felt to imply congestive heart failure. In aortic or mitral regurgitation with the enhanced rapid ventricular filling, the third heart sound may not mean that congestive heart failure has developed. An S_3 is to be expected with a major degree of mitral regurgitation.

EARLY EJECTION CLICK   lubbi DUP

Ejection clicks occur in early systole. They may originate in either of the great vessels or their valves. They are generated by abrupt distention of a dilated great vessel during early ventricular ejection or by upward movement of stiffened deformed valve leaflets. Ejection clicks heard in congenital valvular pulmonic stenosis increase in intensity with expiration and decrease with inspiration. The ejection click of pulmonary hypertension is not as apt to vary with respiration. Pulmonic ejection clicks are best heard at the base and high left sternal border. Aortic ejection clicks do not vary with respiration and are heard equally well at apex and base.

SYSTOLIC CLICK   LUB - i - dup

A *systolic click* is not uncommonly heard in persons presumed not to have heart disease. It is usually single, but two or more clicks may be heard (Figure 10-16). They are usually midsystolic, high-pitched, and sharp. They are extremely variable, changing with respiration or position or spontaneously. Clicks are heard best along the left sternal border or just inside the apex. Previously these clicks were felt to be extracardiac in origin and were regarded as benign. More recent information suggests that many systolic clicks are due to prolapse of a mitral valve leaflet and the "benign" systolic click is now highly suspect.

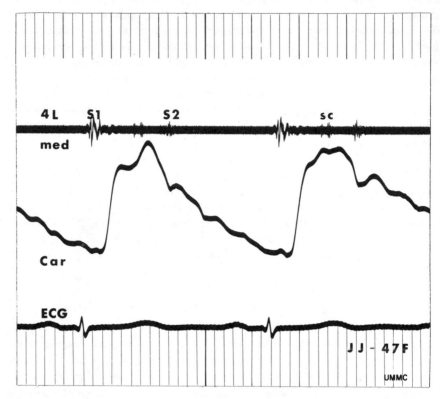

FIGURE 10-16. Mid-late-systolic clicks (sc) from patient with a prolapsing mitral valve leaflet. Although two clicks are seen here, there may be one to several, and their timing in systole may vary as the position of the patient or systemic resistance changes.

A *systolic crunch* or *knock* may be heard in patients with mediastinal emphysema, but it is much more common with small left-sided pneumothorax. In this case it has been referred to as "noisy pneumothorax," frequently being audible to the patient and at times to the physician at some distance from the bedside.

Heart Murmurs. Murmurs are immensely important in cardiac diagnosis. In order to interpret them intelligently, however, you must understand the factors that govern their production and transmission. Further, you must be able to describe them in precise and appropriate terms. The phonocardiogram, although of limited diagnostic value, is helpful as a teaching tool, since it allows time for study and correlation of events which ordinarily are passing by rapidly.

Blood flow within the circulation may be laminar or turbulent. Ideally flow is silent because of its laminar character. With laminar flow through any tube, the layer of fluid adjacent to the wall is stationary while the velocity of flow increases progressively within the inner layers and is maximal at the center. As this velocity exceeds a critical level, turbulence develops which produces vortices or swirls. These in turn emit high-frequency sound vibrations termed murmurs when heard in and around the heart and bruits when heard over peripheral vessels.

Production of murmurs is favored by a number of basic factors which tend to promote turbulence, including: (a) lowering the viscosity of the fluid, (b) increasing the diameter of a tube, (c) changing the caliber of a tube abruptly, and (d) increasing the velocity of flow. The velocity of blood flow is vitally important in governing two important characteristics of a murmur: intensity and pitch. The intensity of a murmur is directly proportional to V^3, that is, the velocity cubed. The faster the flow, the louder the murmur. Thus, exercise, when it speeds blood flow, causes most murmurs to become louder. Pitch is also directly proportional to velocity. The faster the flow, the higher the pitch; the slower the flow, the lower the pitch. There are many other causes of murmur production which are of lesser importance and will not be included here.

Transmission of murmurs also has a profound effect on both pitch and loudness. Sound intensity diminishes in direct proportion to the square of the distance that it must travel. Obviously the closer the murmur source to the chest wall, the louder it will sound. The natural damping effect of interposed tissues is compounded by the tendency of sound to be reflected backward at the interface between media of different densities, such as muscle, lung, bone, skin, and air. In fact, high frequencies tend to be dampened preferentially by this phenomenon; in other words, the pitch will become lower due to the effects of transmission across these interfaces. Another important factor governing pitch is related to the transmission of a murmur within the cardiovascular system itself. As one moves backward against the stream, high frequencies tend to be preserved and lows dampened out. In moving downstream (or forward), high frequencies tend to be lost and lows preserved. This phenomenon is shown in Figure 10-17. It is not surprising, then, that many murmurs sound quite different as the stethoscope is moved over the precordium.

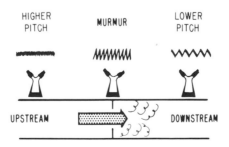

FIGURE 10-17. Changes in pitch of a heart murmur transmitted directly upstream or downstream.

- **Short Duration**

  Early systolic

  Early diastolic

  Late systolic

  Late diastolic
  (Presystolic)

- **Medium Duration**

  Midsystolic

  Middiastolic

- **Long Duration**

  Holosystolic

  Holodiastolic

  Continuous

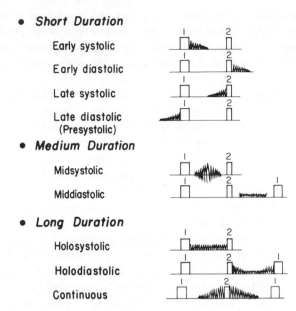

FIGURE 10-18. Duration of heart murmurs.

Description of murmurs is divided into six categories: (1) timing, (2) location, (3) loudness, (4) pitch, (5) duration, and (6) quality. These are defined as follows:

Timing: Systolic, diastolic, continuous.

Location: Point of maximum intensity described in terms of anatomic landmarks: apex; LSB (left sternal border); left base (pulmonic area); right base (aortic area); intermediate zones by exact intercostal space.

Loudness: Graded on a six-point scale. Grade 1/6, barely audible — increasing loudness — Grade 6/6, audible with stethoscope off chest wall.

Pitch: Low, 25—150 cps; medium, 150—350 cps; high, 350—600 cps.

Duration: See Figure 10-18.

Quality: Crescendo: increasing in loudness.

Decrescendo: decreasing in loudness.

Descriptive terms: blowing, harsh, rumbling, musical, cooing, whooping, honking, regurgitant, ejection.

Systolic murmurs. For the sake of simplicity the beginner may group most systolic murmurs into one of two categories: midsystolic (ejection) murmurs and holosystolic (regurgitant) murmurs. This subtyping of systolic murmurs correlates their quality with the underlying pathophysiologic mode of origin and eliminates errors produced by overdependence on the geographic site of maximum intensity, which may be unreliable.

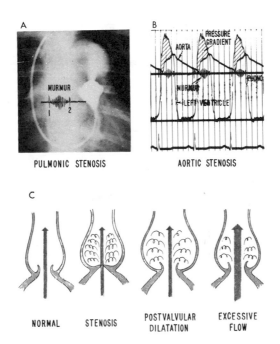

FIGURE 10-19. Mode of origin of the systolic ejection murmur. A. Angiocardiogram show-ing systolic jet. B. Aortic and ventricular pressure curves in aortic stenosis. C. Schematic explanation of murmur.

Systolic Ejection Murmurs. These are produced by the forward outflow of blood through the pulmonary or aortic valves. Because of their hemodynamic basis, ejection murmurs have a characteristic personality by virtue of the fact that they occur in midsystole, are medium-pitched, and rise and fall in a crescendo-decrescendo fashion, ending before the second sound. The four principal causes are:

1. Valvular or subvalvular stenosis.
2. High-velocity rate of ejection through the valves, which may themselves be normal (increased stroke volume).
3. Dilatation of the vessel beyond the valve.
4. A combination of these factors.

These modes of origin are illustrated in Figure 10-19.

Aortic systolic ejection murmurs occur with valvular and subvalvular stenosis (Figure 10-20), primary dilatation of the ascending aorta, and increased left ventricular stroke output. The last may be due simply to hyperkinetic states or may be a compensatory increase in forward flow through the valve because of diastolic backflow with aortic regurgitation. Aortic systolic murmurs are usually

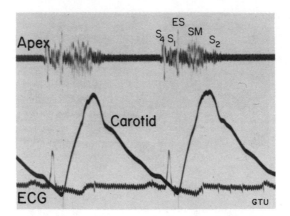

FIGURE 10-20. Graphic record showing many of the features of severe valvular aortic stenosis. The carotid upstroke is very slow, peaking in late systole. A prominent S_4 precedes S_1. An ejection sound (ES) introduces a typical crescendo-decrescendo mid-systolic murmur. S_2 is very soft. (Courtesy Division of Cardiology, Georgetown University Hospital.)

best heard in the aortic area, but they are frequently transmitted to the entire precordium and sometimes are maximal at or inside the apex, where their pitch may seem higher or their sound musical, but they maintain their characteristic ejection quality. They are often transmitted to the carotid arteries.

Pulmonic systolic ejection murmurs occur with pulmonary valvular and subvalvular stenosis, dilatation of the pulmonary artery, and increased pulmonary flow, as with atrial septal defect. They are usually localized to the second and third intercostal spaces and may be referred to the left upper chest.

Functional or innocent systolic murmurs, described in the previous section, are soft ejection murmurs. The turbulence in this case develops in the absence of any structural abnormality. The "hemic" systolic murmur commonly heard with anemia and the basal systolic murmurs frequently associated with thyrotoxicosis, fever, and exercise are functional murmurs, but in these instances the cardiac output is elevated.

Regurgitant Systolic Murmurs. These are produced by backflow of blood from the ventricle to the atrium through an incompetent mitral or tricuspid valve, or by flow through a ventricular septal defect. Since they are due to the escape of blood from a chamber of relatively high pressure into one of relatively low pressure, and since this pressure differential lasts longer, regurgitant murmurs differ from ejection murmurs. They are longer in duration, usually holosystolic, and may engulf the first or second sound or both. They are plateau-shaped and either of constant intensity or, at times, increased in late systole. They do not rise and fall in intensity as an ejection murmur does. Figure 10-21 illustrates the mode of origin of a regurgitant systolic murmur.

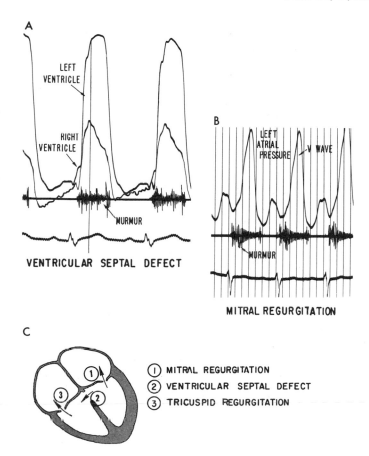

FIGURE 10-21. Mode of origin of the regurgitant systolic murmur. A. Intraventricular pressure curves in ventricular septal defect. B. Left atrial pressure curve in mitral regurgitation. The V wave represents the pressure effect of backflow from the left ventricle to the left atrium through an incompetent mitral valve. C. Schematic representation of three common hemodynamic abnormalities.

Mitral regurgitation usually causes a blowing holosystolic regurgitant murmur. It tends to be loudest at the apex and is transmitted laterally toward the axilla. There may be an associated thrill. Mild congenital forms of mitral regurgitation and certain special abnormalities of function related to weakened papillary muscle contraction may result in a characteristic variant which begins in midsystole. Such a murmur may begin with a prominent systolic click. An S_3 is usually found with mitral regurgitation of significant degree. (See Figure 10-22.)

Tricuspid regurgitation causes a murmur, similar to that of mitral insufficiency, which is best heard over the tricuspid area. It can also be distinguished by the fact that it may become louder with inspiration.

The murmur of *ventricular septal defect* is heard in the third and fourth left

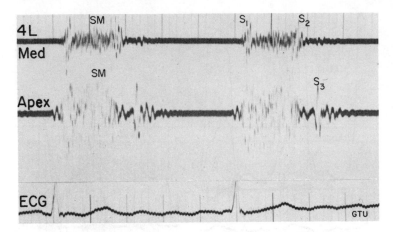

FIGURE 10-22. Pansystolic murmur of mitral regurgitation. The murmur goes all the way to the second sound and is plateau-shaped rather than crescendo-decrescendo. A very prominent S_3 is seen at the apex. (Courtesy Division of Cardiology, Georgetown University Hospital.)

intercostal spaces. It frequently causes a thrill and has a loud, coarse quality which is practically pathognomonic.

<u>Diastolic murmurs.</u> Diastolic murmurs may also be divided into two categories: high-pitched decrescendo murmurs of aortic and pulmonic insufficiency, and lower pitched murmurs of mitral and tricuspid stenosis.

Aortic regurgitation results in a high-pitched murmur which begins immediately with aortic closure and diminishes progressively with diastole. Its intensity varies roughly with the size of the leak (Figure 10-23). It is heard best along the left sternal border with the *patient sitting and holding his breath in expiration.* Because of its high pitch, this murmur is best heard using the diaphragm endpiece. A low-pitched diastolic murmur may be heard at the apex in patients with aortic insufficiency. This murmur, which sounds like the murmur of mitral stenosis, is referred to as an Austin Flint murmur.

Pulmonic regurgitation causes a diastolic murmur that cannot be distinguished from the aortic counterpart simply by auscultation alone. Its pitch, timing, quality, and location are similar, although it tends to be more localized to the pulmonic area. Pulmonic insufficiency is a common finding with severe pulmonary arterial hypertension; then it is called a Graham Steell murmur. Congenital pulmonic insufficiency occurs with normal pressures in the pulmonary artery and results in a diastolic murmur which is low- to medium-pitched and begins at an interval after the second sound.

Mitral stenosis characteristically produces a low-pitched, localized, apical rumble. Often it can be heard *only with the bell* and with the patient *rolled onto his left side.* As is the case with the third and fourth heart sounds, the low-pitched rumble of mitral stenosis must be sought precisely over the apical impulse.

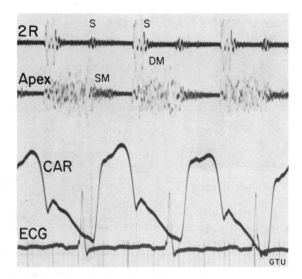

FIGURE 10-23. Phonocardiographic tracing from 30-year-old woman with severe aortic regurgitation secondary to Marfan's syndrome. The second sound is very intense and is followed by a loud diastolic murmur (DM). The short early systolic murmur (SM) represents increased flow across the aortic valve and *not* aortic stenosis. Such a systolic murmur is heard in almost every case of severe aortic regurgitation (it may be loud enough to be associated with a thrill) and should not be interpreted as representing aortic obstruction unless other information (carotid upstroke, aortic valve calcium) is suggestive. (Courtesy Division of Cardiology, Georgetown University Hospital.)

Figure 10-24 illustrates the hemodynamic basis for the mitral diastolic murmur. The murmur is loudest in early and late diastole, at which time the pressure gradient is greatest across the narrowed mitral valve. The early diastolic component is frequently initiated by a sharp click called a *mitral opening snap.* The opening snap can usually be heard at the apex but often is more easily discerned medial to the apex or along the left sternal border. Its separation from the second sound is related to left atrial pressure; the higher the left atrial pressure, the closer the opening snap is to the second sound, and vice versa. This *2-OS interval* may be used to estimate roughly the severity of the stenosis.

The murmur has a decrescendo quality through early and middiastole. During the latter third of diastole the gradient increases sharply due to atrial contraction, causing a *presystolic accentuation* of the murmur. The typical configuration is seen in Figure 10-24B. With atrial fibrillation, coordinated contraction of the atrium ceases and the presystolic accentuation disappears (Figure 10-24C).

The diastolic murmur of *tricuspid stenosis* is similar in timing and quality to the mitral murmur but is frequently higher-pitched and localized near the tricuspid area or along the left sternal border. Inspiration usually makes it louder.

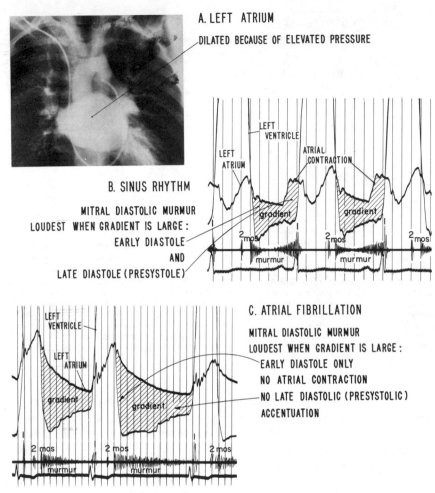

FIGURE 10-24. Mitral stenosis. A. Angiocardiogram showing dilated left atrium. Hemo-dynamic basis for mitral diastolic murmur. B. Normal sinus rhythm. C. Atrial fibrillation.

Continuous murmurs begin in systole and continue into diastole without stopping. Although it is called continuous, such a murmur may not occupy the entire cardiac cycle. Continuous murmurs are found in those situations in which there is flow from a high to a lower pressure chamber uninterrupted by an opening or closing of cardiac valves. The prototype of a continuous murmur is the Gibson murmur of patent ductus arteriosus (Figure 10-25). This murmur begins in systole, peaks around the second sound, and then spills over into diastole, but may not go all the way to the first sound. It is heard maximally under the left clavicle and in the pulmonic area. Other causes of continuous murmurs are pulmonary and coronary A-V fistulae. Even though the combined

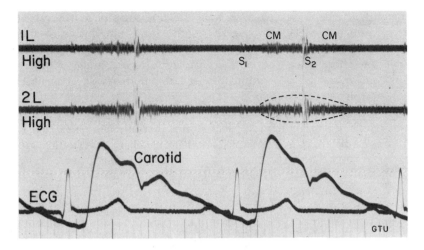

FIGURE 10-25. Phonocardiogram from 15-year-old girl with patent ductus arteriosus demonstrating typical continuous murmur (CM). Note how the murmur starts in early systole, peaks around the second sound, and trails off into diastole. (Courtesy Division of Cardiology, Georgetown University Hospital.)

murmurs of aortic stenosis + aortic insufficiency, mitral regurgitation + aortic regurgitation, or ventricular septal defect + aortic regurgitation may fill the entire cardiac cycle, by definition they are *not* continuous murmurs but combinations of separate systolic and diastolic murmurs.

Pericardial Friction Rub. The pericardial friction rub is a sign of pericardial irritation. It may be heard anywhere over the precordium but is often loudest along the left sternal border or directly over the sternum. Sometimes it is quite loud, but frequently it is faint, high-pitched, and evanescent. Firm pressure with the diaphragm of the stethoscope and/or auscultation with the patient sitting up, leaning forward with breath held in expiration, are maneuvers that will bring out a friction rub. Friction rubs have a superficial, scratchy, to-and-fro quality, often suggesting squeaky leather. They lag somewhat, giving the effect of being slightly out of step with the heart sounds. A fully developed friction rub has three components, corresponding with the systolic, early diastolic, and presystolic phases of the cardiac cycle. One should hear at least two components before diagnosing a rub. The intensity of the rub may vary with respiration, but a true pericardial rub should be audible with respiration halted. Friction rubs may be simulated by movements of the stethoscope's endpiece on the surface of the skin, particularly by hair.

Auscultatory Adjuncts. Several simple bedside interventions which will often enhance one's diagnostic accuracy can be used as adjuncts to routine cardiac examination.

Isometric (handgrip) exercise. (Figure 10-26.) Isometric exercise (simply done by having the patient squeeze tightly with both hands) results in tachycardia and increase in blood pressure and cardiac output. These hemodynamic alterations cause the murmurs of aortic regurgitation, mitral regurgitation, and mitral stenosis to intensify and the murmurs of left ventricular outflow obstruction to diminish.

Amyl nitrite. The inhalation of amyl nitrite over a period of 15 to 20 seconds results in a prompt fall in mean arterial pressure with reflex increase in heart rate, cardiac output, and stroke volume. Characteristic alterations of the

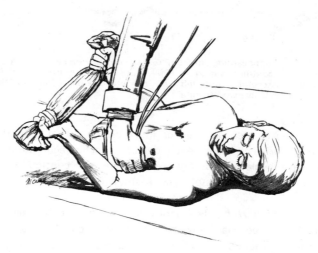

FIGURE 10-26. Isometric exercise can be accomplished simply by having the patient make tight fists. This is easier for the patient if he has something to grip. Here, this is achieved at the bedside by having him squeeze a rolled towel. The patient must be reminded to keep breathing to avoid the added effects of a Valsalva maneuver. The left lateral decubitus position and application of the bell are ideal techniques to bring out diastolic extra sounds.

auscultatory findings ensue. The lower systemic resistance produces either no change or a decrease in regurgitant murmurs, i.e., those of mitral regurgitation, aortic regurgitation, and ventricular septal defect. The increased flow causes increased intensity of outflow murmurs, i.e., aortic stenosis, pulmonic stenosis, and mitral stenosis. This intervention will help distinguish outflow from regurgitant murmurs and the murmur of organic mitral stenosis from the Austin-Flint murmur of aortic regurgitation.

Prompt squatting. (Figure 10-27.) The rapid assumption of a squatting position causes an increase in the mean blood pressure. Trivial murmurs of aortic regurgitation can be brought out or increased in intensity with this maneuver. The increase in arterial blood pressure will result in a decreased obstruction in

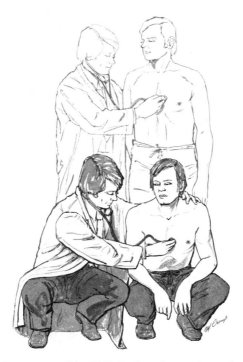

FIGURE 10-27. Prompt squatting. While both patient and examiner are standing, the examiner should find the area of maximum intensity and "tune in" to the auscultatory event in question. Then both the patient and examiner rapidly assume the squatting position. The expected auscultatory changes will occur within the first few cardiac cycles after squatting. It may be necessary to repeat the squat several times to confirm changes. Elderly or weak patients may require support from an assistant.

IHSS with a decrease in murmur, while the murmur of valvular aortic stenosis does not change.

Disorders of Rhythm

The electrocardiogram is indispensable for the precise recognition and identification of arrhythmias, but much can be learned by careful examination at the bedside. Listed below are some of the common arrhythmias with their associated physical findings.

Sinus bradycardia: of sinus origin; rate slower than 60; may be seen in well-conditioned athlete or in elderly patient with disease of conducting system.

Sinus tachycardia: rate 100—150; usually varies with position, exercise, and respiration; carotid sinus pressure may cause gradual slowing followed by gradual resumption of the initial rate.

Prolonged A-V conduction: (first-degree A-V block) faint first heart sound.

Complete A-V block: intermittent cannon waves in the jugular pulse; audible atrial sounds; varying intensity of the first heart sound; slow ventricular rate (less than 40); bounding pulse; aortic systolic ejection murmur. When treated with an artificial pacemaker, the findings are the same except for a faster ventricular rate. Since most pacemakers activate the right ventricle first, the splitting of the second sound is analogous to that seen with left bundle branch block, i.e., reversed or paradoxical.

Premature beats: may be ventricular or atrial in origin; underlying regular rhythm usually discernible; frequently disappear with exercise.

Atrial tachycardia (paroxysmal): abrupt onset and cessation; rate 150—250 and very regular; at times terminated by carotid sinus pressure; not influenced by respiration, exercise, or position.

Ventricular tachycardia: basically regular; rate 150—250; variable intensity of the first sound; occasional cannon waves in jugular pulse; not influenced by respiration, exercise, or position; beat-to-beat variation in systolic blood pressure.

Atrial flutter: ventricular rate variable; unaffected by exertion or position; rhythm regular or irregular; carotid sinus pressure may cause brief period of slowing because of increased A-V block; flutter waves may be visible in jugular pulse.

Atrial fibrillation: ventricular rate variable; increases with exercise; rhythm grossly irregular with variable intensity of the first heart sound; variable amplitude of the arterial pulse; apical-radial pulse deficit.

Ventricular fibrillation: absent blood pressure and pulse; faint vague precordial sounds; dilated pupils; apnea; sudden death.

CLASSIC PHYSICAL FINDINGS IN COMMON
CARDIAC DISORDERS

Mitral Stenosis. Cold hands and feet; small pulse; tapping apex impulse; parasternal lift; presystolic apical thrill; accentuated M_1 and P_2; mitral opening snap; mitral diastolic rumble. Atrial fibrillation (late).

Mitral Regurgitation. Chronic, usually rheumatic. Normal or small, brisk, arterial pulse; apical systolic thrill; sustained apical lift displaced to left; normal or soft M_1; apical regurgitant systolic murmur; early diastolic extra sound. Mild forms may show murmur only. (See Figure 10-22.) *Acute, severe — usually caused by disruption of supporting structure apparatus.* Varying degrees of congestive heart failure, loud S_1; harsh decrescendo holosystolic murmur, S_3 and S_4; right ventricular lift; increased P_2; right-sided filling sounds.

Aortic Stenosis. Cold hands and feet; small, slow-rising pulse; narrow pulse pressure (in severe cases); sustained apical lift displaced to the left; systolic thrill in aortic area; decreased or absent A_2; systolic ejection murmur at base and over carotids; systolic ejection click; presystolic extra sound; reversed split; carotid thrill. (See Figure 10-20.)

IHSS (Idiopathic hypertrophic subaortic stenosis, hypertrophic obstructive cardiomyopathy). Brisk carotid pulse often with two beats (see Figure 10-13); prominent A wave in jugular venous pulse; thrill at apex or left sternal border; triple apical impulse (presystolic and two systolic waves); fourth heart sound; crescendo-decrescendo murmur at apex and left sternal border; aortic regurgitation murmur and ejection sounds not present.

Pulmonary Stenosis. Normal pulse; jugular A wave; parasternal lift; pulmonic thrill; pulmonary component of second sound absent or soft and delayed, causing widely split P_2; systolic ejection murmur in pulmonic area; presystolic extra sound in tricuspid area (severe form); pulmonary ejection click (mild form).

Aortic Regurgitation Prominent carotid pulsations; water-hammer or bifid pulse; capillary pulsations of the nailbeds; diffuse, sustained apex impulse displaced down and left; loud M_1, accentuated A_2; ejection systolic murmur in aortic area; decrescendo diastolic murmur along left sternal border; low-pitched rumbling diastolic murmur at apex (Austin Flint). (See Figure 10-23.)

Systemic Hypertension. Elevated blood pressure; pulse hard, tortuous, and usually full; hypertensive funduscopic changes; sustained double apical lift displaced to the left; normal or accentuated M_1; accentuated A_2; presystolic extra sound; aortic systolic ejection murmur.

Pulmonary Hypertension. Cyanosis (at times); small pulse; narrow pulse pressure; cold extremities; atrial fibrillation (late); giant jugular A wave; parasternal lift; systolic lift in the pulmonic area; accentuated P_2, closely split S_2; presystolic extra sound at tricuspid area; pulmonary ejection click; pulmonic diastolic murmur (Graham Steell); systolic ejection murmur in pulmonic area.

Tricuspid Stenosis. Small pulse; giant A wave in jugular pulse; elevated jugular pressure; quiet precordium; tricuspid diastolic murmur accentuated with respiration; tricuspid opening snap. (Note: Tricuspid stenosis is usually associated with mitral stenosis.)

Tricuspid Regurgitation. Jugular V wave; elevated venous pressure; right ventricular parasternal lift; systolic thrill at tricuspid area; tricuspid regurgitant systolic murmur louder with inspiration; atrial fibrillation; early diastolic extra sound (over right ventricle).

Prosthetic Heart Valves. Mitral: Sharp component to first sound as valve closes; similar sound as valve opens, about .08 to .10 second after the second sound, analogous to the mitral opening snap. Characteristically no murmur is generated over such a valve. *Aortic:* Sharp opening component heard in the first sound; closing sound heard at time of S_3; harsh, short, systolic ejection murmur (sometimes with thrill) typically heard. Sounds of either mitral or aortic prosthetic valves may be heard at a distance from the patient, without a stethoscope.

Myocardial Infarction. Tachycardia; pallor; small arterial pulse; narrow pulse pressure; apical late systolic bulge of ischemic myocardium; soft heart sounds; presystolic or early diastolic extra sounds; pericardial friction rub; apical systolic murmur (papillary muscle dysfunction); any of the arrhythmias.

Atrial Septal Defect. Normal pulse; brisk parasternal lift; lift over pulmonary artery; normal jugular pulse; systolic ejection murmur in pulmonic area; low-pitched diastolic rumble over tricuspid area (at times); persistent wide splitting of P_2. (See Figure 10-15.)

Ventricular Septal Defect. Small pulse; normal jugular pulse; parasternal lift and left ventricular apical lift; systolic thrill and loud systolic regurgitant murmur in third and fourth interspaces along left sternal border; apical diastolic rumble; wide split of P_2 with normal respiratory movement.

Pericarditis. Tachycardia; friction rub; diminished heart sounds and enlarged heart to percussion (with effusion); paradoxic pulse; neck vein distention, narrow pulse pressure and hypotension (with tamponade).

SPECIAL TECHNIQUES

Electrocardiography. This is a routine procedure for evaluating patients with possible or potential cardiovascular disease. It is such a common part of diagnostic and therapeutic management that it can best be considered as part of the clinical examination. Indeed, it has its greatest value when interpreted immediately by the physician as an integral part of his bedside observations. Its chief contribution is in the confirmation of rhythm disturbances and in the immediate diagnosis of acute myocardial infarction. Selective ventricular and atrial enlargement, electrolyte abnormalities, and many other disorders which affect the myocardium result in discernible though less specific pattern changes which are also of diagnostic value.

Roentgenography of the Chest. This is also an integral part of clinical cardiovascular diagnosis. Like the electrocardiogram, it is more valuable when available for immediate correlation with physical findings. It is far more reliable than percussion for determining heart size and contour. The bronchovascular markings are also helpful for estimating pulmonary congestion and excessive pulmonary arterial blood flow. The simple posteroanterior and lateral views may at times be supplemented by oblique projections or special overpenetrated films to bring out certain abnormalities. Many cardiac diagnoses, such as coarctation of the aorta, are possible from the chest film alone.

Fluoroscopy. Performed in conjunction with roentgenography of the chest, this method yields useful information with regard to valvular calcification, motion of the ventricular wall, and pulmonary vasculature which frequently cannot be obtained from routine roentgenograms.

Phonocardiography. This technique involves simultaneous graphic registration of electrocardiogram, heart sounds and murmurs, as well as various pulse waves on paper (as in Figures 10-7, 10-14a, 10-15, 10-16, 10-20, 10-22, 10-23, 10-24, and 10-25). Its major value is in the teaching of cardiac physical diagnosis, although it sometimes helps in timing of heart sounds (splits vs. ejection sounds vs. extra sounds). A phonocardiogram is *not* a substitute for careful physical examination. Additionally, very soft acoustical events such as high-pitched aortic insufficiency murmurs or faint S_3's are technically very difficult to record and can usually not be proven by this method.

Echocardiography. This new method records ultrasound waves which are reflected onto the chest wall from cardiac structures. Clearly established uses of echocardiography include evaluation of mitral valve motion and detection of pericardial fluid. Typical abnormalities of right ventricular size and septal motion in atrial septal defect and septal and mitral valve motion in IHSS are now well documented. More sophisticated analysis of mitral valve motion in various types of mitral regurgitation and left ventricular wall motion is being done with increasing frequency and accuracy.

Exercise Electrocardiography. Monitoring of the electrocardiogram during and immediately after graded exercise (usually with a treadmill or bicycle ergometer) allows one to assess the cardiac response to effort and to determine if there is evidence of myocardial ischemia during exertion. There are noninvasive methods of determining cardiac output available which can be combined with the exercise testing.

Cardiac Catheterization. This is a routine procedure for precise identification and quantification of cardiovascular disorders. It is primarily used for pre-operative confirmation of surgically correctable heart disease. Measurements of intracardiac pressure, oxygen saturation, and flow are often combined with special angiographic and other techniques at the time of study.

Angiocardiography and Cineangiography. These are special studies, usually performed along with cardiac catheterization, which involve x-ray filming during injection of radiopaque contrast media into the various cardiac chambers and great vessels. This technique allows for qualitative assessment of intracardiac and vascular anatomy and ventricular function as well as gross estimation of the degree of valvular regurgitation or intracardiac shunting.

Coronary Arteriography. This specialized type of angiography involves selective injection of contrast media into the coronary arteries. This study may be done for diagnostic purposes in the evaluation of patients with chest pain or for whom surgical revascularization is being considered. Since this study entails more risk than routine cardiac catheterization, it should be applied only in those situations in which information pertinent to diagnosis or treatment can be obtained in no other way.

Central Venous Pressure. This is easily monitored at the bedside using a small indwelling catheter inserted from a peripheral or subclavian vein with the tip in an intrathoracic vein, and a simple water manometer. Since central venous pressure reflects right heart pressure, it may not be at all indicative of the situation in the left heart and is being rapidly replaced by pulmonary artery pressure monitoring for the treatment of the acutely ill cardiac patient.

REFERENCES

1. Adolph, R. J., and Fowler, N. O. The second heart sound: A screening test for heart disease. *Mod. Concepts Cardiovasc. Dis.* 39:91, 1970.
2. Ask-Upmark, E. *Bedside Medicine.* Uppsala: Almqvist-Wiksells, 1963.
3. Benchimol, A., and Tippit, H. C. The clinical value of the jugular and hepatic pulses. *Progr. Cardiovasc. Dis.* 10:159, 1967.
4. Conn, R. D., and Cole, J. S. The cardiac apex impulse: Clinical and angiographic correlations. *Ann. Intern. Med.* 75:185, 1971.
5. Craige, E. Gallop rhythm. *Progr. Cardiovasc. Dis.* 10:246, 1967.
6. Dohan, M. C., and Criscitiello, M. G. Physiological and pharmacological manipulations of heart sounds and murmurs. *Mod. Concepts Cardiovasc. Dis.* 39:121, 1970.
7. Feinstein, A. R. Acoustic distinctions in cardiac auscultation: With emphasis on cardiophonetics, synecphonesis, the analysis of cadence, and the problems of hydraulic distortion. *Arch. Intern. Med.* 121:209, 1968.
* 8. Fowler, N. O. *Examination of the Heart. Inspection and Palpation of Venous and Arterial Pulses.* New York: American Heart Association, 1967.
9. Groom, D. Comparative efficiency of stethoscopes. *Am. Heart J.* 68:220, 1964.
10. Harvey, W. P. Some pertinent physical findings in the clinical evaluation of acute myocardial infarction. *Circulation* (Suppl. IV:175), 1969.
11. Hurst, J. W., and Logue, R. B. *The Heart.* New York: McGraw-Hill, 1970.
*12. Hurst, J. W., and Schlant, R. C. *Examination of the Heart. Inspection and Palpation of the Anterior Chest.* New York: American Heart Association, 1967.
13. Jones, F. L. Frequency, characteristics and importance of the cervical venous hum in adults. *N. Engl. J. Med.* 267:658, 1962.

14. Julius, S., and Stewart, B. H. Diagnostic significance of abdominal murmurs. *N. Engl. J. Med.* 276:1175, 1967.

15. Lange, R. L., Botticelli, J. T., Tsgaris, T. J., Walker, J. A., Gani, M., and Bustamante, R. A. Diagnostic signs in compressive cardiac disorders — constrictive pericarditis, pericardial effusion, and tamponade. *Circulation* 33:763, 1966.

16. Leatham, A. Auscultation of the heart. *Lancet* 2:703, 757, 1958.

*17. Leonard, J. J., and Kroetz, F. W. *Examination of the Heart. Auscultation.* New York: American Heart Association, 1967.

18. Levine, S. A., and Harvey, W. P. *Clinical Auscultation of the Heart.* Philadelphia: Saunders, 1959.

19. Marx, H. J., and Yu, P. N. Clinical examination of the arterial pulse. *Progr. Cardiovasc. Dis.* 10:207, 1967.

20. McCraw, B., Siegel, W., Stonecipher, H. K., Nutter, D. O., Schlant, R. C., and Hurst, J. W. Response of heart murmur intensity to isometric (handgrip) exercise. *Br. Heart J.* 34:605, 1972.

21. Perloff, J. K. *The Clinical Recognition of Congenital Heart Disease.* Philadelphia: Saunders, 1970.

22. Silverman, M. E., and Hurst, J. W. Abnormal physical findings associated with myocardial infarction. *Mod. Concepts Cardiovasc. Dis.* 38:69, 1969.

23. Spodick, D. H. Pericardial friction — characteristics of pericardial rubs in fifty consecutive, prospectively studied patients. *N. Engl. J. Med.* 278:1204, 1968.

24. Spodick, D. H., and Quarry-Pigott, V. M. Fourth heart sound as a normal finding in older persons. *N. Engl. J. Med.* 288:140, 1973.

*25. Sprague, H. B. *Examination of the Heart. History Taking.* New York: American Heart Association, 1967.

26. Tavel, M. E. *Clinical Phonocardiography and External Pulse Recording.* Chicago: Year Book, 1972.

27. Wood, P. *Diseases of the Heart and Circulation.* Philadelphia: Lippincott, 1956.

*These four booklets on examination of the heart, published by the American Heart Association, are available free through your local Heart Association. They are a thorough, yet concise review of cardiac diagnosis written by well-known authorities and we highly recommend them to you.

GASTROINTESTINAL SYSTEM 11

Richard D. Judge
George D. Zuidema

Evaluation of the gastrointestinal system is complete only when it includes a blend of clinical, radiologic, and endoscopic techniques. Even in the hands of a very skilled physician, the history and the physical examination are usually a prelude to selected special studies. It is within this context that the following recommendations for clinical assessment of the gastrointestinal system are made.

GLOSSARY

Anorexia Loss of appetite.

Ascites Free fluid within the peritoneal cavity.

Borborygmi Audible bowel sounds due to active peristalsis. Not necessarily a pathologic sign.

Cachexia Profound malnutrition and weight loss.

Colic Acute, cramping abdominal pain occurring in waves or surges.

Constipation Difficult evacuation of stool because of firm consistency. Should be distinguished from simple infrequency of bowel movements.

Deglutition Swallowing.

Diarrhea Passage of frequent stools of watery consistency.

Dyschezia Painful defecation.

Dyspepsia Indigestion, usually with uncomfortable abdominal fullness, belching, and nausea.

Dysphagia Difficulty in swallowing.

Flatulence Passage of gas from the lower bowel.

Guarding Involuntary spasm of the abdominal muscles, frequently localized to an area of underlying pain and tenderness.

Hematemesis Vomiting of blood.

Ileus Intestinal obstruction with dilation of the bowel and obstipation. May be on a mechanical or paralytic basis.

Jaundice Yellow coloration of the skin, mucous membranes, and sclera, due to accumulation of bilirubin pigments in serum and tissues. Also referred to as *icterus.*

Malabsorption Inadequate and disordered absorption of intestinal nutrients.

Melena Passage of black stool, which is darkened by altered blood. The blood may be deep maroon or black and tarry in appearance.

Nausea An unpleasant sensation of impending vomiting, frequently localized to the epigastrium.

Obstipation Persistent failure to pass any stool.

Preprandial and postprandial Before and after a meal.

Rebound Abbreviated term for rebound tenderness; abdominal discomfort on sudden withdrawal of the palpating hand.

Scaphoid A thin, concave-shaped abdomen. Literally "shiplike," derived from the Greek.

Steatorrhea Excessive fat in the stool.

Stomatitis Inflammation of the mouth.
Tenesmus The sensation of the need to evacuate the bowels, but without result.
Tympanites Distention of the abdomen, due to presence of gas or air in the intestine or peritoneal cavity.
Xerostomia Dryness of the mouth.

TECHNIQUE OF EXAMINATION AND NORMAL FINDINGS

General Inspection

Clinical assessment of the gastrointestinal system begins with a brief general survey of the patient. His facial expression, for example, and his position in bed indicate whether he is relaxed, tense, or in pain. Briefly notice the state of skin and nails and the hair distribution. Outward signs of the patient's nutritional state indirectly reflect gastrointestinal function. (See Chapter 4 for a detailed consideration of nutrition.) Since serious fluid loss may accompany gastrointestinal disease, the state of hydration is also important. It may be assessed by gently squeezing the skin, as on the dorsum of the hand. A glance at the sclera for possible jaundice and similar inspection of the skin for pallor, pigmentation, or yellow discoloration are also of value. Notice whether there is any unusual odor. Normally there is none, although a mild degree of halitosis may at times be detected, particularly in patients with dental disease.

Mouth

A detailed examination of the mouth is considered in Chapters 7 and 8, but it should be emphasized that inspection of the oral cavity is an important part of the gastrointestinal examination. Quickly observe the lips, oral mucosa, teeth and gingiva, and tongue. Is salivation adequate? Although some nutritional deficiencies are reflected in the oral cavity, many nonpathologic changes may be observed. Some of these include (1) cracking of the lips due to exposure to weather and cold, (2) angular stomatitis from poorly fitting dentures, (3) mild gingivitis, a very common finding in otherwise healthy people, (4) prominent fungiform papillae set in a background of white coating, giving the appearance of "strawberry tongue," (5) denudation of the filiform papillae, usually along the midline, giving an appearance called "geographic tongue," and (6) normal furring or coating of the tongue, which is merely dead epithelium combined with yeast and saprophytes. The well-known "coated tongue" may be of some concern to a patient, but it is common in healthy people.

Abdomen

As the first step in the examination of the abdomen, make sure that the patient is completely relaxed and properly positioned. His head should lie comfortably on one or two pillows, and his arms should be at his sides. The knees should be raised slightly in order to relax the abdominal musculature. Draw aside the bedclothes to make sure there is adequate exposure from the costal margin to the symphysis pubis. It is helpful if the bed or examining table is high enough to permit you to work in comfort, and the patient close enough to the edge to permit access to the entire abdomen. Both the room and your hands should be warm, since chilling produces involuntary muscle spasm which hampers satisfactory examination. Lighting, of course, must be adequate.

A systematic plan for abdominal examination follows the usual sequence, that is, inspection, palpation, percussion, and auscultation. By closely following this method, the examiner may avoid unfortunate omissions.

For convenience, we divide the abdomen into topographic segments. This permits precise localization of physical signs and symptoms, and makes it possible to correlate physical signs with the anatomic location of viscera within the abdomen. A number of systems for describing topographic anatomy have been advocated. The most useful method is to divide the abdomen into four quadrants, by means of two intersecting lines (Figure 11-1A). One line extends

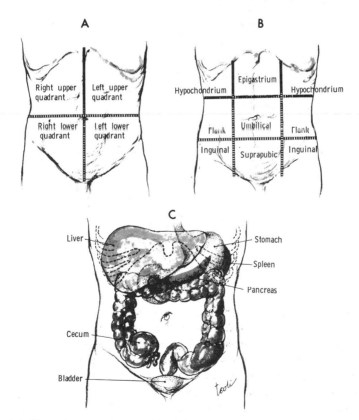

FIGURE 11-1. Topographic anatomy of the abdomen.

vertically from the xiphoid to the pubic symphysis. The other is a horizontal line across the abdomen at the level of the umbilicus. In this way the abdomen is divided into right upper, right lower, left upper, and left lower quadrants. The structures located in each of these areas are listed below.

Right Upper Quadrant	Left Upper Quadrant
liver	stomach
gallbladder	spleen
duodenum	left kidney
pancreas	pancreas
right kidney	splenic pancreas for body and tail
hepatic flexure of colon	splenic flexure of colon

Right Lower Quadrant	Midline Structure	Left Lower Quadrant
cecum	bladder	sigmoid colon
appendix	uterus	left ovary and tube
right ovary and tube		

The plan outlined above is the most widely used guide. Students will, however, probably be familiar with such terms as epigastrium, umbilical, suprapubic, and flank. These terms are descriptive and are often useful. They are derived from a system of topographic anatomy in which the abdomen is divided into nine regions (Figure 11-1B). This division is accomplished by the use of two vertical and two horizontal lines, dividing the abdomen into the right hypochondriac, epigastric, left hypochondriac, right flank, umbilical, left flank, right inguinal, suprapubic, and left inguinal, respectively. Certain of these terms have found general clinical use, largely because they permit very precise localization of a symptom or a sign. The use of this system in its entirety, however, has certain disadvantages, for such extensive subdivision produces regions so small that they contain very few structures. The term hypochondriac or inguinal region is rarely used. On the other hand, it is often convenient to speak of an umbilical or epigastric location, and this terminology is standard.

Figure 11-1C presents the location of the abdominal viscera. It is well to keep these relationships in mind as you perform the physical examination.

Inspection. First of all, observe the skin, its color and its texture. Are there any unusual lesions, striae, or surgical scars? Usually the venous pattern is barely perceptible, and the drainage of the lower two-thirds of the abdomen is downward. Next, observe the general contour. Is it symmetrical? Is there any localized bulging or prominence? Is the umbilicus in the midline and normal? Instruct the patient to cough or bear down, and notice whether this produces any bulging. Does this maneuver cause any pain? Normally it does not. Notice any unusual movements. Normally there is a rhythmic respiratory movement causing a slight protrusion with inspiration due to descent of the diaphragm. Are there any visible peristaltic waves? Finally, are there any vascular pulsations?

Occasionally in the thin individual with a flat or scaphoid abdomen, the normal aortic pulsation may be evident in the epigastrium.

Percussion and Palpation. Percussion and palpation are performed together. No two physicians follow exactly the same approach, and in time you, too, will develop your own method. Whatever it is, adhere to it strictly, for this is the surest way of avoiding costly omissions.

Right-handed examiners nearly always prefer to approach the abdomen from the patient's right side. Since the area is sensitive and not infrequently ticklish, some preparation against the shock of first contact from the examining hand is helpful. Chat with the patient; warm your hands by rubbing them together; touch the patient first on the forearm, very lightly, and do not try to elicit information until he settles down. If he absolutely cannot relax, have him palpate his own epigastrium, then place your hand on top of his, and finally beneath his. Finally, ask him to withdraw his own hand. This maneuver, though rarely necessary, can be helpful with very tense patients and with children.

Keep in mind that this will be one of the most difficult parts of the routine examination. Proceed slowly; do not rush. When difficulty is encountered, have the patient breathe with the mouth open. This automatically causes some relaxation. Such suggestions as, "Now, relax as if you were falling asleep," may be valuable under certain circumstances. If the bed is low, do not hesitate to kneel down beside it or sit down on the edge. The law that prohibits sitting on the patient's bed can be violated under special circumstances if done tactfully. Finally, concentrate all your senses on the examination. In particular, watch the patient's facial expression. A slight wince or almost inaudible gasp may be of major importance as you proceed.

Consider next your objectives. They are simple and limited. You must try to determine the presence or absence of (1) tenderness, superficial or deep, (2) organ enlargement, (3) abdominal mass, (4) spasm or rigidity of the abdominal muscles, (5) ascites, or (6) exaggerated tympanites. There is no universally accepted sequence for routine abdominal examination, and you may begin in any one of the quadrants and proceed in a clockwise manner until all four have been examined. It should be immediately emphasized, however, that modification of your approach (as explained in Chapter 12) will almost certainly be necessary in patients with serious abdominal complaints. With this reservation, the following course is suggested: (a) percussion, (b) light palpation, and (c) deep palpation.

Percussion is of relatively limited diagnostic value, but it is a simple method of relaxing abdominal tension. Percuss the four quadrants briefly while noting the degree of resonance. Next, delineate the upper and lower borders of the liver in the midclavicular line. They should be no more than 10 cm. apart, but liver dullness along the lower border may be partially obliterated by gas in the overlying bowel, making the overall dimension less than 10 cm. Next, outline Traube's space (the gastric air bubble) in the left upper quadrant, and then if possible find the area of splenic dullness lateral to this, remembering that percussion has very limited value in delineating enlargement of the spleen. Move to the suprapubic area and outline the upper border of the urinary bladder if possible. Watch carefully for evidence of tenderness while percussing.

Light palpation is aimed primarily at eliciting minor degrees of tenderness and guarding. It is performed with the flat of the hand, not the fingertips, and sudden increases in pressure should be avoided. When changing position remove the hand rather than dragging it across the surface, which produces a disagreeable sensation resulting in undue muscle spasm. Such voluntary guarding may be suspected when the tightness follows temporary relaxation during the first phase of expiration. It usually passes away gradually as the examination progresses.

Deep palpation requires greater experience and skill. The flat of the right hand is usually used, and at times the left is placed over it for reinforcement. Pressure is *very* gradual and steady; special care is necessary at this point in order not to cause undue discomfort for the patient. The bimanual technique (Figure 11-2A) is always helpful for deep palpation, particularly of the liver and kidneys. The posterior hand is placed between the twelfth rib and the iliac crest, just lateral to the paraspinous muscles. In palpating the liver, place the anterior hand firmly inward and upward in the right upper quadrant and instruct the patient to take a deep breath and hold it. You will want to release your pressure slightly at the height of inspiration, at the same time moving the fingertips gently upward toward the costal margin. When palpable, the liver edge is felt to slip over the fingertips at this moment. Proper placement of the anterior hand is important, as shown in Figure 11-2A. Horizontal placement of the hand (Figure 11-2, inset) tends to force the whole liver backward, making the edge less accessible. Another common error is concerned with the level of placement of the palpating hand. Begin low, below the percussed border of dullness. Liver enlargement can be missed by palpating too close to the costal margin, so that the whole organ is beneath the hand and the edge is not palpated. At times, palpation can be better accomplished by the "hooking" technique (Figure 11-2B).

Palpation of the kidneys and spleen is considered elsewhere (Chapters 13 and 14).

Normal findings on palpation and percussion are highly variable, and depend largely on the degree of obesity and general body build, as well as on the patient's ability to cooperate. The aorta is often palpable in the epigastrium and may be slightly tender. The normal aorta in the elderly, asthenic patient is easily mistaken for an aneurysm. The descending colon and cecum are commonly felt with considerable ease, particularly when they contain feces, and this normal finding may be misinterpreted as neoplasm unless x-ray studies are available. The liver edge may be normally felt, particularly in women and children. At times the liver may extend 4 or 5 cm. below the right costal margin without actually being enlarged. Delineation of the overall size by percussion may help to differentiate this ptosis (dropping down) from true enlargement. Overall size in

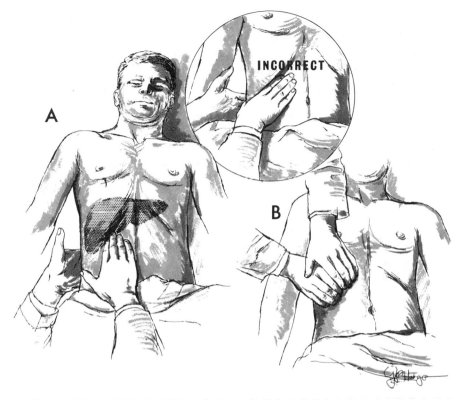

FIGURE 11-2. Technique of palpating liver. A. The anterior hand should parallel the rectus muscle, not the costal margin (inset). B. The hooking technique is occasionally preferable.

the midsternal line is usually less than 15 cm., but the patient's size and habitus are important considerations. Percussable enlargement of the urinary bladder may or may not be of pathologic significance, depending on other factors. Remember always that in the premenopausal woman an enlarged uterus may simply reflect pregnancy.

Auscultation.   Most intestinal sounds originate from the small bowel and have a high-pitched, gurgling quality, better sensed than described. Frequency of intestinal sounds varies in relation to meals, but usually five or more sounds occur each minute. Occasionally it takes more time and patience to determine that normal bowel sounds are indeed present. Vascular bruits and friction rubs are considered elsewhere (Chapters 10 and 13).

Rectal Examination

It is impossible to overemphasize the importance of the rectal examination. This simple and vital procedure is too often passed by because it entails extra effort on the part of the physician and tends to be somewhat disagreeable to the patient. No gastrointestinal evaluation is complete without it.

The skin surrounding the anus should be carefully inspected for signs of inflammation or excoriation. The examination itself consists of digital and endoscopic study. Digital examination (described in further detail in Chapters 14 and 17) is performed with the patient in the supine or knee-chest position, or flexed at the hips and bending over. On occasion it is helpful to examine the patient in the squatting position, for this may bring a high-lying rectal lesion within reach of the examining finger. Insertion of the examining finger should be gentle and gradual, and sufficient time should be allowed for the patient to relax after insertion has been accomplished. Excessive resistance at the anal ring is commonly due to simple spasm caused by nervousness, and this can at times be overcome by asking the patient to strain a little before palpation is begun. If there is considerable spasm and local pain due to anal pathologic changes, a local anesthetic suppository may be used to minimize discomfort. After evaluating sphincter tone and carefully palpating the anal wall to determine the presence or absence of hypertrophic papillae, inflamed crypts, or strictures, investigate the possibility of a rectal mass and the condition of the prostate in the male. In the female patient rectal examination is performed as part of the pelvic examination.

CARDINAL SYMPTOMS AND ABNORMAL FINDINGS

Pain

Pain of gastrointestinal origin varies greatly, depending on its underlying cause. The major pain mechanisms include (1) capsular stretching, as in liver congestion due to heart failure, (2) irritation of the mucosa, as in acute gastritis, (3) severe smooth muscle spasm, as in acute enterocolitis, (4) peritoneal inflammation, as in acute appendicitis, and (5) direct splanchnic nerve stimulation, as in retroperitoneal extension of a neoplasm such as carcinoma of the pancreas.

The character, duration, and frequency of gastrointestinal pain are a function of their mechanism of production; the location and distribution of referred pain are related to the anatomic site of origin. Time of occurrence and factors which aggravate and relieve the discomfort, such as meals, defecation, and sleep, also have special significance which is directly related to the underlying cause.

An outline of major sites of localization usually includes the following:

Esophageal Midline retrosternal with radiation to the back at the level of the lesion.
Gastric Epigastric; radiation occasionally to back, particularly left subscapular.
Duodenal Epigastric: radiation to back, particularly right subscapular.
Gallbladder Right upper quadrant or epigastric; radiation to right subscapular or midback.
Pancreatic Epigastric; radiation to midback or left lumbar area.
Small Intestinal Periumbilical.
Appendicular Periumbilical migrating later to right lower quadrant.

Colonic Hypogastrium, right or left lower quadrant, depending on site of lesion; sigmoid pain may radiate to the sacral region.

Rectal Deep pelvic localization.

Motor Dysfunction

Dysphagia is usually more prominent and severe with solid food than liquids. The sensation is generally localized by the patient to the approximate level of obstruction. It is an important symptom which must *never* be ignored.

Vomiting is often associated with upper gastrointestinal disorders. The patient should be questioned regarding the frequency, time of occurrence, and aggravating factors, as well as the quantity, color, odor, and taste of the vomited material.

Bowel function varies greatly among individuals. Direct observation of a stool specimen by the physician is far more accurate than reliance on a patient's description. Normal frequency of bowel action varies from several times daily to once every three to five days. *Diarrhea* or *constipation* of recent onset requires detailed description. The former is particularly important when it occurs at night.

Additional important gastrointestinal symptoms include hematemesis, melena, anorexia, a sense of abdominal fullness, heartburn (pyrosis), regurgitation, flatulence, and belching.

General Inspection

Since the gastrointestinal tract is highly sensitive to the emotional state, a logical first step is to assess the patient's face and demeanor for signs of depression, agitation, exhaustion, hostility, or fear. Patients with gallbladder or liver disease may show evidence of jaundice. The yellow discoloration of the sclera and skin is frequently much more obvious in daylight than in artificial light, where it may pass unnoticed, even when moderately severe. Other important skin changes include (1) pallor, indicating chronic or acute gastrointestinal blood loss, (2) pigmentation that may reflect malabsorption, regional enteritis, or hemochromatosis, (3) xanthomas resulting from special forms of cirrhosis, (4) erythema nodosum from ulcerative colitis, (5) a flush suggesting carcinoid neoplasm, (6) generalized edema caused by malabsorption, and (7) spider angiomata or petechiae, palmar erythema, and hair loss, which together indicate chronic liver disease.

The state of nutrition is another important indirect sign. Weight loss and emaciation need little emphasis because of their obvious importance. *Nail changes* include (1) koilonychia (spooning) with chronic iron deficiency, and (2) clubbing with intestinal malabsorption, regional enteritis, ulcerative colitis, and hepatic cirrhosis. *Dehydration* may cause reduction of ocular tension, xerostomia, and persistent ridging of the skin on the dorsum of the hand when lightly pinched.

Mouth

Severe halitosis may occur with chronic gastroesophageal disease, particularly neoplasm. Fetor hepaticus is a term describing a characteristic odor associated with severe liver failure; at times it is apparent at some distance from the patient's bed. In peptic disease of the stomach and duodenum, the patient's breath often has an acid odor.

Glossitis and stomatitis often accompany deficiency states caused by chronic gastrointestinal disorders, such as sprue, with associated malabsorption and depleted body stores of iron, B_{12}, folic acid, niacin, thiamine, and riboflavin. Melanin spots about the face and mouth are a sign of intestinal polyposis. They are frequently drab brown or dark bluish-black, and may be mistaken for simple freckles.

Abdomen

Inspection. Superficial abdominal veins may be dilated and tortuous because of vena caval obstruction. With the portal hypertension of hepatic cirrhosis, the veins may appear to radiate from the umbilicus, as a result of back flow through collateral veins within the falciform ligament. A scaphoid abdomen often accompanies cachexia; protuberance may result from gaseous distention, ascites, or neoplasm. With intestinal obstruction, peristaltic waves may be particularly visible passing across the abdomen. A decrease in respiratory movements of the abdomen may suggest intraperitoneal fluid, or it may be associated with acute abdominal pain, such as with peritonitis.

Palpation and Percussion. Areas of tenderness are identified by light palpation. They should be localized as accurately as possible without producing undue discomfort. At times, probing with a single finger may further outline a tender area. As previously explained, it may be necessary to cause additional discomfort during palpation of the abdomen. The need for this is best explained to the patient in advance so that he understands its importance.

A clear understanding of anatomic relationships and their variations is indispensable for the interpretation of tenderness of visceral origin. Sometimes, however, abdominal pain and tenderness do not actually arise from abdominal organs. This is referred to as parietal tenderness, and it is identified by having the patient, in the supine position, contract the abdominal muscles (by raising his head or feet) so as to prevent the transmission of pressure to the underlying viscera. Under these circumstances, persistent tenderness on light palpation or gentle pinching probably arises in the abdominal wall itself. Similarly, a mass which remains palpable under these conditions is probably situated superficially within the abdominal wall. The reactions termed hyperesthesia, rebound, and guarding are explained in Chapter 13.

Palpable masses should always be localized with respect to the previously described landmarks, and they should, if possible, be described in terms of consistency and contour. Frequently, however, they are only vaguely outlined, particularly when they are associated with tenderness or fluid or when the abdomen is obese or tense. Gastric, pancreatic, and colonic neoplasms, pancreatic cysts, and distended gallbladders may be palpable, usually at advanced stages of the disease. Genital and urinary tract disorders, of course, also can cause intra-abdominal masses.

The principal causes of hepatic enlargement are (1) congestion, (2) cirrhosis, (3) neoplasm, and (4) hepatitis. Tenderness is more likely with congestion and inflammation; irregular nodularity with neoplasm; a very hard consistency with cirrhosis. There are many additional causes for liver enlargement which are beyond the scope of this discussion. Splenomegaly is considered in Chapter 13.

Evaluation of the greatly distended abdomen is conducted by palpation and percussion. When the percussion note is high-pitched and tympanitic (drumlike), there is probably an obstruction which has produced gaseous distention of the underlying small intestine, stomach, or colon. A flat or dull percussion note suggests either the presence of fluid in the peritoneal cavity or abdominal fullness associated with ovarian cyst or obesity. The usual method for demonstrating intra-abdominal fluid requires two examiners (Figure 11-3). The assistant places the edge of his hand on the midline of the abdomen in order to limit the transmission of the impulse by the abdominal wall. The examiner then taps one flank while palpating the opposite flank, in order to detect the transmission of a fluid wave. A false-positive sign may result if the patient is very obese or if there is a large ovarian cyst containing a sizeable volume of encapsulated fluid. With ascites the distention is symmetrical and the flanks are

FIGURE 11-3. Method of demonstrating intra-abdominal fluid by eliciting fluid wave.

particularly full. A tympanitic note detected in the midline, anteriorly, reflects associated gaseous distention of the bowel. (The gas-containing bowel floats.) The area of dullness is localized to the flanks and may be noted to shift with a change of position. In contrast, ovarian cysts may produce asymmetrical abdominal swelling; the dullness is located anteriorly with the tympanitic note in the flanks as the gas-containing bowel is displaced laterally. Shifting dullness on change of position is usually not present.

A fluid wave may not be obtained if the volume of ascites is only moderate and abdominal distention is slight. In such a case it may be possible to elicit a fluid wave if the examination is performed during the expiratory phase of a cough or during a Valsalva maneuver. This produces contraction of the abdominal muscles, reduces the volume of the abdominal cavity, and temporarily puts the ascites under enough tension to elicit a wave. This sign may be obtained in the presence of intra-abdominal cysts or fluid-filled intestines, although under these circumstances "shifting dullness" in the flanks will not be obtained [3].

Ascites may be associated with intra-abdominal masses, which are obscured by the presence of the fluid and therefore difficult to palpate. Under these circumstances it is sometimes possible to detect such a mass by *ballottement*. This technique calls for lightly thrusting the fingers into the abdomen in the region of the suspected mass. The thrust will tend to displace the fluid, permitting the mass to bound upward, producing a characteristic tapping sensation against palpating fingers.

Auscultation. Abnormal bowel sounds are distinguishable only when a clear appreciation of the normal variation in peristaltic sound has been developed by the observer. Peristalsis may be increased, diminished, or absent in the presence of intra-abdominal disease. It is necessary to listen for a prolonged period, at least several minutes, before deciding that bowel sounds are entirely absent. This condition usually indicates paralytic ileus, which may be due to generalized peritoneal irritation.

Peristalsis may be increased in cases of acute gastroenteritis or acute mechanical obstruction. In the latter, in particular, the patient is aware of painful cramps which tend to increase and subside in relation to peristaltic activity. The sounds are characteristically high-pitched and of a crescendo nature, with a tinkling quality at the end of the peristaltic run. Partial obstruction may be associated with gurgling and tinkling sounds related to the presence of dilated, fluid-filled loops of the small bowel undergoing episodic contraction. There is no characteristic rhythmic pattern, and cramps may not be a feature (see Chapter 12). With outlet obstruction of the stomach, a succussion splash is at times detectable because of the presence of fluid and gas in the distended organ. This is easily appreciated by placing the stethoscope diaphragm over the epigastrum and shaking the patient vigorously from side to side. A characteristic sloshing and gurgling sound is readily identified. It is important to determine the interval since the previous meal, as it is possible to elicit this sign in a normal person immediately after ingestion of a large quantity of fluid. (See also Chapter 12, The Acute Abdomen.)

SPECIAL TECHNIQUES

Esophagoscopy. Esophagoscopy carries certain risks and should be performed by highly skilled individuals. The accuracy of the examination is fairly high, although the upper one-third of the esophagus is difficult to examine because of the manipulations which must be performed at this point in introducing the esophagoscope. The chief hazard consists of the risk of esophageal perforation during the procedure. This can occur either in the cervical esophagus from pressure by the esophagoscope against the cervical spine, or within the mediastinum in the course of manipulation or biopsy of an esophageal lesion. The principal value of the esophagoscope is to provide accurate diagnosis of esophageal neoplasms, and it is often of value in the detection of esophageal varices. Flexible, fiberoptic esophagoscopes have recently been introduced. Their use has improved the diagnostic accuracy and lessened the hazard of this study.

Gastroscopy. The technique of gastroscopy is quite similar to that used in esophagoscopy. Fiberoptic gastroscopes are available which permit biopsy and photography of gastric lesions. It is easiest to visualize lesions along the lesser curvature, in the distal stomach along the greater curvature, and in the antrum. The flexible gastroscope also permits examination of the fundus with a fair degree of accuracy.

Duodenoscopy. Flexible, fiberoptic duodenoscopes permit examination, biopsy, and photography of duodenal lesions. Special instruments are also available for transpapillary injection of contrast material to the common bile duct or pancreatic duct for x-ray visualization.

Anoscopy. Anoscopy is performed with the patient in the knee-chest or Sims position. The use of a slit anoscope permits inspection of the anal canal for detection of fistula-in-ano, fissure-in-ano, hemorrhoids (internal), and carcinoma. At the time of anoscopy, inspection of the perianal region including perineum, buttocks, genitalia, and thighs may also be carried out.

Sigmoidoscopy. The sigmoidoscope is used for the discovery of lesions which are too small or too soft to be palpable by the examining finger, or which lie above the digital examination but are still too low to be seen by careful barium enema examination. Because of the importance of this examination it is described in detail here. The sigmoidoscope with the obturator in place should be warmed and lubricated. The instrument is inserted along the longitudinal axis of the anal canal. As the tip of the instrument reaches the ampulla of the rectum, the obturator is withdrawn and the instrument is inserted further under direct vision. The sigmoidoscope should never be advanced blindly, for it is possible to perforate the wall of the bowel unless gentleness and care are used. It should never be advanced with force on the mucosa of the bowel ahead of the tip of the instrument. The lumen should always be visualized well before the instrument is advanced. If necessary, the gentle use of air insufflation will keep the lumen open and make examination easier.

About 12 to 15 cm. above the anal orifice the rectosigmoid level is reached. It is necessary to direct the instrument anteriorly in order to pass the sacral promontory. This may be a difficult part of the examination and the patient may experience some lower abdominal cramps. If it is difficult to pass beyond this point, it is advisable not to force the additional examination on the patient. However, once the promontory of the sacrum is passed it is comparatively easy to manipulate the sigmoidoscope to its full length of 25 cm. At times mucosal folds may impede the passage of the sigmoidoscope and the observer will seem to have encountered a "dead end." These mucosal folds may be smoothed by air insufflation, or it may help to withdraw the tip of the sigmoidoscope a short distance to obtain another approach. The best examination is attained as the instrument is being withdrawn. The examiner can then devote his undivided attention to careful inspection of the mucosa. In the distal sigmoid the rugal folds of the mucosa are arranged concentrically. As the tip of the instrument is withdrawn through the rectosigmoid the mucosa becomes smooth except for the spiral folds which make up the rectal valves of Houston. If the sigmoidoscope is withdrawn with a circular motion the rectal valves are flattened and the observer is able to see lesions which lie behind them. Fluid and fecal particles are evacuated by cotton swabs or gentle suction. When pathologic conditions are found at sigmoidoscopy, it is possible to obtain a biopsy for diagnosis. Small polyps may be removed in toto with biopsy forceps or by an electric snare.

Colonoscopy. The ascending, transverse, descending, and sigmoid portions of the colon may be examined by the use of a flexible, fiberoptic colonoscope. The colon is thoroughly cleansed, and the instrument is introduced transanally. Under careful observation, the instrument is advanced under x-ray control. It is possible to visualize, photograph, and take biopsy specimens from intraluminal lesions by this technique. Inflammatory disease of the bowel and intestinal obstruction are contraindications for this type of study.

Radiographic Techniques. Contrast studies using barium sulfate permit examination of esophagus, stomach and duodenum, and small and large bowel. The barium swallow is probably the most reliable method for examination of the esophagus. It is useful in detecting a wide variety of lesions such as diverticula, stricture, carcinoma, achalasia, or hiatus hernia. This special radiographic study should be supplemented by esophagoscopy for direct visualization. Because of its location, direct examination of the esophagus is impossible except in these ways.

The upper gastrointestinal series is a barium contrast study of the stomach and duodenum. If the progress of the barium is followed downward to visualize the small bowel, it is termed a "small bowel follow-through." Barium studies are useful in defining a variety of gastric or duodenal lesions such as ulcer, carcinoma, polyps, gastritis, and gastroesophageal varices. Since in many of these conditions there are few physical signs, these contrast studies are of great diagnostic importance.

The colon is best examined by means of a barium enema, which outlines the large bowel above the level of the rectosigmoid. The method is generally used to supplement proctoscopy, sigmoidoscopy, and rectal examination. If the examiner suspects small lesions such as polyps, it is often helpful to coat the colonic mucosa with thin barium, and inflate the bowel by air insufflation. This is termed an "air-contrast study."

The gallbladder is visualized by another contrast technique, called the cholecystogram or Graham-Cole test. The contrast material is an iodinated organic compound, given orally, which is excreted by the liver in bile. As the contrast material is concentrated within the gallbladder, this structure becomes radio-opaque. A similar preparation may be injected intravenously and used to outline the common bile duct and biliary tree. This method is helpful when the gallbladder has been removed or is obstructed by a stone in the cystic duct.

The portal vein can be visualized by the percutaneous injection of water-soluble contrast material into the spleen. The technique is termed splenoportography and is used to aid in the diagnosis of cirrhosis of the liver and portal hypertension, or thrombosis of the portal vein. Another method of visualizing the portal system is to inject contrast medium into the superior mesenteric artery by selective catheterization. Multiple films are taken during the late phase of injection to capture the venous phase and outline the portal vein and its tributaries.

Gastric Analysis. A nasogastric tube may be used to remove samples of gastric secretion for measurements of volume, pH, chloride content, and free and combined acid. The test may be performed in the fasting state or after stimulation with histamine or insulin. These studies are usually performed in the diagnosis of peptic ulcer. The gastric aspirate may also be examined microscopically for the presence of malignant cells indicating carcinoma of the stomach, or acid-fast bacilli confirming a diagnosis of pulmonary tuberculosis.

Stool Analysis. Various studies may be performed on stool specimens. The benzidine test or guaiac test will detect the presence of occult blood. Bacterial cultures are important in diagnosing bacillary dysentery or staphylococci or leukocytes in suspected cases of staphylococcal enterocolitis, or ova and parasites in amebic dysentery.

REFERENCES

1. Badenoch, J., and Brooke, B. N. *Recent Advances in Gastroenterology.* Boston: Little, Brown, 1965.
2. Bockus, H. L. *Gastroenterology* (2nd Ed.). Philadelphia: Saunders, 1963.
3. Cardon, L. A new sign in the diagnosis of minimal and moderate ascites. *Ill. Med. J.* 92:239, 1947.
4. Chandler, G. N. *A Synopsis of Gastroenterology.* Baltimore: Williams & Wilkins, 1963.
5. Joes, F. A., and Gummer, J. W. P. *Clinical Gastroenterology.* Springfield, Ill.: Thomas, 1960.
6. Naish, J. M., and Read, A. E. A. *Basic Gastroenterology.* Bristol, England: J. Wright, 1965.

THE ACUTE ABDOMEN   **12**

George D. Zuidema

"Correct diagnosis is the essential preliminary
to correct treatment." — Zachary Cope [1]

Successful diagnosis of acute abdominal conditions is dependent upon careful application of anatomic and physiologic principles. This axiom calls for application of knowledge concerning the structural relationships of the abdominal viscera in order to put physical diagnostic techniques on a rational basis. In addition, it is important to appreciate the structure, function, and embryologic derivation of the voluntary muscles bordering the peritoneal cavity and the derivation of their innervation. This detailed knowledge assumes clinical significance when an inflammatory or neoplastic process involves adjacent structures and interferes with their function or produces tenderness and spasm. (Figure 11-1A, B illustrates the topographic anatomy of the anterior abdominal wall. The respective surface areas are identified for descriptive purposes. Figure 11-1C shows the location of the intra-abdominal viscera, and reveals the normal relationships.) Thorough knowledge of the location of the various organs is essential to the accurate diagnosis of intra-abdominal disease, and also makes possible the interpretation of radiologic studies.

The term "acute abdomen" suggests the importance and urgency which often accompanies acute intra-abdominal disease. Because of this urgency one should attempt to clarify the problem and arrive at a precise diagnosis *when the patient is seen for the first time.* The success of the examiner will be dependent upon the thoroughness of his examination and his ability to pursue an orderly approach to the problem. Carelessness will be accompanied by a marked decline in accuracy. Early diagnosis is of crucial importance; the delay of even a few hours may permit peritonitis to develop or perforation of a viscus to occur. Delay also almost inevitably results in unnecessary morbidity and the likelihood of an increased mortality rate. The temptation to temporize is often strong in order to observe the course of the patient. While this may sometimes be justified, in many instances it will only result in the loss of a golden opportunity to treat surgical emergencies early and achieve superior results. It is often stated that abdominal pain that develops in a patient who has previously been well and persists for over six hours is caused by a condition requiring surgical attention. The practice of consistently performing a thorough examination of the acute abdomen will result in making a correct diagnosis early in the course of the disease, permitting prompt surgical treatment if indicated.

Although the patient with acute abdominal distress is usually in pain, it is important to withhold the use of sedative or analgesic drugs until a satisfactory diagnosis has been made and one is fairly certain whether surgical intervention will be required. Analgesics alter the physical signs and symptoms to such an extent that accurate diagnosis often becomes impossible following their administration.

ANATOMIC CONSIDERATIONS

Knowledge of embryologic development is often helpful in interpreting physical signs. This is particularly true in explaining referred pain. For example, the diaphragm has its origin in the region of the fourth cervical nerve, and its nerve supply is derived from the third, fourth, and fifth cervical nerves which make up the phrenic nerve and accompany the muscle fibers in their descent. Referred pain to the shoulder becomes a valuable clinical sign, since upper abdominal or lower thoracic lesions producing irritation of the diaphragm often cause pain or hyperesthesia in the region of distribution of the fourth cervical nerve. Consequently, pain on top of the shoulder may accompany such processes as subdiaphragmatic abscess, perforated ulcer, diaphragmatic pleurisy, acute pancreatitis, or ruptured spleen. The pain is felt in the supraspinous fossa, over the acromion or clavicle, and in the subclavicular fossa. Pain on top of both shoulders indicates either bilateral or midline diaphragmatic irritation.

The body segments which embryologically go to form the pelvis are not represented by nerve supply to the abdominal muscles. Consequently, pelvic peritonitis, which may produce irritation of the pelvic nerves, is not likely to cause abdominal-wall rigidity.

The hollow viscera within the peritoneal cavity have walls composed of smooth muscle fibers. While there is no sensation of touch within the visceral walls, stretching, distention, or marked contraction against resistance produces acute, severe pain. In mild degrees this may be of little or no significance, but when severe obstruction, sufficient to produce local distention, is present, this crampy pain is termed colic; e.g., intestinal, biliary, renal, or uterine colic. The pain is likely to be paroxysmal and excruciating and is usually referred to the segmental distribution of the spinal nerves corresponding to the part of the spinal cord from which the sympathetic nerves took origin. For this reason colic of the small intestine produces pain referred to the epigastrium and umbilical regions. Colic of the large intestine is more likely to be referred to the hypogastrium. Biliary colic may present in the right subscapular region, while renal colic may be manifested by pain in the flank, with possible radiation (in the male) to the corresponding testis. The intensity of the pain is so severe that patients experiencing colic will frequently move about restlessly, twisting and turning and attempting to find a comfortable position in bed. This should be contrasted with the patient with peritonitis who attempts to lie perfectly still lest any movement aggravate the pain.

Another practical application of anatomy is often seen in inflammatory and neoplastic conditions. In appendicitis or other conditions secondarily affecting the psoas muscle, one observes flexion of the thigh due to contraction of the psoas muscle. Moderate degrees of psoas irritation may be tested for by having the patient lie on the opposite side and fully extending the thigh on the affected side. Similarly, pelvic inflammation or abscess may secondarily involve the fascia overlying the obturator internis muscle. This may be detected by putting the muscle through full movement by rotating the flexed thigh inward. Resulting pain is referred to the hypogastrium.

Flank pain is frequently accompanied by radiation to the corresponding testis. This phenomenon is also explained on the basis of embryologic development. The testis originates in the same region as the kidney and descends to the scrotum shortly prior to birth, carrying with it its blood and nerve supply. Consequently, inflammatory or obstructive pain in the region of the kidney or the renal pelvis may have a referred component to the testis.

It is obviously necessary to exclude medical diseases before deciding on the need for surgical intervention. A number of medical conditions may mimic intra-abdominal emergencies. Examples are typhoid fever, pericarditis, lower lobe pneumonia, nephritis, tuberculous peritonitis, and tabes dorsalis. These possibilities emphasize the need for a thorough, objective, and systematic approach to examining entirely the patient who presents with abdominal pain.

HISTORY

The technique of history taking may have to be modified in some respects to be effective. The patient who is suffering from acute, severe pain is not likely to be able to cooperate by furnishing a complete narrative history. In fact, under the urgency of the situation it may be important to depart from the routine and ask direct questions. Historical information derived from the patient himself should be supplemented whenever possible by detailed questions asked of his family and friends. The relatives or friends who accompany the patient to the hospital should be consulted regarding significant points before they are permitted to leave.

Whenever possible a complete and detailed history and physical examination should be performed. Under some circumstances the severity of the illness may make prompt, emergency treatment imperative. Under these circumstances, the diagnostic approach will require abbreviation. It is still necessary, however, to obtain sufficient information to provide an adequate working diagnosis. Shortcuts are likely to be expensive in time, accuracy, or human life. Each symptom must be carefully and thoroughly evaluated in terms of its relationship to physical findings and other symptoms.

The patient's *age* is pertinent. The incidence of certain disease processes is often limited to certain age groups. Recognition of this factor makes it possible to improve one's diagnostic accuracy.

A *clear description of the disease process* is of utmost importance. The patient should be asked to fix the exact time at which the pain began and the manner of its beginning. For example, pain of gradual onset may be associated with appendicitis, while the sudden onset of acute abdominal pain, awakening the patient from sleep, may be associated with perforation of a duodenal ulcer. It should also be noted whether the onset of the pain is related to some injury or exertion, no matter how trivial. The severity of the condition may be estimated by inquiring as to whether the patient collapsed or lost consciousness at the onset of the symptoms. Severe symptoms are more likely to be associated with such intra-abdominal catastrophes as acute pancreatitis, perforated ulcer, ruptured ectopic pregnancy, or stangulation obstruction of the bowel. The character, distribution, and mode of onset of the pain should be carefully evaluated. Even the generalized pain associated with perforated ulcer or hemorrhage from a tubal pregnancy usually begins in a specific location, later spreading and becoming generalized. For example, with perforated duodenal ulcer pain characteristically originates in the epigastrium with severe intensity, but rapidly becomes generalized. For a time the pain may be more acute in the right flank and right lower quadrant, as the irritating gastric fluid passes down the gutter on the right side of the abdomen.

Pain that arises from the small intestine is usually felt primarily in the epigastrium and periumbilical areas. This is true whether it is simple mechanical intestinal obstruction or strangulation obstruction, since the innervation of the small bowel corresponds to the distribution of the ninth through the eleventh thoracic nerves. Since the innervation of the appendix is derived from the same source, this also explains why appendicitis usually begins with onset of pain in the epigastrium, followed by radiation to the right lower quadrant as the peritoneum and the psoas muscle become secondarily involved. Pain associated with large-bowel conditions is usually referred to the hypogastrium or to the actual site of the lesion.

In addition to noting the origin of the pain it is also significant to *follow its change in localization*. The shifting of pain from the upper abdomen to the lower abdomen may be associated with the accumulation of irritating or infected peritoneal fluid in the pelvis. This may occur with a perforated ulcer or acute pancreatitis.

The *nature of the pain* is often a help in diagnosis. Crampy, constricting pain is characteristic of biliary colic, while burning pain is more likely associated with peptic ulceration. Back pain which is severe and constant may be associated with pancreatitis. Appendicitis is usually associated with a constant aching pain, except when a fecalith is present, in which case it may be colicky in nature.

The *radiation of the pain* is often significant. This is particularly true of colic associated with obstruction of hollow viscera, for the pain radiates to the area of distribution of the nerves coming from that segment of spinal cord supplying the affected viscus. For this reason biliary colic is frequently referred to the area beneath the right scapula, and renal colic is frequently referred to the testicle.

The *relationship of pain to respiration* should always be noted. With intra-abdominal sepsis such as peritonitis or abscess, deep inspiration may cause pain. On the other hand, pleuritic pain is made worse by inspiration but often disappears when the patient holds his breath.

A complete history inquires into the possible *relationship of pain to urination.* In addition to the many urinary conditions which may produce this, peritonitis or an abscess lying adjacent to the bladder may cause pain on urination and may even be associated with hematuria.

Vomiting may be associated with the following intra-abdominal conditions: acute gastritis; irritation of the peritoneum or mesentery — e.g., vomiting may appear early in the course of peritonitis associated with appendicitis or perforated ulcer; obstruction of hollow viscera producing smooth muscle spasm — e.g., biliary or ureteral colic or intestinal obstruction may be associated with vomiting; bacterial toxins or certain tissue metabolites may have a direct central action and produce vomiting on a reflex basis. This is often seen in cases of septic peritonitis.

It is important to note the *time relationship between the onset of pain and the exact time of vomiting.* With biliary-tract calculus or sudden, severe peritoneal irritation, vomiting occurs early in the course of the disease and is likely to be violent. Low small-bowel obstruction may be associated with delayed vomiting, while vomiting occurs promptly with high small-bowel lesions. With large-bowel obstruction vomiting may be a very late feature or may not occur at all. With appendicitis the onset of pain almost always antedates vomiting by several hours.

Not all intra-abdominal emergencies are associated with vomiting. Massive intraperitoneal hemorrhage may occur in absence of vomiting, and intussusception may be deceptive because vomiting may occur late or not at all.

The *physical characteristics of the vomitus* should be noted. In acute gastritis the vomitus consists largely of gastric contents occasionally flecked with small amounts of blood. With intestinal obstruction the characteristics of the vomitus show considerable variation. As the condition progresses the character changes from gastric contents to bilious material, becoming yellowish green and finally consisting of brown, feculant-smelling fluid. Feculant vomiting may occur in either dynamic or adynamic intestinal obstruction.

If the patient denies vomiting, it is worth asking him about nausea or anorexia. There is considerable variation in the ease with which people vomit, and the presence of nausea or loss of appetite may in some individuals carry the same significance as vomiting does in others.

The condition of the bowels and nature of the stool should be investigated. An estimate of the *patient's normal bowel habits* should be obtained as well as careful analysis of how the acute illness may cause departure from his normal routine. The *presence of gross or occult blood* should be noted. The combination of blood and mucus in the stool is suggestive of intussusception. Pelvic infections may alter bowel habits and produce lower abdominal pain and diarrhea, and in some instances tenesmus.

A careful *menstrual history* should be recorded, noting the characteristics of the period, including the last date of onset and the nature of the flow. This is necessary if one is to consider the diagnosis of threatened abortion, tubal pregnancy, or other gynecologic problems.

Past history of the pain involved may be extremely valuable in making a diagnosis of such conditions as hiatus hernia, duodenal ulcer, or gastric carcinoma. The possible relationship to previous attacks of jaundice, hematemesis, melena, and so on, may also contribute important information.

TECHNIQUE OF EXAMINATION

Inspection

The patient's facial expression may provide helpful information. A pale face with beads of perspiration may be associated with the shock accompanying acute pancreatitis or strangulation obstruction of the intestine. Simple pallor may be associated with massive intraperitoneal hemorrphage from a ruptured spleen. It should be remembered, however, that intra-abdominal catastrophes may occur without producing characteristic changes in the facies. In the late stages of many acute intra-abdominal processes, profound changes in facial expression may be accompanied by dulling of the eyes; shrinking of the tongue; and a cool skin, reflecting loss of circulating blood volume and impending failure of the circulation.

The position assumed by the patient is worth noting. With biliary or intestinal colic he may be unable to lie quietly; intraperitoneal hemorrhage may produce profound restlessness. This is to be contrasted with the patient suffering from generalized peritonitis who lies quietly with his knees drawn up to relax his abdominal muscles and relieve intra-abdominal tension.

The respiration rate should be noted, for a rapid rate may be associated with intrathoracic disease. Tachypnea may also be associated with generalized peritonitis, intestinal obstruction, intra-abdominal hemorrhage, or anxiety.

Fever is not a constant companion of intra-abdominal disease. In the presence of shock, septicemia, acute pancreatitis, strangulating intestinal obstruction, or perforated ulcer, the temperature may be normal or subnormal at onset. A low-grade fever may accompany the early stages of acute appendicitis, with higher temperatures occurring with peritonitis. High fever is not often associated with the early stages of acute abdominal disease. The movement of the abdominal wall with respiration should be observed. Limitation of movement may be associated with abdominal distention, rigidity of the abdominal musculature, or limited mobility of the diaphragm.

Palpation

Before beginning active palpation the examiner should ask the patient to cough. If peritoneal irritation is present, coughing will elicit a sharp twinge of pain which may then be localized to the involved area. This permits the examiner to carry out the major portion of his abdominal examination without touching the area of maximal tenderness.

The pulse rate and its character should be carefully noted. It is true that a normal pulse does not necessarily mean a normal condition within the abdomen, although it may indicate that the patient is reacting well to his disease. It is worthwhile to follow the pulse rate at intervals, for it is likely to increase in rate as the intra-abdominal infection or hemorrhage progresses. As peritonitis advances, the pulse may show slight irregularity or be somewhat bounding. In advanced peritonitis the pulse is rapid and thready. This is a bad prognostic sign.

Hyperesthesia should be tested for routinely. This may be done by lightly stroking the abdomen with the point of a pin, stroking the abdomen from above downward. The patient is requested to note if the pin stroke feels sharper at a given location. Hyperesthesia suggests the presence of visceral or parietal peritoneal irritation. It may be detected in the segmental distribution of that portion of the spinal cord from which the affected viscus is innervated or along the distribution of the peripheral nerves which may be involved directly by the inflammatory process. This physical sign is helpful when present, but it should certainly not be considered a constant finding in acute abdominal conditions, and its absence does not rule out intra-abdominal disease.

It is important to examine the sites of possible external herniation as a routine measure. Particular attention should be paid to the femoral canal to rule out the presence of a small hernia or a Richter's hernia. Incarcerated or strangulated hernias are so often the cause of or associated with intra-abdominal processes that this observation should be performed without fail in all cases. The femoral artery should be palpated during this part of the examination, since absence of its pulsations or inequality between the two sides may suggest embolic disease or the presence of a ruptured or dissecting aneurysm.

The examiner should test for muscular spasm of the abdominal wall. His hands should be warm; if necessary they may be warmed by washing with warm water prior to this phase of the examination. All areas of the abdomen should be palpated to evaluate the extent of the muscular spasm present. It is helpful to ask the patient to breathe deeply during this part of the examination, for voluntary rectus muscular spasm will give way as the patient exhales. True muscular spasm will not change, and the abdomen will remain rigid and tense during expiration. Extensive rigidity involving both rectus muscles is suggestive of diffuse peritoneal irritation. With localized or early peritonitis the spasm may be limited to a portion of the abdomen. It is helpful to palpate both recti simultaneously in order to evaluate the extent and severity of the muscle spasm.

Palpation of the abdomen includes examination of the costovertebral angles bilaterally. This may be easily accomplished using the index finger. It is also helpful to palpate with one finger to gently outline the areas of tenderness within the abdomen. This serves the dual purpose of achieving accurate localization while producing minimal discomfort to the patient.

The next step in the examination consists of deep palpation of the abdomen in an effort to determine the size of liver, kidneys, and spleen and to discover the presence of any abnormal intra-abdominal masses.

The recognition of masses may be made easier by repeating the abdominal examination after analgesics have been administered or the patient has been anesthetized prior to operation.

Rebound tenderness is elicited by exerting deep pressure into the abdomen in an area away from the suspected acute inflammatory process and then quickly releasing the pressure. If peritoneal irritation is present the patient experiences a twinge of pain either at the site of pressure or in the area of inflammation. This test is more reliable than cough tenderness. If the peritoneal irritation is localized to an area of inflammation, the rebound tenderness will be referred to that area. If generalized peritoneal irritation is present, rebound tenderness will be referred to the area of pressure. This test may be of particular value in obese patients. It is often accompanied by marked discomfort to the patient and should not be employed in the presence of obvious diffuse generalized peritonitis.

Intra-abdominal inflammation which secondarily involves the iliopsoas muscle may be detected by the iliopsoas test. The patient is asked to flex his thigh against the resistance of the examiner's hand (Figure 12-1A). If inflammation is present in this location, the contraction of the psoas muscle will be accompanied by pain. An alternative way of testing this function is to have the patient lie on the opposite side and extend his thigh toward the affected side. This test is not likely to be positive in the presence of subacute infection or if the abdominal wall is rigid.

If the inflammatory process lies adjacent to the obturator internus muscle, as in the presence of pelvic abscess, lower abdominal pain may be elicited by flexing the thigh to a 90-degree angle and rotating it internally and externally. This is known as the obturator test (Figure 12-1B).

It is sometimes difficult to differentiate an acute upper abdominal process from intrathoracic disease. In this situation deep pressure on the opposite side of the abdomen directed toward the affected side will produce pain if the basic process is intra-abdominal, but will not elicit pain if the disease is intrathoracic.

The sign of inspiratory arrest (Murphy's sign) may be seen in the presence of acute cholecystitis. This is elicited by having the patient take a deep breath while the examiner maintains pressure against the abdominal wall in the region of the gallbladder. As the liver descends with inspiration, the gallbladder comes in contact with the examining hand and the patient experiences a sharp pain and inspiration is arrested.

Inflammatory processes involving the liver or gallbladder may be elicited as

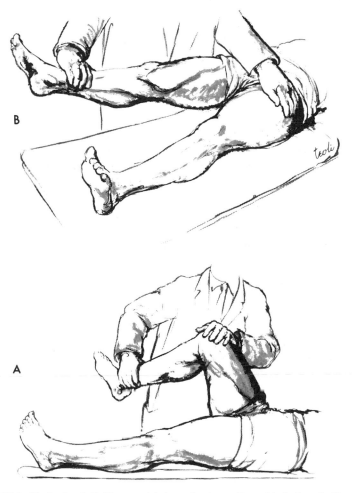

FIGURE 12-1. Methods of eliciting special signs of retroperitoneal irritation. A. Demonstration of iliopsoas sign. B. Demonstration of obturator sign.

tenderness in the right upper quadrant upon fist percussion of the lower anterior chest wall. When this test is negative one should hesitate to diagnose right upper quadrant inflammation.

An important part of the examination by palpation includes digital examination of the rectum. It is desirable to have the patient lying on his back, permitting the rectum to be examined anteriorly, posteriorly, and on both sides with the examining finger. This maneuver may be helpful in detecting such conditions as prostatitis, seminal vesiculitis, pelvic abscess, appendicitis, or tubo-ovarian abscess. Bimanual examination of the rectum and vagina permits careful evaluation of the contents of the pelvis and the cul-de-sac. This is an important part of the examination of female patients and should be included as an integral part of all examinations of such patients with acute abdominal distress.

Percussion

It is sometimes of value to estimate the extent of liver dullness in the right upper quadrant by means of percussion. The liver normally extends from the fifth rib to a point just below the costal margin. If a resonant note is obtained in what should normally be a dull area anteriorly, this is suggestive evidence that there is free air within the peritoneal cavity. In the presence of intestinal obstruction, gas-filled bowel may be pushed up anteriorly, and this test is of no value.

In some instances it may be possible to demonstrate the presence of shifting dullness by means of percussion in patients with acute abdominal disease. This test alone may be of little help, since no information is obtained regarding the character of the fluid present and it is seldom that this would alter one's decision to operate. Needle paracentesis of the abdominal cavity may be of help in obtaining a sample of the free fluid present, permitting its culture and microscopic examination prior to laparotomy. This procedure is of little or no value in the presence of intestinal obstruction where loops of fluid-filled and gas-filled bowel predominate.

The chest should be examined routinely in order to rule out diaphragmatic pleurisy, lower lobe pneumonia, pericarditis, or pleural effusion. Any of these conditions may be confused with intra-abdominal disease.

Auscultation

The examiner should be accustomed to hearing the sounds accompanying normal peristalsis. Absence of bowel sounds suggests paralytic ileus due to diffuse peritoneal irritation. Before bowel sounds can be said to be absent, however, it is necessary to listen for periods of at least five minutes in all portions of the abdomen. Increased peristalsis may be present in acute gastroenteritis. Increased peristalsis will also usually be audible in patients with acute intestinal obstruction. In these patients the abdomen tends to be silent between bouts of colic. As the cramps occur the bowel sounds gradually increase in intensity, rise to a crescendo, then pass away. The patient will often complain of crampy abdominal pain which coincides with the onset of the peristaltic activity. As paralytic ileus subsides, or in the presence of chronic partial small-bowel obstruction, a variety of gurgling and tinkling sounds may be detected. These are produced by peristaltic activity in dilated, fluid-filled loops of bowel. In general, no specific rhythm may be present, and it may or may not be accompanied by cramping pain.

SPECIAL TECHNIQUES

Frequently, simple laboratory tests will be helpful in establishing or confirming a diagnosis. A white blood cell count and differential count should be performed in virtually every case. The differential may be particularly helpful in the evaluation of difficult cases. It is sometimes stated that a differential count need not be performed if the white cell count is less than 9000. This statement is inaccurate and this policy should not be accepted. Aged and very young patients, for example, may have differential counts with a clear shift to the left in the face of a normal white cell count.

A flat film of the abdomen and a chest film should be taken as routine measures. Very few cases are sufficiently emergent to permit deviation from this rule. The potential dividends in terms of useful information to be obtained are considerable. Lower lobe pneumonia may be associated with abdominal pain. Pleural effusion may occur with pancreatitis. Perforation of the esophagus may present initially as upper abdominal pain and usually follows severe vomiting. Just as certain intrathoracic problems may present with abdominal pain, abdominal conditions may also show intrathoracic symptoms. Patients with acute cholecystitis may complain of substernal pain, and it will be necessary to differentiate the condition from angina pectoris or myocardial infarction. Patients with acute symptoms of hiatal hernia usually have substernal burning secondary to reflux esophagitis. Careful examination of the chest is essential to total evaluation of the patient with acute abdominal disease. Many additional examples could be cited. These should be sufficient to prove the point.

Patients admitted with a history compatible with large-bowel obstruction may profitably be evaluated by sigmoidoscopy and judicious barium enema. Obstructing carcinomas, ulcerative colitis, diverticulitis, volvulus, and so on, may be detected in this way when simple physical diagnostic techniques might be inadequate to arrive at a definitive diagnosis.

REFERENCE

1. Cope, Z. *The Early Diagnosis of the Acute Abdomen* (11th Ed.). London: Oxford University Press, 1957.

HEMATOPOIETIC SYSTEM

13

Muriel C. Meyers

GLOSSARY

Agranulocytosis Acute reduction in the number of circulating polymorphonuclear leukocytes.

Anemia Reduction in the number of circulating red blood cells and/or hemoglobin with respect to age and sex.

Epistaxis Nosebleeds.

Erythrocytosis Increase in the number of circulating red blood cells with respect to age and sex.

Hematopoietic Blood-forming.

Hemolysis Accelerated dissolution or destruction of red blood cells in vivo.

Infectious mononucleosis Systemic infection associated with enlargement of lymph nodes and spleen, and eliciting a specific peripheral blood response.

Leukemias A group of disorders of the blood-forming organs characterized by excessive proliferation and/or failure of differentiation of one of the types of white blood cells.

Leukocytosis Increase in the white blood cell count above normal.

Leukopenia Decrease in the white blood cell count below normal.

Lymphadenopathy Any lymph node enlargement.

Lymphoma General term for neoplasms originating from the lymphoid reticulum.

Myeloid Pertaining to the bone marrow.

Petechiae Pinpoint-sized hemorrhages.

Polycythemia Erythrocytosis.

Polycythemia, secondary Erythrocytosis secondary to chronic hypoxia and other rarer causes such as renal tumors.

Polycythemia vera Primary or idiopathic erythrocytosis, in which there is leukocytosis, thrombocytosis, and splenomegaly.

Pruritus Itching.

Purpura Purplish discolorations caused by bleeding into the skin and visible mucous membranes, that is, "black and blue spots."

Splenomegaly Enlargement of the spleen.

Thrombocytopenia Reduction below normal in the number of circulating platelets.

Thrombocytosis Increase above normal in the number of circulating platelets.

In this chapter the major components of the hematopoietic system — the lymph nodes, spleen, bone marrow, and circulating blood — will be considered as a unit. There is a sound basis for considering the physical manifestations of diseases of the lymphoid and myeloid reticulum together. According to the monophyletic theory subscribed to by many authorities, lymphocytes, granulocytes, monocytes, red blood cells, and platelets may be considered to arise from a single pleuripotential mesenchymal precursor.

There are many important physical signs associated with hematologic disorders. Some of these are due directly to changes in the reticulum of the lymph nodes, spleen, or liver; others are indirect manifestations involving the skin, mucous membranes, ocular fundi, or bone. Although the final diagnosis almost invariably requires laboratory confirmation, the physical findings are of great value in directing the physician's attention to the proper special studies.

TECHNIQUE OF EXAMINATION AND NORMAL FINDINGS

Lymph Nodes

Although there are over sixty standard groups of lymph nodes listed by the anatomists, clinicians generally need consider the palpable nodes only in terms of three regional groups: cervicofacial and supraclavicular; axillary and epitrochlear; and inguinal and femoral. Needless to say, many groups of nodes lie beyond the reach of physical examination. Some of these can be evaluated by radiologic techniques, others only surgically.

The standard examination of the lymph nodes requires only simple inspection and palpation. It is always useful to compare sides, using the middle three fingers for palpation. Movements should be slow and gentle; your fingers should oscillate up and down, back and forth, and in a rotary motion.

In recording your findings, five characteristics should routinely be included: location; size, preferably giving the diameter in cemtimeters but at times using descriptive terms such as split pea, bean, almond; tenderness; degree of fixation — movable, matted, fixed; and texture — hard, soft, firm.

Normal lymph nodes are not palpable. However, nodes enlarged as a consequence of prior inflammation are frequently palpable. Cervical nodes up to 1 cm. in diameter are almost always felt in children of up to 12 years of age. If these nodes are biopsied, they commonly show chronic hyperplastic lymphadenitis. Such nodes are usually 0.5 to 1 cm. in diameter and are referred to as "shotty," which has come to mean firm, freely movable, and nontender. They are most common in the occipital region (from old scalp infections), axillae (infections of the hands), and the inguinal regions (old infections of the genitalia and feet).

In adolescents and adults palpable inguinal lymph nodes are so common as to be almost the rule. At times, as a consequence of recurrent infections of the feet, they may be quite large without being of particular clinical significance. For this reason the inguinal region is a poor site for biopsy, since the nodes may be distorted by chronic lymphadenitis. Femoral lymph node enlargement is more commonly of pathologic significance.

Cervicofacial Nodes. Look for asymmetry. Is there any visible lymphadenopathy? Be methodical in your palpation, beginning above and posteriorly (Figure 13-1A) and proceeding downward as follows: (a) occipital and postauricular; (b) submaxillary and submental; (c) anterior triangle (upper end

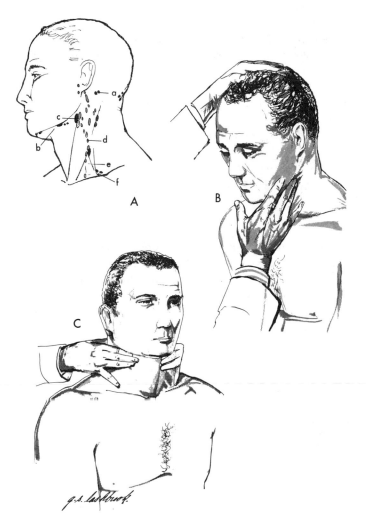

FIGURE 13-1. Technique of examination of cervicofacial lymph nodes. A. Location of major groups of nodes. B. Anterior approach. C. Posterior approach.

of deep cervical chain); (d) downward along the sternocleidomastoid muscle (superficial cervical nodes); (e) posterior triangle (lower end of deep cervical chain); (f) supraclavicular. There are two satisfactory methods of palpation. With the *anterior* approach the hand not used for palpation controls the head (Figure 13-1B). With the *posterior* approach the neck is first flexed to obtain proper relaxation of the muscles, and corresponding zones on both sides are palpated simultaneously (Figure 13-1C). *Palpation must be light* or small nodes will escape notice.

Axillary and Epitrochlear Nodes. The method of palpating the axilla is crucial. The patient may be either sitting or supine. In either case the arm must be supported (Figure 13-2A). Cup your hand slightly and reach as high into the apex of the axilla as possible. Now pull down, exerting gentle pressure against the thorax with the fingertips. Repeat several times, checking, in order, the lateral group (posterior), the central group, and the pectoral group. It is important not to *abduct* the arm too far, for this tenses the skin of the axilla, interfering with deep palpation and at times causing considerable discomfort for the patient. The epitrochlear nodes are palpated as shown in Figure 13-2B.

Inguinal Nodes. Palpate the superficial inguinal and femoral nodes, using the rotary motion described above.

Spleen

The spleen is palpated as a part of the abdominal examination. The patient is supine, arms at side, knees flexed slightly. Outline the area of splenic dullness as a first step. This should be done not so much to delineate the splenic size but rather to loosen the abdomen. Percussion may outline a greatly enlarged spleen, directing initial palpation to the left lower portion of the abdomen. Figure 13-3A shows the position of the examiner's hands. Note that *pressure is light.*

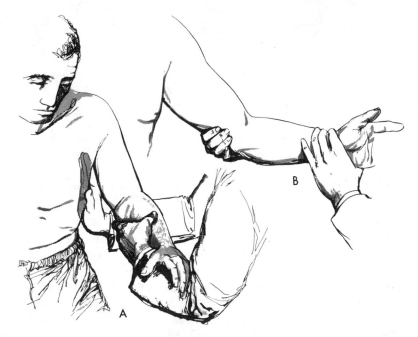

FIGURE 13-2. Technique of examination of (A) axillary lymph nodes and (B) epitrochlear lymph nodes.

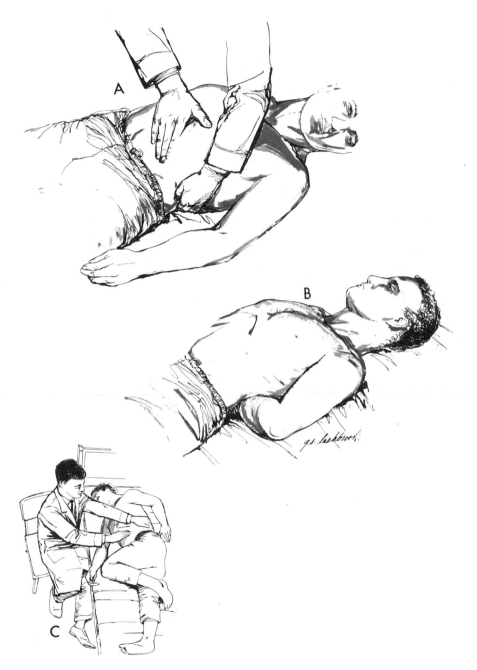

FIGURE 13-3. Palpation of the spleen. A. Positioning of examiner's hands. B. Placement of patient's left forearm to elevate his left flank. C. Examination of patient positioned on his right side to bring spleen forward.

Press the tips of the index and middle fingers of the right hand to a point *just beneath the costal margin.* Then ask the patient to turn his head to the side away from you and take a long, deep breath through his mouth. Do not move the hand as the patient inhales. The edge of an enlarged spleen will then brush against the fingers, lifting them slightly upward. As the patient exhales, probe the left upper quadrant more deeply, moving the fingertips in a slightly rotary motion. If nothing is felt, drop the hand about 1 cm. and repeat. *Do not dig.* This will cause spasm of the muscles, making palpation difficult. Further, slight splenic enlargement can be missed because the fingertips may be below the splenic edge, which will glide over the backs of the fingers.

Two special maneuvers may be helpful: (1) Have the patient slip his left forearm under the small of his back (Figure 13-3B); this will tend to thrust the spleen upward. (2) Roll the patient on his right side with the right leg straight and the left knee flexed. Notice in Figure 13-3C that the left shoulder, supported by the flexed left arm, is rotated posteriorly to the left in order to thrust the spleen forward and medially. Accordingly, the tips of the palpating fingers should be placed 1 or 2 cm. below the costal margin with this maneuver. The keys to satisfactory splenic palpation are proper instructions to the patient with respect to breathing, and *gentleness* by the examiner.

Normally the spleen is not palpable in the adult. It must be two or three times normal size before it becomes palpable.

Liver

The technique for examination of the liver is described in Chapter 11.

CARDINAL SYMPTOMS AND ABNORMAL FINDINGS

Pallor, ease of fatigue, weakness, lassitude, shortness of breath on exertion, faintness, and vertigo are the important general symptoms of anemia of whatever cause. They are due chiefly to increased circulatory effort, in part to deficient oxygenation of the tissues, especially the brain. In an otherwise healthy individual, symptoms of anemia do not become conspicuous until the hemoglobin is lowered to half its normal value, unless there is a coincidental fall in blood volume, as in shock or hemorrhage.

The leukemias and lymphomas commonly give rise to anemia and thrombocytopenia. In turn, epistaxis, gum bleeding, petechiae, purpura, or frank hemorrhage anywhere in the body may occur. Systemic symptoms include weight loss, fatigue, heat intolerance, night sweats, fever, and pruritus. Rapidly enlarging lymph nodes of whatever cause, with acute distention of the capsule, vary from "sore" to exquisitely tender. Slowly enlarging nodes produce no symptoms until they are large enough to result in mechanical difficulties.

Enlarged mediastinal and hilar lymph nodes can compress the trachea, causing respiratory embarrassment with a dry, "brassy" cough, progressive dyspnea, orthopnea, and cyanosis. In addition, obstruction to lymphatic and venous return can cause swelling of the face and neck (superior mediastinal syndrome). Enlarged retroperitoneal, periaortic and perifemoral nodes can cause ascites and edema of the lower extremities. By encroachment on the stomach and intestine, massive splenic enlargement leads to early satiety, constipation, and/or diarrhea.

In both polycythemia vera and secondary polycythemia, total blood volume and blood viscosity are increased. In consequence, congestive heart failure, headache, hemorrhage, and thromboembolic phenomena are common.

Lymph Nodes

It is frequently impossible to determine whether a lymph node is normal or abnormal simply by its texture. Any node, regardless of how it may feel, may show, on microscopic examination, typical *granulomatous, neoplastic,* or other changes.

Localized lymphadenopathy is usually due either to inflammation or to neoplasm. *Acute lymphadenitis* causes enlarged, tender, rather soft nodes, which are sometimes associated with induration and the red streaks of *lymphangitis.* The primary site of infection is usually obvious. *Metastatic lymphadenopathy* is usually stony hard, nontender, and somewhat fixed to the underlying structures. From his knowledge of anatomy, the physician *must* know the various patterns of lymphatic drainage to estimate the most likely site of a primary lesion when metastatic nodes are identified. Localized adenopathy may also result from chronic granulomatous processes such as tuberculosis or Hodgkin's disease.

Generalized lymphadenopathy is usually due to inflammation or neoplasm. Secondary syphilis, viral infections (infectious mononucleosis, measles), and hypersensitivity reactions (serum sickness) are common causes of acute, generalized adenopathy. Such nodes are soft, movable, and usually slightly tender. Many chronic systemic infections may at times produce generalized enlargement. The lymphomas are a common cause of generalized lymph-adenopathy. Initially they may cause painless, progressive, discrete enlargement of the nodes. Involvement may at first be localized, but later it frequently becomes generalized, and the nodes may become firm, matted, and fixed. *Leukemia* may or may not produce generalized lymphadenopathy. With the granulocytic forms, lymphadenopathy is uncommon. With monocytic leukemia, lymphadenopathy is commonly associated with oropharyngeal infection and is most conspicuous in the cervicosubmandibular areas. Chronic lymphocytic leukemia often results in generalized lymph node enlargement, and may be suspected by the feel of the nodes, which are usually 1 to 3 cm. in diameter, nontender, elastic (rubbery), and freely movable. Other varieties of leukemia may produce variable degrees of lymph node enlargement. In acute leukemia, because of rapid enlargement, the nodes are frequently tender.

Many other disorders are at times associated with generalized lymphadenopathy. It would be impractical to attempt to enumerate them here.

Spleen

Splenomegaly is common to many different and unrelated types of disease. It may be due to hyperplasia, congestion, or infiltrative replacement of the splenic pulp by neoplasm, myeloid elements, lipid, or amyloid.

Splenic hyperplasia occurs as a response to many systemic bacterial, parasitic, viral, or mycotic infections. Acute enlargement occurs with hematogenous dissemination of the infectious organisms, as, for example, in bacterial endocarditis, septicemia, or miliary tuberculosis. Chronic enlargement of the spleen occurs with malaria, rheumatoid arthritis, and other relapsing or progressive inflammatory diseases. Splenic hyperplasia is common to many of the chronic anemias, whether due to conditioned deficiencies, hemolysis, or inherited defects in erythropoiesis. It is also the cause of splenomegaly in polycythemia vera.

Splenic congestion as a consequence of portal hypertension may be secondary to chronic hepatic disease, chronic congestive heart failure, or occlusion of the splenic or portal veins. The most common cause of congestive splenomegaly is cirrhosis of the liver.

Splenic infiltration by neoplastic cells results in the marked enlargement associated with the leukemias and lymphomas. Occasionally splenomegaly may also be produced by replacement of the splenic pulp by amyloid, lipid-filled reticuloendothelial cells (e.g., Gaucher's disease), or myeloid elements (extramedullary hematopoiesis).

A classification of splenomegaly according to the degree of enlargement is listed below. The designations depend on the distance (in centimeters) of the splenic edge below the left costal margin, on deep inspiration, as follows: slight enlargement, 1 to 4 cm.; moderate enlargement, 4 to 8 cm.; great enlargement, more than 8 cm.

Slight Enlargement	*Moderate Enlargement*
Subacute bacterial endocarditis	Cirrhosis of the liver
Miliary tuberculosis	Acute leukemia
Septicemia	Chronic lymphocytic leukemia
Rheumatoid arthritis	Lymphoblastoma
Syphilis	Infectious mononucleosis
Typhoid	Polycythemia vera
Brucellosis	Hemolytic anemia
Congestive heart failure	Sarcoidosis
Acute hepatitis	Rickets
Acute malaria	
Pernicious anemia	

Great Enlargement

Chronic granulocytic leukemia	Agnogenic myeloid metaplasia
Chronic malaria	Rare diseases: Gaucher's disease,
Congenital syphilis in the infant	Niemann-Pick disease, kala-azar,
Amyloidosis	tropical eosinophilia

If the spleen is greatly enlarged, it may be missed on routine palpation. This pitfall can be avoided by (1) careful preliminary inspection during which the splenic edge may actually be visible in the abdomen with respiration, (2) preliminary percussion of the area of splenic dullness, and (3) repeated palpation at ever lower levels of the abdomen until the pelvic brim is reached. When it is suspected that the splenic capsule has been acutely distended by rapid enlargement of the spleen such as can occur in infectious mononucleosis, splenic infarction, or intrasplenic hemorrhage, great caution must be exercised lest by *excessive examination or manipulation splenic rupture be provoked.*

When there is tenderness to palpation over an enlarged spleen, it is always worthwhile to listen for a *splenic friction rub.* This is a common finding with splenic infarction.

Skin

Many hematologic disorders have cutaneous manifestations. *Pallor* and coldness of the skin may be present in patients with chronic anemia. The nail beds and conjunctivae are particularly helpful in the estimation of the significance of pallor, since many Caucasians are normally fair-skinned. *Rubor,* particularly of the face and neck, may be present in patients with polycythemia vera. There is often associated dilatation of the superficial veins and venules, as well as suffusion of the bulbar conjunctivae (bloodshot eyes). *Cyanosis* may be associated with secondary polycythemia. *Icterus* (jaundice) may be a sign of rapid hemolysis. *Purpura* may be due to any of a number of deficiencies of the hemostatic mechanism. Purpuric lesions are classified either as petechiae or ecchymoses. *Petechiae* are small, superficial, cutaneous or mucosal hemorrhages, less than 5 mm. in size. Ecchymoses are larger (greater than 5 mm.); they are purplish in color and have irregular borders. The common bruise is an ecchymosis. *Pruritus* (with certain lymphomas) may be intense, resulting in extensive excoriation of the skin. There may be primary invasion of the skin by lymphoma or leukemia, and almost any type of secondary dermatitis may develop with these two disorders (see Chapter 5).

Mucosa

Glossitis and stomatitis are common with the deficiency anemias. When they are longstanding, atrophy of the glossal papillae occurs, making the tongue pale and smooth. The gums may bleed whenever there is a hemorrhagic tendency associated with a blood disorder. *Gingival* and *mucosal ulcerations* of the mouth and pharynx occur with leukemia, particularly with the acute forms. A characteristic plum-colored swelling of the gingivae may be the most prominent physical finding in monocytic leukemia; it can be almost pathognomonic. Necrotic mucosal ulcers and pharyngitis are important signs of agranulocytosis.

Bone

Bone tenderness and bone pain commonly accompany disorders of the blood-forming organs. Although the sternum is commonly tender to point pressure, exquisite tenderness may develop as a consequence of increased intramedullary proliferation of blood cells, as in the leukemias and regenerative anemias, such as hemolytic anemia. *Localized bone tenderness* may be due to localized invasion of bone by leukemia or other malignancies of hematopoietic tissue (e.g., multiple myeloma, Hodgkin's disease).

Ocular Fundi

Many hematologic disorders produce funduscopic signs, including retinal edema, hemorrhages, exudates, and dilatation and tortuosity of the veins. These changes may be on the basis of hypoxemia, thrombocytopenia, stasis, capillary injury, or metabolic deficiencies. Abnormalities of the retina are further discussed in Chapter 6.

SPECIAL TECHNIQUES

Examination of the Peripheral Blood.   The peripheral blood is the most easily biopsied organ of the body. The ordinary complete blood count is a standard, routine laboratory procedure that should be performed on all patients. Its value in the recognition of most of the disorders of the blood-forming tissues is obvious.

Examination of the Bone Marrow.   Aspiration and needle biopsy of the bone marrow are simple and relatively painless, but interpretation of bone-marrow films requires considerable experience.

Lymph Node Biopsy.   Excisional biopsy of a peripheral node is frequently the simplest method of establishing a tissue diagnosis in diseases involving the blood-forming system. Whenever it is possible, a cervical, supraclavicular, or axillary node should be selected. Inguinal nodes are rarely satisfactory, for reasons previously stated.

Positive-Pressure Test for Capillary Fragility (Rumpel-Leede Phenomenon). Place a blood pressure cuff around the upper arm and determine the systolic and diastolic pressures. Inflate the cuff midway between the systolic and diastolic levels (but not to exceed 100 mm. Hg) and observe for five minutes for petechiae. Normally, no more than one petechia forms. Grade 0 to 4+, depending on the number of petechiae formed and the speed with which they develop. If petechiae begin to develop promptly, *discontinue the test immediately.* Severe purpuric skin damage can be precipitated in patients with pronounced capillary fragility by maintaining venous occlusion for the prescribed period.

X-ray. Such examination of the chest, including lateral views, is essential in evaluating intrathoracic lymphadenopathy.

Lymphangiography. This refers to the serial filming of the trunk, particularly the pelvis and abdomen, after the injection of a radio-opaque medium into peripheral lymphatics. Information with respect to size, contour, and texture of deep femoral, iliac, and periaortic nodes can be obtained by this technique. *The mesenteric nodes are not visualized.*

REFERENCES

1. Dameshek, W., and Gunz, F. *Leukemia* (2nd Ed.). New York: Grune & Stratton, 1964.
2. Mengel, C. H., Frei, I., and Nachman, R. *Hematology: Principles and Practice.* Chicago: Year Book, 1972.
3. Moseley, J. E. *Bone Changes in Hematologic Disorders.* New York: Grune & Stratton, 1963.
4. Wintrobe, M. M. *Clinical Hematology* (6th Ed.). Philadelphia: Lea & Febiger, 1967.

GENITOURINARY SYSTEM
Ralph A. Straffon

14

GLOSSARY

Anuria Complete suppression of urinary output.

Bacteriuria Presence of bacteria in the urine.

Balanoposthitis Inflammation of the glans penis and prepuce.

Bulbocavernosus reflex Reflex elicited by squeezing the glans penis, which results in constriction of the bulbocavernosus muscle and the anal sphincter.

Chordee Bowing of the penis during erection secondary to fibrotic plaques in the corpora cavernosa, or to congenital absence of the distal portion of the urethra.

Cryptorchism Failure of one testis or both testes to descend into the scrotum.

Cystitis Inflammation of the urinary bladder.

Dysuria Painful micturition.

Enuresis Involuntary voiding during sleep.

Hematuria Blood in the urine; when blood is detected visually it is called *gross hematuria*, and when only microscopically, *microscopic hematuria*.

Hydrocele Cystic scrotal mass containing clear, amber-colored fluid.

Nephrocalcinosis Deposition of calcium in the renal parenchyma.

Neurogenic bladder Dysfunction of bladder secondary to abnormality of its innervation.

Oliguria Urinary output less than 400 ml. per 24 hours.

Paradoxical incontinence Involuntary dribbling of urine secondary to chronic urinary retention.

Phimosis Tightness of the prepuce so that it cannot be retracted to uncover the glans penis.

Pneumaturia Voiding of urine containing free gas.

Polycystic kidney Congenitally abnormal kidney that contains numerous cysts of various sizes.

Precipitous micturition Involuntary loss of urine caused by sudden urge to void.

Priapism Prolonged erection of the penis.

Proteinuria Presence of protein in the urine; *orthostatic proteinuria,* presence of protein in a urine specimen taken while the patient is upright, which is absent from specimen taken while patient is supine.

Pyuria Presence of leukocytes in the urine.

Residual urine Urine that remains in the bladder after micturition.

Spermatocele A cystic scrotal mass containing whitish, milky fluid with spermatozoa.

Stress incontinence Involuntary loss of urine during period of increased intravesical pressure such as can be produced by coughing, straining, or lifting.

Varicocele Dilated veins of the pampiniform plexus in the scrotum.

TECHNIQUE OF EXAMINATION AND NORMAL FINDINGS

Kidneys and Ureters

The upper urinary tract is assessed as a part of the abdominal examination. The patient is supine with the knees slightly raised. Begin by scrutinizing the upper abdomen for obvious asymmetry or bulging, particularly in the flanks. Close observation during the patient's deep inspiration and expiration may give important clues as to the site of pathologic change.

In palpating the renal areas, place one hand posteriorly beneath the costal margin and press directly upward (Figure 14-1). The other hand palpates for the kidney and is placed below the costal margin at about the midclavicular line. The patient is then asked to take a deep breath, a maneuver that depresses the diaphragm and pushes the kidney downward. As the patient inhales, the hand is pressed inward and upward toward the costal margin. The technique is similar to that used for palpation of the liver and spleen except that the hand is gradually pressed more deeply into the abdomen. The right kidney is palpated from the right side, while the left kidney is examined by reaching across the abdomen or preferably by moving around the patient to his left side. When pathology is suspected, it may be valuable to have the patient lie on his side. The uppermost kidney tends to fall downward and medially in this position, making it somewhat more accessible to palpation.

Percussion is not used routinely, but it may be valuable at times in generally outlining masses in the renal area which are unusually large. Auscultation is valuable for detecting bruits that may originate in the renal arteries. It is carried out over the costovertebral angles posteriorly and in both upper quadrants of the abdomen. Transillumination is a particularly valuable technique in children. The room is darkened and the light source is pressed into the costovertebral angle posteriorly.

Since the right kidney normally is somewhat lower than the left, it is occasionally palpable, particularly in asthenic patients. Nothing of note is normally detected by transillumination, percussion, or auscultation. The ureters are not accessible to physical examination.

FIGURE 14-1. Palpation of the kidney.

The examiner should look also for possible tenderness originating in and around the kidney. Renal pain is elicited by pressure at the costovertebral angle — the angle formed by the junction of the twelfth rib and the paraspinous muscles. In this region the kidney is nearest to the skin surface, and deep pressure by the examiner's fingers may elicit pain due to intrinsic renal parenchymal disease. Costovertebral angle pain must be differentiated from pain produced by muscular spasm. Muscle pain can be demonstrated by deep palpation directly over the back muscles that lie medial to the costovertebral angle.

It is important to have a clear understanding of the innervation of the kidneys and ureters. Afferent autonomic nerves carrying sensation from the kidneys reach the spinal cord through the tenth, eleventh, and twelfth thoracic nerves. Referred pain of renal origin is therefore interpreted by the patient over the somatic distribution of these nerves in the abdominal wall. The fibers supplying the ureter enter the spinal cord from the twelfth thoracic nerve and the first three lumbar nerves. Pain referred from the ureter is distributed over the somatic distribution of the subcostal, iliohypogastric, ilioinguinal, and genitofemoral nerves, depending upon the portion of the ureter which is diseased. Since both the ilioinguinal nerve and the genital branch of the genitofemoral nerves supply the scrotum, pain referred from the ureter often radiates into the testicular and scrotal areas.

The Urinary Bladder

The bladder is examined by inspection, percussion, and palpation. When distended, it produces a bulging mass in the lower part of the abdomen over which dullness may be elicited by percussion. At times this dullness may extend up as far as the umbilicus. The region of the symphysis should be carefully palpated. Bladder pain is usually elicited by direct palpation over the suprapubic area. Bimanual examination may be performed at the time of rectal examination with the male in the lithotomy position; the examiner places one finger in the rectum pressing upward, the opposite hand on the lower abdominal wall (Figure 14-2). The best results are obtained with the patient under anesthesia. In the female, the bladder is easily palpated bimanually at the time of pelvic examination.

The empty bladder is not accessible to physical examination, but when distended with urine it can be mistaken for a lower abdominal tumor unless this possibility is kept in mind. On bimanual examination, the empty bladder feels much like a thick-walled, collapsed balloon. It is not tender and is freely movable, with no lateral extensions or palpable discrete masses.

Male Genitalia

In the normal uncircumcised male the foreskin should be easily retractable. At the time of retraction the external meatus is examined by separating it with the

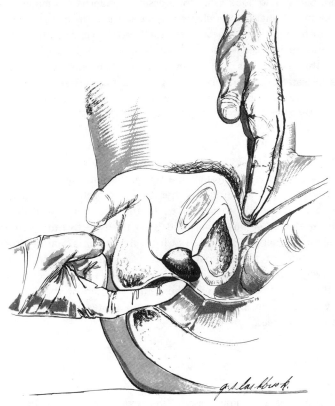

FIGURE 14-2. Bimanual palpation of the urinary bladder.

thumbs placed on either side of the distal glans penis. The shaft is then carefully palpated while searching for areas of tenderness or induration, and the urethra is milked downward to express any secretions present. The skin of the scrotum is inspected, and each testis is palpated between the thumb and the first two fingers. The comma-shaped structure bulging on the posterolateral surface of each testis is the epididymis, and it is palpated in the same manner. The spermatic cord extends upward from the epididymis to the external ring. The vas deferens can be easily palpated as a small, solid cord between the thumb and index finger, using the opposite hand to exert gentle downward traction on the testis.

Normally, no masses or areas of tenderness are palpable along the shaft of the penis. The testes lie freely in the scrotum. In the average man, they are 4 by 3 by 2.5 cm. in size, being correspondingly smaller in boys. The epididymis and vas deferens are discretely palpable but not tender. Transillumination is a commonly employed technique for examining the scrotal contents. It is performed in a darkened room. Normally, the testes are not translucent.

Prostate Gland

As explained previously, the prostate is assessed during rectal examination. The technique of digital examination of the rectum is considered in Chapter 11. The prostate gland is best examined while the patient is standing and bending over the examining table (Figure 14-3) or, when he is unable to stand or is in bed, in the Sims's position. Ample lubrication of the examining finger and perianal region facilitates the procedure. The index finger is introduced pointing toward the umbilicus, as this approximates the direction of the anal canal. The patient is asked to bear down slightly, which helps to relax the anal sphincter, and with gentle pressure the finger is easily introduced into the anal canal.

The muscle tone of the anal sphincter is estimated, and the bulbocavernosus reflex is tested. The patient is asked to relax the sphincter as much as possible, and the glans penis is squeezed with the opposite hand. Normally this produces involuntary contraction of the anal sphincter. Voluntary contraction may produce a false-positive test. The presence of a bulbocavernosus reflex signifies an intact reflex arc in the region of the sacral cord which also innervates the urinary bladder.

The prostate gland is examined by gentle palpation of the anterior wall of the rectum. The upper limits, lateral margins, and medial sulcus should be outlined.

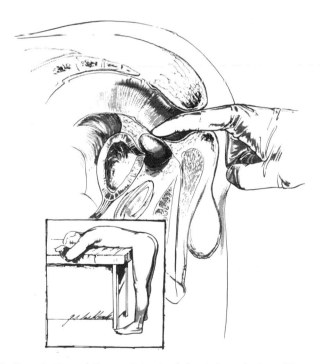

FIGURE 14-3. Examination of the prostate gland. Inset shows best position, with patient bending over examining table.

Each lobe is carefully palpated while the examiner searches for areas of irregularity or enlargement. The region of the seminal vesicles extends upward and laterally along the upper margin of the prostate gland.

Normally the sphincter tone is good, and the bulbocavernosus reflex is present. The prostate varies greatly in size, usually increasing with age. It is smooth and rubbery in consistency and normally not tender. The lateral borders are usually well defined. The seminal vesicles are normally not palpable.

CARDINAL SYMPTOMS AND ABNORMAL FINDINGS

Irritative Bladder Symptoms. Frequency, urgency, and dysuria are symptoms that often occur together and are usually secondary to inflammatory disease of the bladder, the prostate, or the urethra.

Prostatism. Symptoms produced by enlargement of the prostate gland can be divided into two groups: (1) irritative bladder symptoms as described above, and (2) obstructive symptoms characterized by a urinary stream decreased in size and force, with hesitancy and interruption of the stream during voiding. Obstructive symptoms may also be produced by a urethral stricture, a bladder neck contracture, or a urethral valve.

Incontinence. Careful history taking will identify the type of incontinence and lead to the appropriate treatment. *Stress incontinence,* or the involuntary loss of urine caused by straining, coughing, or lifting, occurs most frequently in the multiparous woman, and often is associated with cystourethrocele. The incontinence can be corrected surgically. *Urge incontinence* (precipitous micturition) is the involuntary loss of urine caused by the sudden urge to void. It may occur with inflammatory disease of the bladder and urethra, but also with a neurogenic bladder with uninhibited contractions. *Dribbling incontinence* is the constant loss of urine in various amounts with or without stress. It may be produced by a vesicovaginal fistula, ectopic ureter, or after prostatectomy when the sphincters of the bladder have been damaged. *Paradoxical incontinence,* the involuntary dribbling of urine, is due to chronic urinary retention. This may be produced either by obstruction of the urethra in the male, as in benign prostatic hypertrophy, or secondary to a neurogenic bladder, as in sensory paralytic bladder, produced by tabes dorsalis.

Pain. Renal pain usually is present in the costovertebral angle and may radiate anteriorly. It may be due to pyelonephritis, calculi, perinephric abscess, tumor, glomerulonephritis, or intermittent hydronephrosis. *Vesical pain* is usually present in the suprapubic region. Pain may be severe, with bladder distention, and it may be relieved by voiding, as occurs with interstitial cystitis; or pain may be continuous when associated with urinary retention or acute cystitis.

Testicular pain usually is due to neoplasm, infection, or local trauma, but pain may occasionally be referred to this region. *Prostatic* and *urethral pain* may be referred on occasion to the low back area.

Mass. A palpable mass in the renal area may be produced by neoplasm, hydronephrosis, polycystic kidney, or perinephric abscess. When a suprapubic mass can be palpated, it usually is due to distention of the bladder, but it may be a neoplasm in this region. A scrotal mass may be inflammatory, neoplastic, or secondary to trauma. If it can be transilluminated, it probably is either a spermatocele or a hydrocele.

Pneumaturia. This may develop with urinary tract infection due to gas-forming bacteria, or it may be produced by an enterovesical fistula due to inflammatory or neoplastic disease of the gastrointestinal tract.

Hematuria. Hematuria, either microscopic or gross, must be considered to be due to neoplasm until proved otherwise. Initial or terminal hematuria is usually associated with disease of the lower urinary tract. Blood throughout urination may come from the kidney, the ureter, or the bladder. Calculi, infection, trauma, and acute glomerulonephritis are frequently associated with hematuria.

Nocturia. Nocturia is usually a significant symptom and is seen in association with benign prostatic hypertrophy, diabetes mellitus, urinary tract infections, and with reversed diurnal rhythm as occurs with renal and circulatory insufficiency.

Gastrointestinal Symptoms. Nausea, vomiting, and abdominal distention may be associated with renal or ureteral calculi. Hydronephrosis is frequently a silent lesion that may produce symptoms that suggest gallbladder disease or duodenal ulcer when the right kidney is affected, and a lesion of the colon when the left kidney is affected. The presenting symptoms of chronic renal insufficiency (azotemia) frequently are nausea and vomiting.

Kidneys and Ureters

Acute pyelonephritis usually produces fever, and the patient may be extremely ill. There is tenderness to deep palpation or percussion in the costovertebral angle, and the entire flank region may be tender. When the peritoneum overlying the kidneys is affected by the inflammatory reaction, signs of peritonitis are present, with abdominal distention, muscle spasm, rebound tenderness, and hypoactive bowel sounds.

    Perinephric abscess may be associated with a low-grade or septic elevation in temperature. There usually is exquisite tenderness on the affected side, and frequently a bulging mass may be felt in the flank. Scoliosis of the spine with the concavity pointing toward the affected side occurs because of irritation of the psoas major and quadratus lumborum muscles. The diaphragm is elevated and somewhat fixed on the affected side; because of inflammatory reaction, basilar rales may be present. Edema of the skin may occur over the abscess.

Obstructive uropathy, regardless of its location, produces an increase in the hydrostatic pressure within the renal collecting system. In general, the higher in the urinary tract the lesion is located, the greater the effect upon the kidneys. Initially there is hypertrophy of the musculature of the renal pelvis, but as the obstruction persists or progresses, decompensation and dilatation occur. The resultant enlarged kidney may be palpated on bimanual examination. In small children, transillumination of the renal areas may help to differentiate cystic mass from solid tumor. As the back pressure increases, the hydronephrotic process progresses; the renal blood supply is compromised, producing ischemia. Eventually the renal parenchyma is destroyed, leaving a thin-walled cystic mass.

Benign tumor of the kidneys is rare and usually is too small to be palpated. The *embryoma* (Wilms's tumor) is malignant. It usually occurs in children under the age of 5 years. The presenting sign often is a palpable mass in one or both renal areas. *Renal-cell carcinoma* (hypernephroma), the most common renal malignant neoplasm, if of sufficient size may produce a palpable mass in the flank. Extension of the tumor into the renal vein and inferior vena cava will produce dilated veins in the abdominal wall. The left spermatic vein empties into the left renal vein and may be obstructed by a tumor growing into the renal vein. This produces on the left side of the scrotum a varicocele that does not decompress when the patient is supine.

Renal calculi may produce costovertebral-angle tenderness, particularly when there is associated pyelonephritis. Acute renal colic frequently produces abdominal distention and either hypoactive or absence of bowel sounds.

Hypertension of renal origin caused by a stenotic lesion of the renal artery may produce a *systolic bruit* that can be heard anteriorly in the upper quadrants of the abdomen.

Polycystic kidneys are usually bilateral and contain multiple cysts. As the cysts enlarge, palpable masses are produced in the renal areas. Unless infected, the renal masses are usually not tender.

Renal ectopia will produce a palpable mass in the lower part of the abdomen. In *crossed renal ectopia,* both kidneys are on the same side and are often fused, giving rise to a rather large mass which suggests neoplasm. A *horseshoe kidney* is produced by the fusion of the lower pole of each kidney, producing an isthmus of renal tissue across the midline, which may be palpable in a thin patient.

Ureterocele, retrocaval ureter, and *congenital stricture* of the ureter produce physical findings only when hydronephrosis results from the lesion. Ectopic ureter in the male always opens within the external sphincter and produces no characteristic physical findings. In the female, the ectopic ureteral orifice may empty into the uterus, cervix, vagina, or vestibule, and dribbling incontinence may result. Inspection of these areas will frequently disclose the ectopic ureteral orifice, particularly after the intravenous injection of indigo-carmine to color the urine blue. The ureter usually is involved in the inflammatory process of *pyelonephritis* and *tuberculosis.* In chronic ureteritis the ureter becomes fibrotic and shortened and may be palpated on rectal examination in the male or on vaginal examination in the female.

Ureteral calculi originate in the kidney and then pass into the ureter. There are three narrowed areas where calculi commonly become arrested: the ureteropelvic junction, the pelvic brim where the ureter crosses the iliac vessels, and the ureteral vesicle junction. Ureteral colic produces severe pain; the patient is usually agitated and unable to remain still; there is great tenderness in the costovertebral angle and often abdominal distention with hypoactive bowel sounds.

Bladder

Exstrophy of the bladder is easily identified from inspection of the lower abdominal wall. Rami of the symphysis pubis are separated and the patient's gait is a "duck waddle." Inguinal hernias and epispadias accompany this congenital anomaly.

Cystitis is the most common disorder of the bladder, particularly in females. There is often tenderness elicited by palpation over the suprapubic region. *Obstruction* below the bladder may be due to vesicle neck contraction, hypertrophy of the prostate, or hypertrophy of the urethral valves. Chronic obstruction causes trabeculation, cellules, and diverticula of the bladder. As residual urine increases, bladder capacity may increase, producing a decompensated bladder. The bladder may then be palpable in the suprapublic region as a midline mass. When large diverticula occur, these may also be palpated. *Tumors* and *calculi* of the bladder are common, particularly in males, but usually produce no physical findings unless the lesions are large.

Neurogenic Bladder. Physical Findings are pathognomonic of various types of neurogenic bladder:

Sensory paralytic bladder is produced by a lesion on the sensory side of the reflex arch, as in tabes dorsalis. There is no bulbocavernosus reflex, and saddle anesthesia is present. When the bladder becomes decompensated, a suprapubic mass may be palpated.

Motor paralytic bladder results from a lesion affecting the motor side of the reflex arch, as in poliomyelitis. There is no bulbocavernosus reflex but saddle anesthesia is absent. The patient is unable to initiate micturition, and a large distended bladder is palpable in the suprapubic region.

Autonomous neurogenic bladder is produced by a lesion affecting sacral segments 2, 3, and 4, such as occurs in myelomeningocele or myelodysplasia. There is no bulbocavernosus reflex, and saddle anesthesia is present. Urine can be forced from the bladder by pressure in the suprapubic region.

Reflex neurogenic bladder due to a transverse myelitis of the spinal cord, as in trauma, characteristically produces a hyperactive bulbocavernosus reflex and saddle anesthesia. Associated neurologic findings due to the paraplegia facilitate this diagnosis.

Uninhibited neurogenic bladder is seen in normal infants before myelinization of the spinal cord and after cerebrovascular accidents. The bulbocavernosus reflex is normal or hyperactive and there is no saddle anesthesia.

Penis and Urethra

Congenital anomalies of the penis are uncommon. *Balanoposthitis* is seen in uncircumcised males, because of recurrent infection of the prepuce and glans penis. There is erythema, local discomfort, and sometimes a purulent discharge. *Phimosis* occurs when it is impossible to retract the prepuce, and is usually secondary to recurrent balanoposthitis. There may be local signs of infection. *Paraphimosis* results when the prepuce is retracted behind the glans penis and cannot be returned to its normal position.

The *primary lesion of syphilis,* which appears two to four weeks after infected sexual contact, is a painless ulcer with indurated borders and a relatively clear base. Palpable inguinal lymph nodes are often present. *Lymphopathia venereum* begins with a small penile lesion that may be papular or vesicular. Painful enlarged inguinal lymph nodes called buboes may ulcerate and drain. *Granuloma inguinale* results in a painful superficial ulceration that is erythematous and velvety in appearance.

Epidermoid carcinoma is usually found in uncircumcised males as a painless ulceration that fails to heal. Growth frequently begins beneath the prepuce. Palpable lymph nodes may indicate metastatic extension of the neoplasm.

Stenosis of the external urethral meatus produces a serious obstructive lesion. There is often meatal ulceration and crusting. *Hypospadias* may be discovered on close inspection of the ventral surface of the penis. Its classification is dependent on the location of the external urethral meatus. An associated *chordee* is produced by fibrosis in the area of the malformed urethra, producing a downward curvature of the penis on erection. *Epispadias* is less common than hypospadias and is often associated with exstrophy of the bladder. Urinary incontinence is often an associated finding. *Urethritis* produces a few abnormal physical findings. The urethra may be tender to palpation, and often there is a purulent urethral discharge.

Scrotal Contents

Acute epididymitis results in a painful mass in the scrotum. Initially it may be possible to distinguish the enlarged, tender epididymis from the testis, but later the testis and epididymis become an inseparable mass. The spermatic cord is often thickened and indurated. *Chronic epididymitis* results from recurrent bouts of epididymitis. The epididymis is enlarged and indurated. *Tuberculous epididymitis* may mimic acute and chronic epididymitis. The vas deferens often contains a group of enlargements that resemble a string of beads, giving rise to the term of "beading" of the vas deferens.

Acute orchitis may occur from any infectious disease process, but most often is associated with mumps parotitis. The testis is enlarged and painful, and the overlying scrotal skin is erythematous.

A *testicular tumor* usually results in an enlarged testis that is not translucent. Any painless nodular area in the testis must be regarded as a tumor. *Hydroceles may develop secondary to tumors;* when the testis cannot be palpated because of a hydrocele, aspiration of the fluid will facilitate palpation. Simple hydroceles are common. They transmit light readily.

Torsion of the spermatic cord occurs spontaneously, most often in prepubertal boys, resulting in acute ischemia to the distal parts of the epididymis and testis. Examination of the scrotum reveals a painful mass that is usually elevated. In this condition, elevating the testis may increase the pain, in contradistinction to epididymitis, in which elevation of the testis will somewhat relieve the pain.

Prostate and Seminal Vesicles

Acute prostatitis results in fever, urethral discharge, and an exceedingly tender, enlarged prostate gland on rectal examination. An abscess may develop and can be demonstrated as a fluctuant mass in the prostate gland. The seminal vesicles often are involved in inflammatory reaction and may be dilated and extremely tender. *Chronic prostatitis* is usually asymptomatic. At times rectal examination shows the prostate to be boggy or irregular. Areas of fibrous tissue may be palpated, thus simulating neoplasm. *Prostatic calculi* are seldom of clinical importance. They may often be palpated at rectal examination and mistaken for carcinoma.

Benign prostatic hypertrophy is extremely common in men over 50 years of age. The prostate is of variable size. Small glands may be as obstructive as the large glands, because it is the intraurethral protrusion that produces obstruction, not extraurethral enlargement. The gland is usually symmetric and has a smooth, rubbery consistency. The median sulcus can be identified and the lateral borders are well defined.

Carcinoma of the prostate occurs in approximately 20 percent of men over 60 years of age. *The initial lesion usually involves the posterior lobe and is readily palpable at rectal examination.* The early lesion feels like a small nodule on the posterior surface of the gland. Similar nodules can also be caused by calculi, chronic infection, or benign adenoma. More advanced malignant lesions usually are stony hard, irregular, and painless on palpation.

SPECIAL TECHNIQUES

Urinalysis. This is a most important study. Too often it is made by the laboratory technician who is unaware of the clinical status of the patient and gives only a cursory examination. Much information is available from the routine urinalysis when it is properly performed.

Both the male and the female urethra harbor bacteria and leukocytes. After careful cleansing of the urethral meatus and with the foreskin retracted in males and the labia separated in females, a clean midstream-voided specimen is collected in a sterile container. Prompt examination of the urine specimen is highly important. Catheterized specimens may be collected from the female with little risk of introducing infection. Catheterization is inadvisable in the male unless the urine is infected and it is done to check the residual urine.

Quantitative Urine Cultures. The number of bacteria in the urine specimen may be quantitatively estimated by making a pour plate, allowing incubation to take place, and counting the colonies. Fewer than 1000 organisms per milliliter of urine signifies only contamination; more than 100,000 organisms per milliliter is nearly always significant.

Plain Film of Abdomen. This is often helpful in the diagnosis of genitourinary disease. The position, shape, and size of the kidneys can often be determined on the plain film. Evidence of bony abnormalities and soft-tissue densities may be identified, as may calcification in the region of the adrenals, kidneys, ureter, bladder, and prostate.

Intravenous Pyelogram. This visualizes the upper urinary tract in the most physiologic manner. It is extremely valuable in the diagnosis of a wide variety of renal diseases, provided that renal function is not too seriously impaired.

Cystogram. This is made by instilling radio-opaque fluid into the bladder by way of a catheter. Voiding cystourethrograms allow observation of the vesical neck and urethra during micturition. The presence or absence of reflux into one or both ureters may be observed on either the cystogram or voiding cystourethrogram.

Cystoscopy and Panendoscopy. The lens system in the panendoscope allows the examiner to look forward and slightly upward, and hence the instrument is particularly useful for observing the posterior wall of the bladder, the vesical neck, and the urethra. The cystoscopic lens system is devised for inspection of the bladder; the view is nearly at right angles to the shaft of the instrument. Accessory instruments are available for ureteral catheterization, biopsy, fulguration, and minor operative procedures.

Retrograde Pyelogram. This is obtained by passing a ureteral catheter up to the renal pelvis at the time of cystoscopy. Contrast medium is injected by way of the catheter to outline the collecting system of the kidneys. This procedure is particularly useful when sufficient outline of the collecting system is not obtained by intravenous pyelography, or when renal function is too poor to allow excretion of the contrast medium.

Selective Renal Angiography. This procedure is performed by making a percutaneous puncture of the femoral artery just below the inguinal ligament and passing a specially constructed catheter upward into the aorta to the area where the renal arteries take off. An aortic injection of contrast material is first made to show the position and number of renal arteries on each side and to evaluate the orifice of each artery as it leaves the aorta. Each renal artery is then selectively catheterized and a direct injection of contrast material made into each one to visualize the blood supply to the renal parenchyma. This procedure is particularly useful in the diagnosis of renal mass lesions or of renal vascular disease which may be producing hypertension, and in evaluating the status of an obstructed kidney which is shown to function poorly on intravenous pyelography.

REFERENCES

1. Campbell, M. F. *Urology.* Philadelphia: Saunders, 1963.
2. Emmitt, J. L. *Clinical Urography.* Philadelphia: Saunders, 1964.
3. Glenn, J. F. (Ed.) *Diagnostic Urology.* New York: Hoeber Med. Div., Harper & Row, 1964.
4. Smith, D. R. *General Urology.* Los Altos, Calif.: Lange Medical Publications, 1961.
5. Welt, L. G. *Clinical Disorders of Hydration and Acid-Base Equilibrium* (2nd Ed.). Boston: Little, Brown, 1959.

HERNIA 15

George D. Zuidema

GLOSSARY

Femoral hernia Type of hernia in which the abdominal wall defect lies along the femoral canal with the hernial sac presenting deep to the inguinal ligament.

Hydrocele Fluid-containing cystic mass, usually arising as a result of incomplete obliteration of the processus vaginalis; common in infancy and often bilateral; may occur in the scrotum as a hydrocele of the testis or appear anywhere along the inguinal canal as a hydrocele of the cord.

Incarcerated hernia One in which the contents cannot be returned to the abdominal cavity; no inflammation or interference with blood supply has, however, resulted.

Inguinal hernia, direct Hernia through weakness in the abdominal musculature with protrusion through the region of Hesselbach's triangle.

Inguinal hernia, indirect Hernia through the internal inguinal ring, with the hernial sac descending beside the spermatic cord toward or into the scrotum.

Reducible hernia One in which the contents of the hernial sac can be returned to the abdominal cavity.

Richter's hernia Type of strangulated hernia in which only a small portion of the wall of intestine is caught in the ring of the defect; as a result, local gangrene may appear without the production of the signs or symptoms of intestinal obstruction; usually seen with a femoral hernia.

Scrotal hernia Inguinal hernia, almost always of the indirect type, of sufficient size to permit the hernial sac and its contents to enter the scrotum together with the contents of the spermatic cord.

Strangulated hernia One in which the blood supply to the viscera within the hernial sac has become obstructed, resulting in necrosis of tissue.

TECHNIQUE OF EXAMINATION

Abdominal hernias have in common a sac lined with peritoneum which protrudes through some defect in the abdominal wall. The contents of inguinal and femoral hernias are variable. Omentum, small bowel, large bowel, or bladder may be encountered within the hernial sac.

Examination of the inguinal canal is not difficult (Figure 15-1). The examining finger is inserted in the lower part of the scrotum and the scrotum is inverted so that the finger passes along the inguinal canal to palpate the external ring. If performed slowly and carefully this examination can be done with minimum discomfort to the patient. The examining finger should always identify the following normal structures: the extent of the os pubis, the spermatic cord as it lies within the inguinal canal, the size and perimeter of the external inguinal ring, and the area of Hesselbach's triangle medial to the deep epigastric vessels.

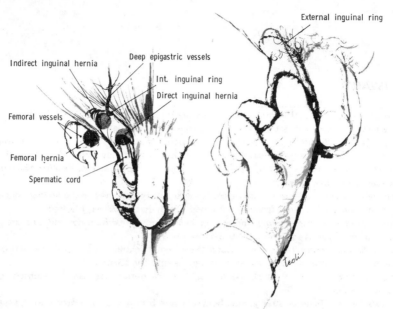

FIGURE 15-1. Anatomic relationships in inguinal and femoral areas. Technique of examining inguinal canal.

The indirect inguinal hernial sac emerges through the internal ring, traverses the inguinal canal with the contents of the cord, and appears at the external ring. If it extends into the scrotum it is termed a *scrotal hernia*. The neck of the indirect hernial sac lies lateral to the deep epigastric artery.

Indirect hernias are the most common inguinal hernias. They are thought to be congenital, and may result from failure of the processus vaginalis to obliterate. Hernias in children and young adults are usually of the indirect type. Hernias limited to the inguinal canal are termed *incomplete,* while those that emerge from the external ring are *complete.* In the female a complete hernia may enter the labium majus as a labial hernia.

The direct inguinal hernial sac protrudes through Hesselbach's triangle medial to the deep epigastric artery and appears at the external ring without passing through the inguinal canal. It is apparent, therefore, that the direct hernial sac does not lie in close relationship with the spermatic cord. *Direct hernias* usually result from weakness of the transversalis fascia in the region of Hesselbach's triangle and present as a rounded swelling. They are almost always reducible and rarely enter the scrotum. Most direct hernias occur in individuals over 40 years of age.

On physical examination the external ring may sometimes appear to be enlarged. Relaxation of the external ring is, however, insufficient evidence on which to make a diagnosis of indirect inguinal hernia. If the examining finger encounters a mass in the inguinal canal, the presumptive diagnosis of hernia may be made. This may be confirmed by palpating an impulse in the external ring when the patient coughs. The differentiation between an indirect inguinal and a direct inguinal hernia should not be difficult. If the hernia can be completely reduced, the examining finger may be inserted into the external inguinal ring, and when the patient coughs the leading edge of the hernia may be palpated with the tip of the finger. On the other hand, if the finger is inserted into Hesselbach's triangle, the sac of an indirect hernia will be felt striking the side of the finger.

It may sometimes be difficult to establish the diagnosis of inguinal hernia in the female. It should be possible to identify the inguinal ligament and os pubis, and from these anatomic points locate the external inguinal ring. If a sac is palpated when the patient coughs, the diagnosis of inguinal hernia may be made. It may be of help to place the palmar surface of the hand over the area of the internal inguinal ring in an effort to feel an impulse with cough. It is occasionally possible to see a small indirect inguinal hernia as a bulge that appears on coughing. Examination in both standing and supine positions may help to bring these points out.

Examination of the femoral region is more difficult than is the study of the inguinal area. The external opening of the femoral canal may be located anatomically just medial to the femoral artery and deep to the inguinal ligament. A simple rule to remember in examining the patient's right femoral area is that if the examiner's right index finger is placed on the patient's right femoral artery the middle finger will overlie the femoral vein and the ring finger will overlie the femoral canal. A swelling lying within the femoral canal which transmits an impulse upon coughing may be diagnosed as a *femoral hernia*. This must be distinguished from psoas abscess, lymphadenitis, or saphenous varix.

Femoral hernia is the commonest hernia in the female, and the points raised in the above discussion of examination of the femoral region are applicable in examining the female patient as well.

As a general rule, the incidence of strangulation is high in femoral hernias. Consequently, early operation is desirable.

ABNORMAL FINDINGS

A *sliding hernia* is a special type that deserves mention. The large bowel (or bladder) slips retroperitoneally between the leaves of its mesentery to herniate or protrude through the defect in the abdominal wall. On the right side the cecum may be the presenting part, while on the left the presenting part may be the sigmoid colon. In either event it is important to recognize this entity, since the wall of the bowel or bladder rather than a peritoneal sac makes up the leading edge of the hernia. Failure to make the proper diagnosis may lead the surgeon to open the bowel accidentally, thinking he is incising a hernial sac. It is often difficult to reduce a sliding hernia, and irreducibility should raise the examiner's suspicions as to this possibility.

When a hernia can no longer be reduced and the contents of the hernial sac cannot be returned to the peritoneal cavity, it is said to be *incarcerated*. If the blood supply to the viscera lying within the hernial sac has been cut off, it is said to be a *strangulated hernia*. It is often difficult and sometimes impossible to tell with certainty whether a hernia is simply incarcerated or whether it is strangulated. It is reasonable to attempt to reduce a hernia which is incarcerated if the incarceration is recent and one can be certain that the contents are completely viable. Strangulated hernias usually show local signs of inflammation, although this is not invariable. When inflammatory signs are present it is unwise to attempt vigorous reduction lest strangulated bowel be returned to the peritoneal cavity.

If it is decided to attempt nonoperative reduction of an incarcerated hernia, this should be performed with the patient in the recumbent position. This may be accomplished by exerting constant gentle pressure over the sac. The patient will often be experienced in reducing the hernia himself and may be able to accomplish this with ease. In difficult cases it may be necessary to lower the head of the bed, flex the leg on the affected side to relax the abdominal muscles, and attempt to gently guide the contents of the hernial sac through the inguinal ring. If local pain and tenderness are present, presumptive diagnosis of strangulation may be made and operative reduction and repair of the hernia is the procedure of choice. It should be noted that it is sometimes possible to reduce the entire hernia, together with the internal ring, into the abdominal cavity without actually freeing the contents of the hernial sac from the constricting internal ring. Under these circumstances continued pain or tenderness in the region indicates the need for urgent surgical attention.

Large femoral hernias may be difficult to diagnose correctly because of their tendency to leave the abdominal cavity by way of the femoral canal and then be directed upward to overlie the inguinal ligament. They may be differentiated from inguinal hernias, however, if it is remembered that the sac of the femoral hernia lies lateral to and below the level of the symphysis pubis, while the sac of the inguinal hernia lies medial to it and above.

It is sometimes difficult to detect early strangulation in femoral hernias, since such local signs as pain or tenderness may be minimal. The femoral canal is the most likely site for the development of a Richter's hernia. The systemic signs of strangulation — such as tachypnea, leukocytosis, and fever — should invariably lead one to inspect this area with great care. The localized gangrene of the bowel wall without intestinal obstruction may lead to perforation and abscess formation just below the inguinal ligament. This should be differentiated from psoas abscess and femoral lymphadenitis.

If the processus vaginalis is incompletely obliterated, fluid may accumulate within the involved segment to produce a cystic mass. In the male this may occur in the scrotum as a hydrocele of the testis. If it occurs within the inguinal canal it is termed a *hydrocele of the cord.* It is necessary to differentiate the hydrocele from an incarcerated hernia or a tumor of the testis. Transillumination in a darkened room may be helpful in establishing the diagnosis of hydrocele. These lesions are particularly common in infancy and are often bilateral.

Hydrocele of the canal of Nuck may occur in the female. In the course of embryologic development, the round ligament leaves the retroperitoneal area to traverse the inguinal canal and insert itself on the labium majus. A processus vaginalis of peritoneum descends with the round ligament, and if the process is incompletely fused and obliterated, a hydrocele may result. This may lie anywhere between the internal inguinal ring and the labium majus. Hydrocele of the canal of Nuck may be difficult to demonstrate but is characterized by its cystic, irreducible, translucent appearance.

Occasionally patients will have indirect and direct hernias simultaneously. These are termed *saddlebag hernias.*

An *umbilical hernia* protrudes through the umbilical ring. In the newborn a congenital umbilical hernia may result from improper closure of the abdominal wall; a hernia of the umbilical cord, also termed an *omphalocele,* is produced. The peritoneal sac is not covered by the skin of the abdominal wall.

True umbilical hernias are common during the first year of life. In this type the peritoneal sac is covered by skin. Increased intra-abdominal pressure due to trauma, cough, or constipation may contribute to their formation.

Umbilical hernias in adults are more common in women. Obesity, pregnancy, ascites, or congenital defect may be contributing factors. They may show wide variation in size, but the neck of the sac is often small. This type of hernia usually contains omentum, but may contain large or small bowel and other viscera as well. Strangulation is a frequent occurrence. Diastasis or separation of the rectus muscles is often associated.

An *epigastric hernia* occurs through a weakness in the linea alba between the xiphoid and the umbilicus, usually due to a developmental defect. Pregnancy, obesity, trauma, or constipation may be contributing factors. These hernias are most common in young adult males. The hernial sac is usually small and may contain omentum, rarely intestine. Strangulation rarely occurs.

Incisional hernias, as the name indicates, occur through surgical incisions. Infection, poor wound healing, faulty wound closure, postoperative vomiting, ileus, partial wound disruption, and obesity may be contributing factors. This type of hernia often reaches large size, and the intestines are usually adherent to the underside of the peritoneum. Strangulation is uncommon but may occasionally occur.

A *spigelian hernia* occurs at some point in the semilunar line at the lateral margin of the rectus muscle, usually in the lower abdomen at the linea semi-circularis where the posterior rectus sheath is absent.

Diaphragmatic hernias occur through defects in the diaphragm, permitting the protrusion of abdominal viscera into the thoracic cavity. Several varieties exist: *congenital,* in which some portion of diaphragm may be absent; *traumatic;*

and *acquired,* of which the most common types are the *hiatus hernia* (sliding, because it is retroperitoneal) and the *paraesophageal* (upside-down stomach).

Several rare varieties of hernia occur. These include *sciatic hernia,* which emerges through the greater or lesser sacrosciatic foramen; *perineal hernias,* which protrude between the muscles or fascia which form the pelvic floor; *lumbar hernias* through the inferior lumbar triangle of Petit; *obturator hernias,* presenting through the obturator foramen; and *internal hernias* within the peritoneal cavity.

BREAST

George D. Zuidema

Careful examination of the breast should be part of every complete physical examination regardless of whether or not the patient has noted any particular signs or symptoms. Breast cancer is the commonest malignancy occurring in women. It is a tumor that offers reasonable chance of cure if it is recognized early and adequate therapy is carried out. Early detection is the key to successful treatment, and early detection of breast carcinoma is completely dependent upon careful performance of this "routine" part of the physical examination.

GLOSSARY

Gynecomastia Hypertrophy of breast tissue in the male patient, which causes resemblance to a female breast.
Mastitis Inflammation of the breast, usually due to pyogenic infection.
Mastopathia cystica (chronic cystic mastitis) Inflammatory condition of the breast characterized by diffuse nodularity and cystic changes.
Paget's disease Excoriating or scaling lesion of the nipple associated with an underlying carcinoma.

TECHNIQUE OF EXAMINATION

The female patient should always be examined in the presence of a chaperon. Complete examination of the male or female breast demands a thorough and systematic approach. Adequate exposure is important. The patient should disrobe to the waist, although the breasts should be kept covered by a towel except during the actual examination. The patient should first be examined in the supine position, then while seated. The examination begins with inspection (Figure 16-1).

Inspection

Some degree of asymmetry is not uncommon and is usually the result of a difference in breast development. Increase in size of one breast may, however, indicate the development of cyst, inflammation, or tumor. Asymmetry is most easily observed while the patient is in the sitting position.

The skin overlying the breast should be carefully observed. The edema associated with inflammatory carcinoma or the ulcerative involvement of the

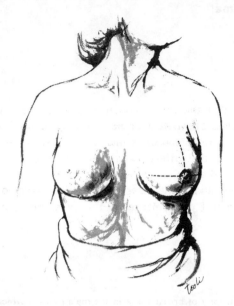

FIGURE 16-1. Division of breast into quadrants for purposes of precise localization.

nipple seen in Paget's disease should be noted. Local areas of redness may indicate underlying inflammation and are important in the detection of early breast infection.

The nipple should be carefully examined for evidence of bleeding, discharge, retraction, or ulceration.

Skin retraction is usually an indication of carcinoma, although it may result from traumatic fat necrosis. It is, however, a sign of malignancy and should be carefully searched for. This is best done by having the patient assume a position that will exert a pull on the suspensory ligaments of the breast. She should be examined while sitting erect and with her arms raised directly overhead. Elevating the arms should result in equal elevation of both breasts. A lesion producing shortening of the suspensory ligaments is likely to produce some retraction or deviation of the nipple (Figure 16-2A). Another method of bringing out retraction is to produce contraction of the pectoral muscles which in turn results in general traction on the breast tissue and tends to exaggerate any retraction which may be present. The patient should place the palms of both hands together and on command push the hands against each other (Figure 16-2B). This may also be accomplished by placing her hands on her hips and pushing forcibly against them (Figure 16-2C). It may be necessary to repeat these maneuvers several times so that all parts of the breasts can be adequately inspected. Another method of demonstrating retraction is to have the patient

FIGURE 16-2. Demonstration of skin retraction. A. Diagram showing how shortening of Cooper's ligaments may produce skin retraction. B. Method of bringing out skin retraction by pressing hands together. C. Method of bringing out skin retraction by pressing against the hip.

lean forward at the waist with her hands placed on the back of a chair. This demonstrates whether the breasts, as they fall away from the thorax, produce equal traction on the suspensory ligaments bilaterally. These maneuvers must be employed to detect early lesions. Obviously, with a grossly detectable lesion, these steps are unnecessary.

Inspection of the breasts should include careful observation of axillary and supraclavicular regions for evidence of bulging, retraction, discoloration, or edema, since these are the important lymphatic drainage areas from the breasts.

Palpation

The next portion of the examination consists of palpation. There is a wide variation in the consistency of normal breast tissue. This variation will depend upon such factors as age, obesity, stage of the menstrual cycle, and pregnancy. As experience is gained in the art of physical examination, the range of normal will become apparent.

Palpation of the breast is best carried out by means of a definite system of examination. Regardless of the patient's presenting complaint it is important to completely examine both breasts and their lymphatic drainage areas lest some serious lesion be overlooked. It is convenient to begin the examination in the upper lateral aspect of each breast. The left breast is usually examined first and palpation is carried out using the fingertips (Figure 16-3A). Gentle, light palpation should first be used, following which deeper exploration may be indicated if there is considerable breast substance. Palpation should be carried out in a clockwise direction until the entire breast has been examined. The nipple should be palpated for the presence of induration or a subareolar mass, and gentle pressure or a stripping action should be used to see whether discharge can be detected. Upon completion of the examination of the left breast, the right breast is examined in a similar manner, again beginning in the upper lateral area and proceeding in a clockwise direction.

Palpation should be performed with the patient in both supine and seated positions. In both instances, the breasts should be examined with the patient's arms at her side and then with her arms overhead (Figure 16-3B, C).

The examiner should note the texture of the skin and the consistency and elasticity of the breast tissue. An increase in firmness may suggest infiltration or neoplasia. Tenderness to palpation usually indicates underlying inflammation. Malignant lesions by themselves are seldom tender but may coincide with widespread chronic cystic mastitis, in which case tenderness may be present.

If a mass is palpated, its size and location should be carefully recorded. This is usually done by considering the breast as the face of a clock with the nipple at the central point. The mass may be precisely located in regard to its distance from the nipple, and its size should be carefully estimated and recorded. It is often helpful to make a sketch of the area, describing the exact location and consistency. The surface of the mass should be described, since malignant tumors may have irregular infiltrating margins while benign tumors may be sharply demarcated and smooth. The consistency of the mass may be helpful, since soft cystic lesions are more likely to be benign whereas firm and irregular masses are more likely to be malignant.

The examiner should observe whether the lesion is freely movable or fixed in position. Benign tumors are usually movable. Inflammatory lesions may be moderately fixed; advanced malignant lesions are often fixed to other structures as they become invasive. In general, benign lesions tend to have discrete margins whereas malignancies tend to have boundaries that are difficult to define.

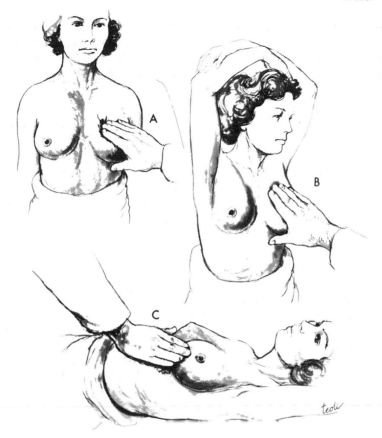

FIGURE 16-3. Positions used for palpation of the breast. A. Erect with arms at side. B. Erect with arms raised overhead. C. Supine.

Both axillae should be systematically examined. This is best performed by examination with one hand while the examiner's opposite hand holds the patient's arm. The boundaries of the axilla should be carefully palpated, and it is helpful to have the patient's arm go through a full range of motion during the examination in order to uncover any lesions which might otherwise be hidden beneath the pectoral muscle or subcutaneous fat.

The supraclavicular areas should be examined in a similar fashion; the neck should be palpated, since the deep jugular nodes may be involved in metastatic spread. Since hepatic metastases are commonly found in patients with advanced disease, the position and character of the liver edge should be noted.

Additional Diagnostic Techniques

Approximately 90 percent of breast malignancies are first detected by the patients themselves prior to seeking medical advice. The physical examination remains the most accurate method of identifying breast lesions. Three additional techniques have been developed, however, for either screening patients for breast masses or providing confirmatory information. These techniques should be regarded as complementary to the physical examination, and should not replace it or substitute for a careful appraisal.

Mammography   This technique consists of standard radiologic examination of both breasts in two planes. Breast carcinomas show areas of calcification and distortion of adjacent connective tissue. It is of greatest value in examination of the postmenopausal woman with large, pendulous breasts. Its greatest inaccuracy is in premenopausal women with chronic cystic mastitis. The overall accuracy of the technique is approximately 80 percent, with false negatives occurring in roughly 20 percent of patients. This must be borne in mind when the decision regarding biopsy is made.

Thermography.   This technique is based on the observation that skin temperature is elevated overlying a breast cancer. An electronic machine, utilizing an infrared camera, is used to scan the breasts and produce an image of the infrared heat pattern on Polaroid film. It is being evaluated as a screening technique.

Xeroradiography.   This is a modification of film mammography in which the image of the breast is recorded on a photoconductive surface with the use of x-rays. It is faster, requires less x-ray exposure, and is more convenient to produce and interpret. The image shows the soft-tissue structures, ducts, and underlying masses within the breast. While still experimental, it may prove useful for screening and adjunctive diagnostic studies.

NORMAL FINDINGS

The normal female breast shows considerable variation in size, shape, and consistency. In the obese patient the breast may be large and pendulous; in the slender person it may be thin and small. In young patients the breast tends to be firm, somewhat elastic in consistency, and cone-shaped. The borders of the breast tissue are clearly delineated, and it is possible to move the entire breast freely over the anterior chest wall. It is often very sensitive to palpation. This may be particularly marked just prior to the menstrual period. In older patients the breast develops an irregular consistency and the sharply delineated border tends to be obscured. This is particularly true after pregnancy and lactation.

The breasts tend to undergo cyclic changes that are reflected in alterations in fullness and thickness of the normal breast tissue. These changes accompany menstruation and are associated with epithelial hyperplasia and fibrosis. When such alterations are marked one sees the clinical picture of mastopathia cystica. Normal breast tissue may be characterized by a faint but distinct generalized nodularity. In some individuals this may be pronounced. This may make the recognition of a distinct tumor nodule difficult.

ABNORMAL FINDINGS

On the basis of a physical examination the examiner should be able to note the following diagnostic points: (1) Describe accurately the location of any detected lesion. (2) Is the lesion solitary or multiple? (3) What is the consistency and extent of the mass? (4) Is the mass tender? (5) Is the mass movable or is it fixed to the chest wall? (6) Is the nipple displaced or retracted? (7) Is there retraction or dimpling of the skin overlying the mass? (8) Are any regional lymph nodes, axillary or supraclavicular, palpable? (9) Can the lesion be transilluminated in a dark room?

Breast *carcinomas* are extremely variable in appearance, and a high index of suspicion is necessary if early diagnosis is to be achieved. Accurate diagnosis may only be obtained by biopsy, and this step should be promptly suggested if any question exists in the mind of the examiner.

Some general features of carcinoma include a firm or hard consistency and lack of tenderness on palpation. Lesions of moderate size shorten the supporting ligaments of the breast, producing dimpling of the skin or nipple. A bloody or purulent discharge from the nipple may occur if there is neoplastic involvement of the ducts or if the carcinoma is intraductal in origin. Advanced lesions may interfere with lymphatic drainage, producing an edematous thickening of the skin termed *peau d'orange.* Advanced carcinomas may also be associated with palpable regional lymph nodes.

Inflammatory carcinoma is a special variety of cancer characterized by infiltration in the skin which produces a red, raised margin resembling acute cellulitis. This may simulate an inflammatory lesion and be associated with pain, fever, and tenderness. This type of carcinoma is more common in premenopausal women, is sometimes seen during pregnancy, and has a bad prognostic outlook.

Intraductal papilloma is characterized by a bloody or dark discharge from the nipple. Gentle pressure in the quadrants of the breast will often permit one to identify the involved duct system. Careful palpation or stripping in this region will often result in production of the characteristic discharge.

Fibroadenoma (adenofibroma) is a nontender, firm lesion which is often multilobulated. It is most often found in young women and is difficult to differentiate from carcinoma by physical findings.

Mastopathia cystica or *chronic cystic mastitis* is an exceedingly common breast lesion characterized by multiple nodules diffusely located in both breasts. The breast tissue is usually thickened and is often tender to palpation. The breasts frequently vary in size and degree of tenderness in association with the menstrual cycle. Occasionally the process may be fairly localized, or discrete nodules may be located in the presence of diffuse cystic changes. In either event biopsy is required for diagnosis.

Traumatic fat necrosis is a lesion which may simulate carcinoma, producing dimpling of the skin and retraction of the nipple. The consistency of the lesion is often firm, adding to the confusion. In some instances no history of trauma may be elicited. This lesion is more often encountered in large breasts containing increased amounts of fatty tissue. Absolute diagnosis can be established only by biopsy.

Paget's disease of the nipple is characterized by an excoriation or dry scaling lesion of the nipple. It may extend to involve the entire areola, tends to bleed easily on contact, and is *always* associated with an underlying carcinoma. It should be noted, however, that the underlying malignancy may not be palpable. Diagnosis may only be achieved by biopsy.

Mastitis is a generalized inflammation of breast tissue, usually occurring during lactation. It is often associated with chills and fever. The involved breast tends to be red, edematous, and tender. Axillary lymphadenitis may occur but fluctuation develops late or not at all. The soft, fatty nature of breast tissue tends to produce a spreading infection with little tendency to localization and abscess formation. This condition is usually associated with pyogenic infection.

Carcinoma of the male breast usually occurs as an irregular hard nodule underlying the areola. Because of the relative paucity of breast tissue, fixation to the chest wall occurs early. Metastasis is common by the time the carcinoma is detected. In general the prognosis is poor.

Gynecomastia is, by definition, a female type of breast occurring in a male patient. It is usually but not always unilateral. It must be distinguished from fatty breast occurring in a normal male. In the young patient the breast tends to assume a conical shape and be glandular in consistency, resembling the breast of a pubertal female. In elderly males the nodularity may be more irregular, and it may be difficult to rule out neoplasm. Biopsy is, of course, required under these circumstances. When gynecomastia is bilateral it may be related to some systemic disease. For example, in patients with cirrhosis of the liver decreased detoxification of estrogens may lead to gynecomastia. Certain testicular tumors may be difficult to rule out neoplasm. Biopsy is, of course, required under these bilateral gynecomastia. Sophisticated biochemical studies may be necessary to complete the diagnostic work-up in these patients.

TECHNIQUE OF SELF-EXAMINATION
OF THE FEMALE BREAST

The technique of self-examination of the breast has been widely advocated as a means of early detection of malignant disease. Patients frequently seek advice regarding the method and frequency of its use. This brief description may be of help in advising patients on this matter.

The patient should establish a regular schedule for monthly breast examination. This is ideally carried out immediately following the end of her menstrual period. An examination carried out during the period may be unsatisfactory because of the temporary changes in consistency and tenderness that so often occur. It is well for the woman to use the menstrual period, however, as a reminder to inspect her breasts. After the menopause monthly examination should, of course, be continued. The patient should be instructed to consult her physician immediately upon detecting a lump of any kind. The judgment and course of action should then be her physician's responsibility. She should also be warned to be alert for dimpling or puckering of the skin, retraction of either nipple, any thickening or change in consistency or any alteration in symmetry, size, contour, or position. A discharge from the nipple may be significant. Pain, swelling, or inflammation may indicate advanced cancer, although more often they are associated with nonmalignant conditions. Having thus been warned about what to look for, she should be instructed in the following steps:

1. *Observation.* The patient should place herself before a mirror with her arms at her sides. She should carefully examine her breasts in the mirror for symmetry, size, and shape, searching for any evidence of puckering, dimpling of the skin, or retraction of the nipple. She should then raise her arms above her head and again study her breasts in the mirror, looking for the same physical signs. She should also be alert for any evidence of fixation of the breast tissue to the chest wall. This may be displayed as she moves her arms and shoulders.

2. *Palpation.* This should be performed in the reclining position. This position permits the breasts to spread over a greater area and thins the breast tissue, making accurate palpation easier. A small pillow or folded towel should be placed beneath the shoulder on the side of the breast to be examined. This raises that side of the body and distributes the weight of the breast tissue more evenly over the chest wall. The arm on the side to be first examined is placed at her side, and the breast is gently examined with the flat surface of the fingers of her opposite hand. The technique calls for gentle palpation of the breast tissue against the chest wall, usually beginning on the outer half of the breast and systematically covering the entire half of the breast, paying particular attention to the upper outer quadrant where the axillary tail of breast tissue is thickest and where most tumors occur.

She should then raise the arm above her head and thoroughly examine the inner half of the breast beginning at the sternum. When the entire breast has been carefully palpated, the pillow is placed beneath the opposite shoulder and the woman investigates the second breast in exactly the same manner.

Palpation of the breast should be thorough and unhurried. Every portion of the breast must be deliberately and carefully examined if small lesions are to be detected.

The patient should be instructed to place the greatest emphasis on the regions where most breast cancers develop, namely in the axillary tail of the breast and beneath the nipple. If the technique is to have any meaning she must establish a definite habit pattern and conduct a thorough examination at monthly intervals. The method will only be effective if it is used regularly.

FEMALE REPRODUCTIVE SYSTEM 17

Theodore M. King

GLOSSARY

Abortion Interruption of pregnancy before the fetus becomes viable (26 weeks).
Adnexa Ovaries and adjacent fallopian tubes.
Amenorrhea Absence of menses.
Cystocele Hernial protrusion of the urinary bladder through the vaginal wall.
Dysmenorrhea Painful menstruation.
Hypermenorrhea (menorrhagia) Abnormally increased volume of menstrual flow.
Menarche Beginning of the menstrual function.
Menopause Termination of menses at the end of a normal reproductive span of years.
Miscarriage Nonmedical term for premature expulsion of a fetus.
Multipara Woman who has borne more than one child.
Polymenorrhea (metrorrhagia) Abnormally increased frequency of menstrual flow.
Primigravida Woman pregnant for the first time.
Primipara Woman who has borne one viable infant.
Rectocele Hernial protrusion of part of the rectum into the vagina.

INTRODUCTION

The objective of this portion of the physical examination is to assure the normality or diagnose abnormalities of the reproductive organs of the woman. More specifically, the objective is to determine the size, shape, and mobility of the reproductive organs, and to locate any source of intrapelvic pain or evidence of inflammation, discharge, or structural abnormality in the genitalia.

In view of this objective, it is well for the student to acquire the habit of referring to the procedure as a pelvic examination rather than as a vaginal examination. The latter, of course, implies an examination only of the vagina itself, rather than the entire pelvic contents.

This examination is best conducted in a room designed and equipped for that specific purpose. Ideally, separate toilet and dressing facilities should be a part of the room. A curtained cubicle is no substitute for four walls and a solid door.

The bladder and rectum should be emptied immediately before the examination, and the patient should disrobe completely.

Equipment

There are certain basic items of equipment which should be available for any routine pelvic examination:

Examination table with stirrups
Instrument table
Sink or basin
Portable, easily adjustable light source
Stool

In addition to the instruments required for the general physical examination, there should also be the following:

Assortment of bivalve vaginal specula of various sizes
Sponge forceps and cotton sponges
Lubricant
Rubber gloves
Cotton applicators
Glass microscope slides
Uterine sounds and probes (sterile)
Cervical biopsy forceps (sterile)
Topical antiseptic solution
Culture tubes
Specimen bottles and fixative solution

General Considerations

To understand the importance of acquiring competence in this particular aspect of physical diagnosis, it might be well to consider two questions:

1. Who should have such an examination? First, of course, all hospitalized women should have the benefit of careful pelvic examination. In addition, any woman coming to the physician because of pelvic complaints (or with the request for routine prophylactic pelvic checkup) should be examined, as well as women with systemic complaints of any kind. The old axiom "If the nature of the story is such that the physician palpates a woman's abdomen, then he is obligated to examine her pelvis" is, if anything, too restrictive.

There is often a great reluctance on the part of physicians (and parents) to examine the very young. This attitude can be very detrimental. In a recently reported series of cases of ovarian malignancy, those patients who were teen-age or younger had a significantly worsened prognosis, chiefly because of the much greater amount of time lost through postponement of the examination necessary for correct diagnosis. Actually, gentle rectal examination in even very young girls can yield a great deal of helpful information; and if there is any suspicion of a pelvic lesion, examination under light anesthesia should be carried out. Youth does not convey an immunity to pelvic disease, and the physician should not deny his patients the best possible care simply because they are young. In general, if a mother thinks the problem serious enough to bring her daughter to the doctor, the doctor should consider the complaints serious enough to examine the patient.

2. Who should carry out the examination? Pelvic examination of the female is a specialized gynecologic procedure only in the sense that auscultation of the

heart is a specialized procedure reserved for the cardiologist. In other words, any physician who retains among his basic skills the ability to percuss the cardiac border or listen to the heart sounds should also retain the ability to use the vaginal speculum and carry out careful bimanual pelvic examination.

In either instance — for either the noncardiologist or the nongynecologist — the purpose of such examinations is to identify those patients who should be referred for specialized diagnosis and treatment. Such referral is also an obligation when the examination has been unsatisfactory or incomplete. Thus, to reassure a woman that "everything is all right" when both ovaries have not been clearly felt could be signing her death warrant. In summary, all physicians — internists, surgeons, pediatricians, as well as gynecologists — should perform pelvic examination on their patients. If the examination is inadequate or the results inconclusive, careful referral rather than bland reassurance is required.

Special Considerations

None of us consider our reproductive organs in exactly the same sense that we regard other parts of our anatomy. Women in particular are unable to consider the gallbladder and the uterus in the same light. Trained from childhood to keep her perineum covered, the woman is required at the time of pelvic examination to expose herself to a comparatively unknown person. This gives the entire process an aura of unpleasantness which a male can only appreciate by imagining that all urologists are women.

The embarrassment which a woman anticipates in relation to this type of examination leads to a natural tendency on her part to postpone it. Furthermore, in her own personal hygiene she will often fail to inspect or palpate lesions which on other parts of her body she would examine minutely. As a physician, therefore, one must proceed with this embarrassment clearly in mind, and do one's best to place the patient at ease. This has practical as well as humane aspects — the more relaxed the patient, the more easily the examination will be accomplished. Patient cooperation is essential, and an undue tightening of the abdominal muscles or the levator ani sling can effectively bar the physician from accomplishing the objective of this examination.

Chaperonage. It should be self-evident that this examination should *never* be performed without a female nurse or aide present. This is not only reassuring to the patient, it is an imperative protection for the physician. Furthermore, most patients need assistance getting onto and off of the examining table.

Draping. It should be evident from the rising and falling of necklines and hemlines that what women consider to be "indecent exposure" varies almost annually. Nevertheless, there are commonly held criteria of "decency" which should be recognized. Thus, in the examination of the breast, the woman does not feel naked unless the areola is exposed, no matter what percentage of the remainder of the breast is undraped.

For the woman in the lithotomy position on the examining table, this "criterion of coverage," interestingly enough, is most often the knee. There are said to be two principles of British Law: first, the trial must be fair; and secondly, the trial must give the illusion of being fair. The same might be said of draping the patient for gynecologic examination; she should be decently covered, and she should have the illusion of being decently covered. As long as she does not look up to see her knee exposed, this illusion is largely maintained.

The most satisfactory draping is a square sheet. The patient holds one corner over her xyphoid, the corners adjacent to this are placed one over each knee, and the fourth corner hangs between her legs and over her perineum.

Matter-of-Factness. The routine nature of the office ritual helps put the patient at ease. The relative impersonality of the physician's behavior, and the manner of arranging the draped sheet all heighten the reassuring effect. In elevating the corner of the sheet which hangs over the perineum and tucking it under the sheet on her abdomen, one is well advised to be looking not at her perineum but directly into her eyes, and to be commenting on something far afield from the complaints of the moment.

Gentleness. The perineum is as tender and sensitive as any area examined in the course of physical diagnosis. The doctor who hurts a patient instantly makes her an opponent rather than an ally in the job of accomplishing a satisfactory examination.

Anyone — male or female — placed in the lithotomy position and then touched abruptly on the perineum will experience the anal sphincter reflex. The external anal sphincter (and portions of the levator ani) will involuntarily contract, creating a barrier to a comfortable examination. Accordingly, the examiner's initial contact with the perineum should be at some distance from the labia and should be firm but gentle.

All of the tender areas of the introitus are anterior (clitoris, labia minora, and urethral meatus) and insertion of the intravaginal fingers should be posterior, with adequate lubrication and an avoidance of jamming or shoving motions. While subsequent portions of the examination are inevitably somewhat uncomfortable, the entire initiation of the procedure can be such that optimum patient comfort (and hence patient cooperation) is achieved and maintained.

EXTERNAL GENITALIA

Inspection precedes palpation here as in other realms of physical examination (Figure 17-1). However, the inspection can be brief and unembarrassing to the patient. One is seeking superficial skin lesions of the groin and perineum, evidence of erythema on the labia, and signs of an abnormal vaginal discharge.

Palpation can likewise be effected briefly and can yield evidence of the

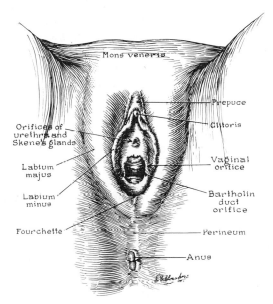

FIGURE 17-1. External genitalia.

strength of the pelvic floor, the presence of Bartholin's cysts, or the presence of pus in Skene's glands. Evaluate the general strength of perineal support by asking the patient to bear down and then observing any ballooning which may develop anteriorly (cystocele) or posteriorly (rectocele). In addition, the perineal body may be assayed by depressing the posterior fourchette with two fingers of the gloved hand and testing the general resistance. The presence of pus in Skene's glands can be determined by stripping these areas (gently!) upward and watching for the appearance of purulent material at the urethral meatus.

With the exception of instances in which you must pause for an unexpected finding (vulvar or labial lesions), the inspection and palpation of the external genitalia can be expertly completed in one to two minutes.

Saline Drop. Before introducing the intravaginal fingers it is often well to obtain material for microscopic examination. This is achieved by inserting a saline-moistened swab into the vagina to collect a specimen and then placing this swab in a small test tube containing 1 or at most 2 ml. of physiologic saline solution. It is advantageous to keep this slightly warm, perhaps by having the nurse hold it in her clenched fist. The examination of this specimen as simply a drop on a regular slide, without stain or fixation of any kind, will yield considerable information. The value of the procedure — as is true of so many diagnostic tests — is usually in direct proportion to the frequency with which the physician carries it out.

Under most circumstances the principal cellular element noted will be the vaginal epithelial cells. The bacterial content will be largely Döderlein's bacillus and *Escherichia coli*. A significant number of leukocytes implies inflammation, and erythrocytes indicate a bleeding lesion. The *Trichomonas vaginalis* parasite is recognized by its flagellate motion and *Candida albicans* by its configuration. For the few moments spent in examining this preparation, its diagnostic value is great.

SPECULUM EXAMINATION

Inspection of the cervix and upper vagina precedes palpation, and is accomplished by means of the vaginal speculum. This is perhaps the most valuable single diagnostic instrument there is, and the student is well advised to spend a few moments learning its simple mechanism.

The Graves or duckbill speculum is made up of two blades. The posterior blade is fixed and the anterior movable. The two blades are held together by a thumbscrew on the handle. Loosening this permits the blades to be separated, which increases the aperture at the proximal end of the speculum. Removing the thumbscrew permits separation of the blades, and in some circumstances it is desirable to use the posterior blade (or two posterior blades) as hand-held retractors. This thumbscrew should always be tightly fastened and the blades in close approximation when the speculum is in ordinary use, or the chances of pinching the labia are increased.

The anterior blade is hinged and carries the thumbpiece on the side, which permits it to be elevated, separating the two blades when the speculum is assembled. The thumbscrew here should be all the way back, permitting the blades to be in close approximation on introducing the speculum into the vagina.

To avoid interference with cytologic studies and minimize discomfort, the speculum should have no lubrication other than warm water. It is best held with the handle grasped loosely and the blades firmly held between the index and middle fingers. On introduction of the speculum, the pressure should be largely against the posterior fourchette and the blades should be oblique. If the flat surfaces of the blades are horizontal, the introitus is often overly stretched; if the blades are vertical, the suburethral area can be hurt.

As soon as the broad portions of the blades have passed the introitus, the speculum is rotated so the blades are horizontal and the handle is elevated so that the speculum is advancing at a 45-degree angle toward the examining table. The blades should not be separated until the speculum is fully inserted. As the thumb presses the lateral thumbpiece to elevate the top blade, the hand should lift on the handle to lower the fixed posterior blade. In this way the two blades move away from each other, and the cervix and vaginal walls are exposed (Figure 17-2).

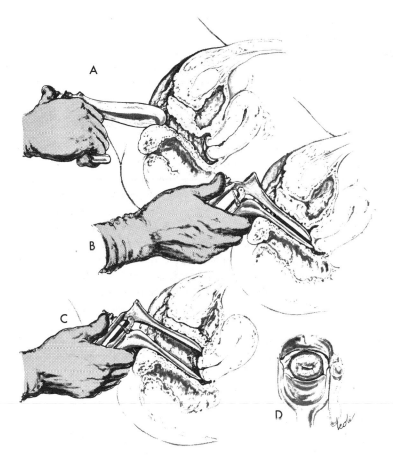

FIGURE 17-2. Insertion of vaginal speculum. A. Blades held obliquely on entering vagina. B. Blades rotated to the horizontal position as they pass introitus. C. Blades separated by depressing thumbpiece and elevating handle. D. Normal parous cervix.

Special Tests

Papanicolaou Cell Spread for Cervical Cytology. There are many ways to obtain the cells for cytodiagnosis. Perhaps the most important single requirement is to obtain an adequate representation of the endocervical sample. The procedure described here is the Fast technique, designed to combine samples from the posterior vaginal pool and from the cervical canal and cervical face. Since this test is for the identification of epithelial lesions, the specimen should be obtained before the upper vagina and cervical face are wiped off.

The slide and the bottle of fixative should be labeled and prepared in advance to facilitate prompt fixing after spreading the cellular specimen. By using the handle of the Ayres spatula or a similar spatula, a thick sample of cells from the posterior cul-de-sac is obtained and placed on the slide at the frosted end. The spatula is immediately reversed and the longer arm of the tip placed in the cervical os and the blade rotated a full circle. This portion of the sample is spread the length of the slide, catching and distributing some of the original cul-de-sac specimen. Then, with the tip of the spatula the entire specimen is spread evenly and the slide dropped immediately into the ether-alcohol.

Maturation Index.   The cornification count of the vaginal cells and the maturation index (the ratio of basal, intermediate, and superficial cells) are used as measurements of the patient's estrogen level. The same procedure is used, but the cell sample is obtained from the vaginal wall.

In the absence of laboratory facilities, the same information can be obtained in the office by the Lugol vapor stain. The same vaginal wall sample is obtained but no fixative is employed. The slide is air-dried and inverted over a small shallow dish of Lugol's solution. The rising vapors stain the cells with a color proportionate to their glycogen content, which in turn is proportionate to the patient's estrogen level. If more than 20 percent of the cells are mahogany color, in contrast to light yellow, the patient is certainly not estrogen-deficient. Estrogenated cells are large and flat, with a flagstone appearance and dark, pyknotic nuclei; crinkled cells with vesicular nuclei are atrophic.

Biopsy.   The cervix is insensitive to cutting and can be liberally biopsied without discomfort to the patient. Various punch biopsy forceps which will obtain samples from the cervical face and endocervix without distortion are available.

It should be remembered that cervical cytology has not replaced the biopsy. Cytologic study has its greatest value for the normal-appearing cervix, as an aid in screening, and in raising the examiner's index of suspicion. Where there is an evident lesion, the examiner's suspicion should already have been alerted, and a biopsy should be taken. A negative cytologic report in the presence of a lesion does not remove the need for biopsy.

Bimanual Pelvic Examination

After the specimens are obtained and the speculum removed, bimanual pelvic examination is carried out. The age-old debate as to whether the intrapelvic examination should be done with the right or the left hand is fruitless. The point is to achieve the stated objectives of the pelvic examination without concern about which hand is used. Indeed, it is often useful to change the intrapelvic hand, since the adnexal regions can often be felt best with the hand of the same side (for example, right hand for right adnexal region).

In general, the abdominal hand brings the pelvic structures within reach of the intravaginal fingers, for palpation. It is well to remember that the area which can be covered depends largely on the position of the abdominal hand. The patient's abdominal wall forms the altitude of a right-angle triangle, the perineum the base, and the intravaginal hand the hypotenuse. It can be seen that moving the abdominal hand higher on the patient's abdomen achieves a greater area "enclosed" in the examination than the mere inches it might have moved. The beginner tends to place the abdominal hand too close to the pubic bone, and he should be encouraged to move it at least three-quarters of the way toward the umbilicus (Figure 17-3).

The examining fingers should be well lubricated and gently inserted over the posterior fourchette. It is often most comfortable to rest one foot on a low stool with the elbow of the intrapelvic hand resting on the knee.

FIGURE 17-3. Bimanual pelvic examination. The abdominal hand brings the pelvic contents to the intravaginal fingers. A. Palpation of the uterus. B. Palpation of right ovary. C. Palpation of right parametrial tissues.

The cervix will be most easily palpated, and as a general rule it points the opposite way from the fundus. If the cervix points posteriorly, the fundus of the uterus will usually be found anteriorly; if the cervix is in the axis of the vagina, the fundus will more often be retroverted and found in the cul-de-sac. Motion of the cervix should not produce pain. However, it is just as well not to ask a direct question on this score, since most patients will answer in the affirmative simply on the basis that the entire examination is uncomfortable. The pain one is seeking (as in ectopic pregnancy or adnexal inflammation) is severe enough that the patient will usually indicate its presence without questioning.

The body of the uterus is then held between the examining hands, and its size, contour, and mobility are determined. Irregularities in its surface (as with myomata) can usually be determined with ease. The cul-de-sac area should also be examined for bulging, tenderness, or masses.

The broad ligament structures beside the uterus do not generally yield palpatory findings. The vaginal fingers go posteriorly under the broad ligament and the abdominal fingers beside the uterus approach them, so that the tissues of the broad ligament and the Fallopian tube can be run between the fingers from bottom to top. In a normal patient no masses should be encountered.

The intravaginal fingers now move downward and backward again at the lateral wall, the abdominal hand moves over just inside the anterior superior spine of the ilium, and the same process is repeated. Here, however, the examiner should encounter the ovary, where again size, shape, and mobility can be determined. The gonad of the female is as sensitive to pressure as the gonad of the male; again, direct questions on this score are meaningless, and gentleness is imperative.

The examination is completed by rectovaginal abdominal palpation (Figure 17-4). Entry into the rectum is facilitated by liberal use of lubrication and having the patient bear down. The middle finger is gently inserted in the rectum and the index finger in the vagina. This immediately gives much greater access to the adnexal regions and the posterior surface of the broad ligament as well as revealing the presence or absence of rectal lesions.

PELVIC INFLAMMATORY DISEASE

Pelvic inflammatory disease has been used to the detriment of patients as a diagnostic term for pelvic pain of unknown etiology. Its use should be limited to the occurrence of pelvic peritonitis as a result of an infection involving the uterus, tubes, or ovaries, caused by pathogenic organisms.

The clinical diagnosis of acute pelvic inflammatory disease has an accuracy of less than 75 percent. The remainder of the individuals with the commonly accepted clinical signs and symptoms of acute pelvic infection have either other surgical problems or no discernible pelvic abnormalities.

The increasing incidence of gonorrheal infection combined with recognition of asymptomatic male carriers of gonorrhea requires that young physicians be knowledgeable of the symptomatology and clinical findings in pelvic infections of women.

FIGURE 17-4. Rectovaginal abdominal examination. Indicates how the perineal body may be depressed to allow deeper entry into the pelvis. Note that one finger is inserted into the vagina and one into the rectum.

Acute Pelvic Peritonitis

Patients with pelvic peritonitis present with a complaint of lower abdominal pelvic pain that is associated with a fever ranging from 100° to 104°. They frequently date the onset of symptoms to immediately following the last menstrual period. Usually patients are without gastrointestinal complaints and their appetites are usually not impaired.

Positive physical findings are limited to the lower abdomen and pelvis. On abdominal examination, the lower abdomen and suprapubic area are tender with the suggestion or presence of rebound. Intestinal sounds are most commonly active. On pelvic examination there may or may not be a vaginal discharge. There is a history of a malodorous discharge before the last menstrual flow. Bimanual examination will demonstrate bilateral adnexal and uterine tenderness. This tenderness will be accompanied by guarding, and one frequently will be unable to delineate the adnexal structures clearly. Moderate-sized ovarian or tubal masses may be missed. Rectal examination is mandatory because it allows demonstration of the presence of posterior pelvic masses characteristic of tubo-ovarian or pelvic abscesses.

Unfortunately the outlined physical findings may be the result of a variety of clinical entities. Such findings are observed in appendicitis associated with perforation or an inflamed appendix that is adherent to pelvic viscera. Other gastrointestinal inflammatory states may also result in pelvic peritonitis. These include the uncommon Meckel's diverticulum and diverticulitis. Leaking ovarian cysts or twisting of either ovarian tumor or adnexa may result in similar complaints. Benign uterine tumors such as degenerating myomata can also cause similar physical findings.

The patient's history may be of assistance in suggesting an exposure to gonorrheal infection. In such cases the collection of cervical cultures may demonstrate the gonococcus without difficulty and the patient may respond rapidly to antibiotics such as penicillin. The duration of the clinical course is frequently related to the number of pelvic examinations; these should be minimized.

Recurrent Pelvic Infection

The continued destruction of ovarian and tubal architecture as a result of recurrent and inadequately treated infections may result in the evolution of tubo-ovarian abscesses or ovarian abscesses. The major problem with such abscesses is rupture with intraperitoneal leakage of their contents. This syndrome is associated with inordinate mortality if operative intervention is not prompt. Those patients who have intraperitoneal leakage develop progressive peritonitis with disappearance of intestinal function and development of ileus associated with nausea and vomiting. Their heightened temperature and tachycardia do not respond to antibiotic therapy. On pelvic examination, pelvic masses will be found in at least 50 percent of the cases. These patients require immediate exploration and cannot be treated in a nonsurgical manner. Because of the high incidence of anaerobic organisms, particularly bacteroides, and other secondary invaders, these patients must have a combination of antibiotics that includes either penicillin and Chloromycetin or Keflin and kanamycin. At time of surgery, bilateral salpingo-oophorectomy and hysterectomy is required. Incomplete pelvic surgery, prompted because of a patient's youth, is associated with an unacceptable incidence of reoperation frequently within the subsequent three to six months.

PREGNANCY

Diagnosis of Pregnancy

The most prominent changes in the pelvis during early pregnancy revolve around the vascular congestion that takes place. The cervix softens and the uterus grows slightly larger and softer. Subsequently, cyanosis of the upper vaginal tissues and the cervix becomes prominent (Chadwick's sign).

The initial enlargement of the uterus is not usually symmetrical, but is more pronounced on one horn, apparently the side of conception. Subsequently, globularity of the uterus becomes evident, in contrast to the anteroposterior flattening normally present. A softened area 2 or 3 cm. in diameter appears on the anterior wall (Ladin's sign).

There is also fullness and congestion of the breasts and increased prominence of the superficial veins.

Diagnosis of Previous Pregnancy

The nipples and areolae retain the brown color acquired during pregnancy and do not resume the former pink color. The linea alba remains brown. The cervical os never resumes the round and symmetrical configuration, but will be irregular and slightly enlarged. The perineum may show a loss of tone or there may be scarring of the posterior fourchette (episiotomy).

Examination of the Abdomen During Pregnancy

The initial prenatal visit calls for a complete examination together with an estimation of the capacity of the pelvis. Subsequent examinations in the average case call only for a brief interval history and a limited examination. The patient's weight and blood pressure should be obtained, and the urine examined for albumin and sugar.

Abdominal examination should be performed on every visit. The height of the fundus of the uterus above the pubic bone should be measured to estimate continued and appropriate uterine growth. The number of centimeters divided by 4 equals the months of pregnancy. This rule is only approximate, but marked deviations from it suggest the possibility of twins or hydramnios if the measurement is great, or of fetal abnormality or death if the fundal height fails to increase.

The palpation of the fetus within the uterus usually begins with an attempt to locate the fetal head by grasping it above the pubic bone with the thumb and middle finger. If the pole of the fetus in this area gives the impression of being pointed rather than firm and rounded, the head should be sought in the fundus, with both hands placed flat on the abdomen and parallel to the uterine axis. The back of the fetus can also be located with the hands flat on either side and the fingers parallel to the uterus. The fetal small parts feel knobby and irregular in comparison with the back, on which the hand can be fitted flat.

The fetal heart is heard best over the anterior shoulder. It is often helpful also to obtain the rump-to-shoulder measurement. If this is over 25 cm., the baby's weight will usually exceed 2500 gm.

SPECIAL TECHNIQUE

Culdocentesis. Needle biopsy of the posterior cul-de-sac is a valuable diagnostic procedure which should be used more often. The bivalve speculum should be spread to the limits of patient comfort, and the cervix grasped with a tenaculum on the posterior lip. Lifting the cervix upward and drawing it down will expose the posterior cul-de-sac. By means of a 20-gauge 3-inch needle on a 10-ml. syringe, the vaginal mucosa is entered in the midline just back of the cervix and the needle advanced with gentle suction applied from the syringe. The finding of pus is usually an indication for further study by posterior colpotomy; the presence of nonclotting blood obviously indicates the need for laparotomy. Clear fluid can be smeared and sent for cytologic diagnosis. The diagnostic value of this procedure is such that it should be used in all cases where the clinical picture is not completely clear.

REFERENCES

1. Benson, R. C. Gynecologic History and Examination. In *Handbook of Obstetrics and Gynecology* (2nd Ed.). Los Altos, Calif.: Lange Medical Publications, 1966.
2. Frost, J. K. Gynecological and Obstetrical Cytopathology. In Novak, E., and Novak, E. R. (Eds.). *Gynecological and Obstetrical Pathology* (4th Ed.). Philadelphia: Saunders, 1958.

MUSCULOSKELETAL SYSTEM 18

Lee H. Riley, Jr.

GLOSSARY

Abduction Motion of a part away from the midline.

Active range of motion Limits of motion through which a joint may be moved by those muscles which cross the joint.

Adduction Motion of a part toward the midline.

Ankylosis Complete loss of motion of a joint.

Arthritis Inflammation of a joint.

Bursa Potential space often filled with fluid between two soft tissue layers which move upon each other.

Calcific tendinitis of the shoulder Inflammatory condition of a tendon about the shoulder, one stage of which is the deposition of crystals containing calcium within the structure of the tendon.

Capsule The fibrous tissue sheath about a joint.

Carpal tunnel The potential space on the volar (anterior) aspect of the wrist, the floor of which is formed by the carpal bones, the roof by the transverse carpal ligamenta. Within the carpal tunnel travel the deep and superficial flexors of the fingers, the long flexor to the thumb, and the median nerve.

Cavus foot Foot deformity in which a very high arch is present.

Club foot (talipes equinovarus) Foot deformity consisting of varus of the heel, equinus of the ankle, and adduction and supination of the forefoot.

Contracture Shortening of soft tissues about a joint which limits normal motion of the joint.

Contralateral On the opposite side.

Crepitation Grating or cracking sensation produced by motion.

Eversion Position achieved by turning a part away from the midline of the body.

Inversion Position achieved by turning a part toward the midline of the body.

Joint effusion Excessive fluid within a joint.

Kyphosis Angular curvature of the spine, the convexity of which is posterior.

Meniscus Fibrocartilaginous structure found between the articular surfaces of certain joints.

Paresthesia Sensation of burning, crawling, or tingling.

Passive range of motion Limits of motion through which a joint may be moved without use of the muscles which cross the joint.

Pronation Position of the forearm achieved by turning the palm down; position of the foot achieved by turning the sole down.

Pseudoarthrosis Lack of bony continuity after the process of bone repair has ceased.

Rotator cuff of the shoulder Common insertion of the subscapularis, supraspinatus, infraspinatus, and teres minor muscles into the proximal humerus.

Spondylolisthesis Forward displacement of a vertebra upon the one below as a result of bilateral defects in the vertebral arch.

Subluxation Lack of a completely normal relationship between the articular surfaces of two bones which comprise a joint, such that, however, the articular surfaces are still in contact (that is, not dislocated).

Supination Position of the forearm achieved by turning the palm up; position of the foot achieved by turning the sole up.

285

Supine Position of lying on the back, face upward.
Synovitis Inflammation of the lining of a joint.
Tendinitis Inflammation of a tendon.
Thoracic outlet Anatomic region between the base of the neck and the axilla through which pass the brachial plexus and subclavian vessels.
Valgus Angulation of a part of an extremity away from the midline.
Varus Angulation of a part of an extremity toward the midline.

TECHNIQUE OF EXAMINATION

The musculoskeletal examination of an individual who has not been acutely injured differs greatly from the examination of an acutely injured patient. For example, the active and passive range of motion of the cervical spine should be determined when the patient shows symptoms of a nontraumatic disorder of the neck and upper extremity. However, this determination is not performed in an individual who has been acutely injured until the mechanical stability of the cervical spine has been demonstrated by roentgenograms. The importance of this distinction cannot be overemphasized.

The physical examination of an adult differs in some respects from the physical examination of a child and especially from that of an infant. The method detailed here is that used for the examination of an adult.

The method of measuring and recording joint motion will not be described in detail. For these techniques the reader is advised to consult the booklet *Joint Motion: Method of Measuring and Recording* [2] published by the American Academy of Orthopaedic Surgeons.

Begin the examination by observing posture and gait as the patient enters the examining area. Gait is divided into two broad phases — stance and swing. The stance phase includes heel strike (as the heel strikes the ground), midstance (as body weight is transferred from the heel to the ball of the foot), and push-off (as the heel leaves the ground). The swing phase includes acceleration, swing-through (as the foot travels ahead of the opposite foot), and deceleration (as the foot slows in preparation for heel strike). Observe each phase and note any awkwardness or change in rhythm. For example, a patient with a painful callus on the ball of the foot will not bear weight normally upon that part of the foot during midstance and push-off.

Look for gross deformities, areas of swelling, and areas of discoloration. The presence of ecchymoses will suggest previous trauma. A red swollen area which is warm and tender to palpation will suggest inflammation. If such an area is about a joint and associated with fluid within the joint, acute synovitides such as rheumatoid arthritis, gouty arthritis, or infectious arthritis are suspected. If such an area is over the diaphyseal region of an extremity, infectious lesion of the underlying bone or soft tissues, or noninfectious inflammatory lesions such as thrombophlebitis, are suspected.

Examination of a Joint

Observe the joint and note any gross deformity and swelling. Swelling within a joint may represent either fluid or thickened synovial tissue. In the former instance, a fluid wave can be demonstrated; in the latter, the firm and somewhat boggy tissue can be palpated and no fluid wave is noted unless, of course, excessive fluid is also present. Palpate about the joint for masses and for points of tenderness which may indicate a torn ligament, an area of osteoarthritis or synovitis, or the torn attachment of a meniscus. Determine the active and passive range of motion of the joints, comparing those on opposite sides. Palpate the joint during active and passive motion to detect crepitation. Test the ligaments which help stabilize the joint and the muscles which control it.

Examination of a Muscle

In the examination of a muscle or a muscle group ascertain the status of the muscle fibers as well as the nerves which supply them. Inspect the muscle for gross hypertrophy or atrophy and for fasciculations, which are isolated contractions of a portion of the fibers. Look for areas of muscle spasm which can be easily palpated and often are accentuated as the joint they span is moved passively. Areas of muscle spasm are tender to palpation. Measure the circumference of an extremity at a given point above and below the patella or above and below the olecranon, and compare this measurement with that of the opposite side. Test the strength of the muscle according to the criteria given in Figure 18-1. Note the consistency of the muscle to palpation and the presence of

100%	5	N	Normal	Complete range of motion against gravity with full resistance.
75%	4	G	Good	Complete range of motion against gravity with some resistance.
50%	3	F	Fair	Complete range of motion against gravity.
25%	2	P	Poor	Complete range of motion with gravity eliminated.
10%	1	T	Trace	Evidence of slight contractility. No joint motion.
0%	0	0	Zero	No evidence of contractility.

FIGURE 18-1. Criteria for grading muscle strength.

tenderness over the muscle or its tendon. Test the deep tendon reflex. The spinal segments and the peripheral nerves which generally innervate major muscle groups of the extremities are listed in Figure 18-2.

Examination of a Bone

Observe the soft tissues covering a bone for obvious deformity, such as bowing, angulation, or tumor. Palpate the bone for areas of tenderness and for masses.

UPPER LIMB MUSCLE	NERVE	C 2	C 3	C 4	C 5	C 6	C 7	C 8	T 1
Sternocleidomastoid; trapezius	Spinal accessory	X	X	X					
Diaphragm	Phrenic		X	X	X				
Deltoid	Axillary				X				
Supraspinatus	Suprascapular				X				
Infraspinatus	Inferior scapular				X	X			
Teres minor	Axillary				X	X			
Subscapularis; teres major	Subscapular				X	X			
Serratus anterior	Long thoracic				X	X	X		
Rhomboideus	Dorsal scapular				X				
Clavicular pectoralis major	Anterior thoracic				X	X	X		
Biceps; brachialis	Musculocutaneous				X	X			
Brachioradialis	Radial				X	X			
Latissimus dorsi	Thoracodorsal					X	X	X	
Sternopectoralis major	Anterior thoracic					X	X	X	X
Flexor carpi radialis; pronator teres	Median					X			
Extensor carpi radialis, longus & brevis; extensor digitorum communis; extensor indicis proprius; extensor carpi ulnaris; extensor pollicis, longus & brevis; abductor pollicis longus; triceps	Radial					X	X	X	
Flexor digitorum sublimis	Median						X	X	X
Flexor digitorum profundis	Volar interosseous; ulnar							X	X
Flexor carpi ulnaris	Ulnar							X	
Pronator quadratus	Volar interosseous							X	X
Dorsal interosseous; volar interosseous	Ulnar							X	
Lumbricals; flexor pollicis brevis	Median; ulnar							X	X
Adductor pollicis brevis; opponens	Ulnar							X	X
Biceps tendon reflex	Musculocutaneous					X	X		
Extensor pollicis tendon reflex	Radial							X	
Triceps tendon reflex	Radial							X	X

FIGURE 18-2. Peripheral nerve and spinal segment innervation of extremity muscles.

LOWER LIMB MUSCLES		NERVE	L 1	L 2	L 3	L 4	L 5	S 1	S 2	S 4	S 5
Hip flexion	Iliopsoas; satorius; rectis femoris; tensor fasciae latae	Lumbar plexus; femoral; superior gluteal; obturator	X	X	X	X	X				
Hip adduction	Adductor major; adductor brevis; adductor longus	Obturator		X	X	X					
Knee extension	Quadratus femoris	Femoral			X	X					
Hip abduction	Gluteus medias; gluteus minimus; tensor fascia femoris	Superior gluteal				X	X	X			
Foot inversion & dorsi-flexion	Tibialis anterior	Peroneal				X	X				
Toe extension	Extensor digitorum, longus & brevis	Peroneal				X	X	X			
Great toe extension	Extensor hallucis, longus & brevis	Peroneal					X	X			
Foot eversion	Peroneus, longus & brevis	Peroneal					X	X			
Foot inversion & plantar flexion	Tibialis posterior	Tibial					X	X			
Toe flexion	Flexor digitorum, longus & brevis	Tibial					X	X			
Great toe flexion	Flexor hallucis longus	Tibial					X	X	X		
Hip extension	Gluteus maximus	Inferior gluteal					X	X	X		
Knee flexion	Biceps femoris; semimembrinous; semitendonosis	Peroneal; tibial					X	X	X		
Foot plantar flexion	Gastrocnemius; soleus	Tibial						X	X		
	Cremasteric reflex	Genital femoral	X								
	Patellar tendon reflex	Femoral			X	X					
	Achilles tendon reflex	Tibial						X	X		
	Anal reflex	Pudendal								X	X

Tenderness of a bone suggests underlying tumor, inflammation, or the sequelae of trauma. This impression is strengthened when percussion of the bone at a site distant from the site of tenderness produces pain not at the site of percussion but at the point of tenderness to palpation. Test the gross structural integrity of a bone by noting its ability to resist a deforming force. Lack of resistance suggests a fracture or the result of a fracture — pseudoarthrosis.

CARDINAL SYMPTOMS

Pain, deformity, and limitation of function are the major symptoms of musculoskeletal abnormality. Limitation of function in itself may be due to pain associated with movement, to bone or joint instability, or to the restriction of joint motion. The joint restriction in turn may be due to muscle weakness from neurologic disease or trauma, muscle contractures from previous injury or disease, bony fusion, or a mechanical block by bone fragments or torn cartilage within the joint.

Pain is a broad and significant symptom. It is important to note its characteristics, location, and relation to the patient's activity. The pain of bone erosion caused by tumor or aneurysm is usually described as deep, constant, and boring. It is apt to be more noticeable and more intense at night, and it may not be relieved by rest or by position.

The pain of degenerative arthritis and muscle disorders is an aching type which is often accentuated by activity and lessened by rest. The discomfort may be increased by certain positions or motions. For instance, the pain resulting from degenerative changes in the cervical spine is often accentuated by maintaining the neck in extension. Subjective paresthesias which do not follow a dermatome distribution are often noted by the patient with degenerative changes involving the cervical or the lumbar spine. These are often described as a "sandy" feeling or as if the foot (or arm) were "going to sleep."

The pain of fracture and infection of bone is severe and throbbing, and is increased by any motion of the part. Acute nerve compression causes a sharp, severe pain radiating along the distribution of the nerve. It is often associated with weakness of muscles supplied by the nerve and sensory changes over the area supplied by it.

Referred pain is that perceived by the patient in an anatomic location removed from the site of the lesion. Pain resulting from a disorder of the hip is often first noted by the patient in the anterior and lateral aspect of the thigh. Pain from a shoulder lesion may be felt at the insertion of the deltoid muscle on the lateral aspect of the proximal portion of humerus. Pain from the lower cervical spine is often referred to the interscapular region of the back, along the vertebral aspect of the scapula, or to the tips of the shoulders and lateral aspects of the arms.

NORMAL AND ABNORMAL FINDINGS

Cervical Spine

In the cervical part of the spine, evaluate the vertebrae, the articulations between them (the facet joints posteriorly and the intervertebral discs anteriorly), and the structures totally or partially contained within these bone structures (the spinal cord, the cervical nerve roots, and the vertebral artery).

Inspect the spine for gross deformities and visible muscle spasm. Among the possible deformities are absence of the normal cervical lordosis, or abnormal shortness of the neck, which may be associated with congenital malformations of the cervical vertebrae. Determine the active and passive range of motion of the cervical spine in flexion, extension, lateral bending, and rotation. Transient giddiness, or even unconsciousness, induced by a particular head or neck position suggests temporary occlusion of a vertebral artery within the neck. This diagnosis may be suspected on physical examination but must be confirmed by arteriography.

Palpate the tips of the spinous processes for general alignment and for points of tenderness, and the cervical muscles — which include the posterior paraspinal muscle group, the trapezius muscle, the sternocleidomastoid muscle, and the scalene muscles — for tenderness and muscle spasm. Decreased motion of the cervical spine associated with points of tenderness to palpation over the spinous processes is commonly seen in tumors and infections of the cervical vetebrae, as well as in degenerative, prolapsed, and herniated cervical intervertebral discs. Palpate the cervical spine anteriorly, with the palpating finger passing medial to the carotid vessels and lateral to the trachea and oesophagus. Tenderness is noted in the presence of disc lesions, infections, and certain tumors.

Maintain the neck in extension by having the patient look at the ceiling for 60 seconds. This test will produce discomfort in the base of the neck, the interscapular region of the back, or the lateral aspect of the shoulders or arms when there are degenerative changes involving either the cervical intervertebral discs (anteriorly) or the facet joints (posteriorly). Vertically compress the extended cervical spine by pressing on the head. This will intensify discomfort in the same areas in the presence of degenerative, prolapsed, or herniated intervertebral discs. Increase in the symptoms upon vertical compression of the extended cervical spine with the head tilted to the side indicates degenerative changes in the facet joints or the uncovertebral articulations (so-called joints of Luschka). Perform Lhermitte's test, with the patient sitting, by flexing the neck and the hips simultaneously with the knees in full extension. Sharp pain which radiates down the spine and into the upper or lower extremities suggests irritation of the spinal dura either by tumor or by a protruded cervical disc.

Examine the thoracic outlet routinely as part of the examination of the cervical spine. Palpate the supraclavicular fossa for muscle spasm, vascular thrills, masses, and points of tenderness. Perform the Adson test with the patient sitting with the forearms in supination and resting on the thighs. Palpate the radial pulse on the side to be tested. Instruct the patient to extend the neck and turn the chin to the side to be tested. The transient disappearance of the radial pulse during inspiration signifies temporary occlusion of the subclavian artery as the anterior scalene muscle is tensed (by extension of the neck and rotation of the skull) while the "floor" of the thoracic outlet rises during inspiration. Compression of the brachial plexus in the supraclavicular fossa is suggested by intensification of pain and paresthesia in the arm as the Adson maneuver is performed, by tenderness to palpation over the brachial plexus, and by neurologic changes in arm and hand (particularly a decrease in the ability to appreciate light touch over the fourth and fifth fingers of the hand). These signs may be intensified as the shoulder is abducted and externally rotated. Measure the circumference of the upper and lower arms and compare with the opposite side. Test muscle strength, the sensory modalities, and the deep tendon reflexes in the upper extremity.

Thoracic Spine

Palpate the tips of the spinous processes for general alignment and points of tenderness. Test the motion of the thoracic spine during flexion, extension, and lateral bending, and observe the symmetry of the ribs. Small rotational deformities of the thoracic spine which produce asymmetry of the rib cage are best appreciated by inspecting the flexed thoracic spine and rib cage from the rear. Perform a neurologic examination of the trunk and the lower extremities. Perform Lhermitte's test when mechanical irritation of the spinal dura in the thoracic area is suspected.

Lumbar Spine

Note the contour of the lumbar spine with the patient standing. A lumbar lordosis is normally present; its absence suggests a spinal abnormality. Palpate the paraspinal muscles for tenderness and spasm and the tips of the spinous processes, noting their general alignment and any points of tenderness that may be present. Carefully differentiate spinal tenderness from flank tenderness, which would indicate an abnormality in the retroperitoneal space. When a spondylolisthesis is present, anterior displacement of the superior spinal segment produces a "step" deformity in which the spinous processes of the superior (cephalad) segment can be palpated anterior to those of the inferior (caudal) segment. Test the range of lumbar motion in flexion, extension, lateral bending, and rotation. Limitation of flexion exerted by tightness of the hamstring muscles in the posterior aspect of the thigh is associated with spondylolisthesis and less commonly with tumors, arachnoiditis, and herniated intervertebral discs in the lumbar and lower thoracic regions.

Tenderness to palpation over the lumbar spinous processes associated with limited motion of the spine may indicate disc lesions, osteoarthritis, spondy-lolisthesis, tumor, or infection. Laxity of the sacroiliac joints may be detected by palpation while the patient, standing erect, lifts first one knee and then the other ("marching in place"). A click can be felt if subluxation of the sacroiliac joint occurs. With the patient supine, stress the articulation between the fifth lumbar vertebra and the sacrum by acutely flexing the patient's hips and knees. Resultant pain in the lower back suggests a mechanical abnormality at the lumbosacral articulation. Tilting the pelvis in this position by rocking the knees to the left and right causes pain when abnormalities of the sacroiliac joint are present.

Measure the leg length from the anterior superior spine to the medial malleolus. Measure the circumference of the thighs and calves and compare opposite legs. Test muscle power of the dorsiflexors and plantar flexors of the foot and ankle by having the patient support the body weight on the toes and on the heels. Test deep tendon reflexes at the knee and ankle and the sensory modalities of light touch, pain, and position sense. Perform the straight-leg raising test with the patient lying supine on the examining table. Flex the hip with the knee in full extension by raising the foot. A sharp pain traveling from the lower back or buttock down the posterior aspect of the leg indicates irritation of the sciatic nerve or its roots of origin within the spine. In the absence of such irritation, this maneuver may be limited by muscle tightness at the back of the thigh, but not by sharp pain.

Perform a rectal examination, a brief evaluation of the urinary tract, an examination of the abdomen, and an evaluation of the femoral pulses. Prostatic disease in the male, pelvic abnormalities in the female, and diseases of the kidneys and ureters may produce discomfort in the lower back. Occlusion of the distal aorta or of the hypogastric arteries as well as an abdominal aortic aneurysm may produce pain in the lower back or in the buttocks.

The Shoulder

Inspect the shoulder anteriorly and posteriorly and look for loss of normal contour and muscle atrophy. Three major muscles are easily visible about the shoulder: deltoid (which covers the shoulder anteriorly, laterally, and poste-riorly), supraspinatus, and infraspinatus, the last two of which originate on the posterior border of the scapula above and below the scapular spine. Visible atrophy of the deltoid muscle follows disuse, injury to the axillary nerve, and diseases of the nervous system such as poliomyelitis. Visible atrophy of the supraspinatus and infraspinatus muscles is seen following nerve injury, diseases of the nervous system, tears of the insertion of these muscles into the rotator cuff of the shoulder, and calcific tendinitis involving that portion of the rotator cuff that represents their insertion. Determine function of the anterior serratus muscle by having the patient push with both hands against a wall. The medial border of the scapula is not held firmly against the chest wall but is displaced posteriorly ("wings") when weakness of the anterior serratus muscle is present. Weakness of this muscle frequently follows injury to the long thoracic nerve of Bell either in the neck or in the axilla.

Note the bony contour of the shoulder and the position of the humeral head. Flattening of the lateral portion of the shoulder is seen when the humeral head is not in its proper position but is displaced as in a subcoracoid dislocation of the shoulder. Prominence of the distal end of the clavicle indicates a dislocation of the acromioclavicular joint or a tumor of the distal end of the clavicle.

Carefully palpate the structures about the shoulder, particularly the rotator cuff, for points of tenderness. The rotator cuff of the shoulder represents the insertion of the subscapularis, supraspinatus, infraspinatus, and teres minor muscles into the proximal humerus. Tenderness to palpation over a segment of the rotator cuff indicates a rotator cuff tear or calcific tendinitis. Diffuse tenderness over the entire rotator cuff suggests pericapsulitis (frozen shoulder) or synovitis of the joint. Tenderness over the long head of the biceps tendon as it lies in the bicipital groove between the greater and lesser tuberosities of the humerus suggests tendinitis of the biceps tendon. This diagnosis is further suggested if shoulder pain is accentuated when the forearm is flexed and supinated against resistance.

Determine active and passive motion of the shoulder and compare with the opposite side. If passive motion of the shoulder is normal but active motion is limited, a tear of the rotator cuff or muscle weakness about the shoulder should be suspected. If active and passive motion are both limited to an equal degree, contractures, arthritis, or a mechanical block (such as calcific tendinitis) should be suspected. Determine sensation over the lateral portion of the shoulder, which is supplied by the axillary nerve.

The Elbow

Note the contour and the carrying angle of the elbow. The carrying angle is the angle formed by the upper and lower arm when the elbow is observed from the front with the forearm in full extension and supination. Change in this angle may be seen following damage to the elbow from trauma, rheumatoid arthritis, or osteoarthritis. Determine the relationship between the distal humerus and the olecranon and the head of the radius. Note points of tenderness and the presence of palpable effusion or synovitis within the joint. Palpate the head of the radius and the tissue just anterior to it for points of tenderness as the forearm is supinated and pronated. Tenderness lateral or anterior to the radial head may indicate lateral epicondylitis (tennis elbow) and may be accentuated as the patient pronates the forearm and extends the wrist against resistance. Palpate the ulnar nerve in the groove in the posterior aspect of the medial condyle of the humerus. Localized tenderness over the ulnar nerve at this spot suggests irritation of the nerve, which frequently follows fractures about the elbow. It may be associated with pain and paresthesia in the fourth and fifth fingers and weakness of those muscles in the hand supplied by the ulnar nerve (adductor pollicis, third and fourth lumbricalis, and interossei). Test the active and passive range of motion of the elbow in flexion, extension, supination, and pronation, and compare with the opposite side. Palpate the subcutaneous surfaces of the proximal ulna for nodules which are often found with rheumatoid arthritis, even when rheumatoid involvement of the elbow is not present.

The Wrist

Observe the contour of the wrist and note any fullness about the joint. Palpate the bony and tendinous structures about the wrist for tenderness and nodularity. Tenderness over the anatomic snuffbox on the lateral aspect of the wrist may be associated with abnormality of the navicular or greater multangular bones. Tenderness over the lateral aspect of the distal radius suggests inflammation of the tendon sheaths of the extensor pollicis brevis and the abductor pollicis longus — deQuervain's disease. The diagnosis is confirmed when a tender nodule can be felt within one or both of the tendons over the lateral aspect of the distal radius and when a positive Finkelstein's test is noted. Perform Finkelstein's test by moving the wrist rapidly into ulnar deviation as the patient holds the thumb flexed in the palm. The test is positive if it induces sudden pain which extends to the thumb.

Compression of the median nerve at the wrist causes tenderness over the nerve, loss of normal sensation in the portion of the hand supplied by this nerve — the flexor surfaces of the thumb, the index and middle fingers, and the lateral half of the ring finger — and weakness of those muscles of the thumb supplied by the median nerve. In addition, percussion over the median nerve at the wrist will produce pain radiating distally into the hand — a positive Tinel's sign.

The Hand

Inspect the hand for deformities. Test active and passive ranges of motion of the fingers for evidence of nerve injury, tendon rupture, muscle fibrosis, joint contracture, or arthritis. Test the function of each muscle to the fingers and wrist. The presence of nodules within the palm associated with firm fibrous bands which limit extension of the fingers suggests Dupuytren's contracture. A palpable nodule within a flexor tendon overlying a metacarpal-phalangeal joint, associated with a palpable click as the finger is flexed or extended, suggests a trigger finger, in which a nodular enlargement of a flexor tendon snaps into or out of the fibrous tunnel through which it passes.

The Hip

Examination of the hip begins with the initial evaluation of the patient's gait. Leg-length inequality, muscle weakness, habit patterns, fusion, contractures, and pain-producing lesions of the hip, knee, ankle, and foot all may produce abnormalities of gait. There are two disturbances, however, which are fairly specific for abnormality of the hip; the Trendelenburg gait and the antalgic gait. The first term denotes a gait marked by a fall of the pelvis, rather than the normal rise, on the side opposite the involved hip when it is weight-bearing. This indicates weakness of the abductor muscles due to intrinsic muscle disease, lack of normal innervation of the muscle, or an unstable hip joint. An antalgic ("antipain") gait reflects the patient's attempt to minimize the force borne through the hip by leaning over the involved hip as weight is borne through it. This moves the body's center of gravity toward the hip joint, decreasing the force exerted on the hip and thus decreasing the discomfort (Figure 18-3).

Palpate the hip anteriorly for points of tenderness and fullness within the joint. The location of the femoral head is approximately one inch distal and one

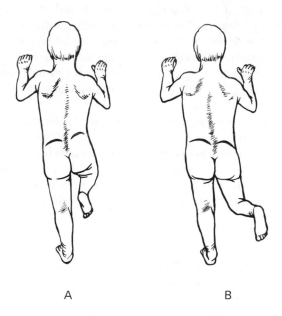

A   B

FIGURE 18-3. The Trendelenburg sign. A. Normal: The right side of the pelvis rises when weight is borne on the left leg. B. Positive Trendelenburg sign: left hip. The right side of the pelvis falls when weight is borne on the left leg.

inch lateral to the point at which the femoral artery crosses the inguinal ligament. In addition, palpate the greater trochanter laterally and the ischial tuberosity posteriorly. Inflamed bursae in either of these locations may account for pain in the hip region. Test the range of motion of the hip in flexion, extension, abduction, adduction, and rotation. Test for flexion contracture of the hip by maximally flexing the opposite hip with the patient supine on the examining table (Figure 18-4). In this maneuver the pelvis is flexed upon the

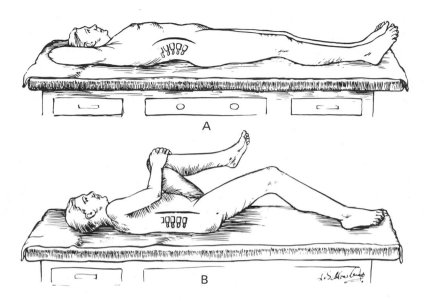

FIGURE 18-4. The Thomas Test. A. With the patient supine on the examining table a flexion contracture of the hip may be masked by lordosis of lumbar spine. B. When the lumbar lordosis is eliminated by maximally flexing the left hip, the angle which the right thigh makes with the surface of the examining table designates the degree of flexion contracture in the right hip.

lumbar spine, and any lumbar lordosis present is eliminated. If a flexion contracture is present, the involved thigh will flex. The angle which the involved thigh forms with the surface of the examining table indicates the degree of contracture. Determine the strength of the flexor, extensor, abductor, and adductor muscle groups by having the patient move the thigh against the resistance of the examiner's hand.

The Knee

Inspect the knee for joint swelling and gross deformity, and the thigh for atrophy of the quadriceps muscle group, which may indicate a significant knee abnormality. Test for excessive fluid within the knee joint by pressing the patella against the femur with the knee in full extension and gently tapping one side of the joint while feeling for a fluid wave on the other side of the joint. In addition, note whether ballottement of the patella is possible. Palpate about the joint line, about the circumference of the patella, over the attachments of the medial and lateral collateral ligaments, over the anserine bursa (on the medial posterior aspect of the proximal tibia about one inch distal to the joint line), and in the popliteal fossa. When tenderness is noted over the medial or lateral joint line, a torn meniscus is suspected. Tenderness about the patella suggests osteoarthritis or chondromalacia. The diagnosis is confirmed if crepitation and pain are elicited when the patella is moved medially, laterally, proximally, and distally, while it is firmly pressed against the underlying femur. Tenderness over the medial or lateral collateral ligament suggests a tear of the ligament or an inflamed bursa between the ligament and the bone. Tenderness over the anserine bursa or in the popliteal fossa suggests symptomatic bursae in these regions.

Determine the range of motion and compare with the opposite side. Full extension and at least 120 degrees of flexion from full extension are normally present. Palpable crepitation within the knee joint during flexion and extension suggests a mechanical incongruity of the joint which may be the result of arthritic changes, a tear of the medial or lateral meniscus, or a loose fragment of cartilage or bone within the joint. Locking of the joint during flexion or extension definitely signifies a mechanical block within the joint, most commonly a torn meniscus or a bone fragment. A mechanical block preventing full extension or a contracture of the posterior capsule is suggested when the knee cannot be fully extended passively. Test for a positive spring sign, which indicates a mechanical block with the consistency of cartilage, by maximally extending the knee passively and then forcibly extending the knee further. The joint will extend and then quickly snap back into flexion when the sign is positive. A torn meniscus is further suggested by a positive McMurray sign. To perform the McMurray maneuver, flex the knee fully with the patient recumbent, steadying the knee with one hand, and slowly extend the knee while holding the tibia in internal rotation. A palpable or audible snap associated with momentary discomfort as the knee is extended suggests a tear in that portion of the lateral meniscus which was between the femoral condyle and the tibial plateau when the snap occurred. This test is repeated with the tibia held in external rotation, which tests the posterior aspect of the medial meniscus. This test is generally not positive when the tear involves the anterior one-third of either meniscus.

Test the integrity of the medial collateral ligament by attempting to force the knee into valgus (knock-knee) deformity with the knee in full extension. If more than 5 to 10 degrees of deformity can be produced, instability of the medial collateral ligament is suspected. Test the lateral collateral ligament by attempting to force the knee into varus (bowleg) deformity with the knee in full extension. Again, if more than 5 to 10 degrees of deformity can be produced, instability of the ligament is suspected. Test the cruciate ligaments with the patient sitting with the knees flexed to 90 degrees. If the tibia can be pulled anteriorly from under the femur (a positive "drawer" sign), laxity of the anterior cruciate ligament is present. If the tibia can be pushed posteriorly under the femur, laxity of the posterior cruciate ligament is present.

The Ankle

Inspect the ankle for gross deformity or swelling, and palpate for points of tenderness or fullness within the joint. Tenderness over the medial or lateral malleolus suggests previous injury to these bony structures. Tenderness distal to the tip of the medial or lateral malleoli but over the medial or lateral ligaments of the ankle suggests previous injury to these ligaments. Tenderness just anterior to the Achilles tendon at its insertion into the calcaneus suggests an inflamed bursa, often noted in patients with rheumatoid arthritis. Tenderness over the posterior tibial tendon as it lies just posterior to the medial malleolus, or over the peroneal tendons just behind the lateral malleolus, suggests inflammation of these structures. Occasionally a tenovaginitis of the posterior tibial tendon or the peroneal tendons, similar to deQuervain's disease of the wrist, is seen. In such cases tenderness is present over the tendon at the point of the constriction posterior to the malleolus; passive motion of the foot will intensify the discomfort; and a palpable nodule may be present within the tendon itself.

Test the active and passive motion of the ankle joint. At least 30 degrees of plantar flexion and 15 degrees of dorsiflexion should be present. Test the stability of the joint, particularly in inversion and eversion. Normal ankle motion is dorsiflexion and plantar flexion. If tilting of the talus can be demonstrated as the ankle is forced into eversion, instability of the medial collateral ligament is present.

Determine the strength of the muscles crossing the ankle joint by having the patient support body weight first on the heel and then on the ball of the foot.

The Foot

Inspect the foot for obvious deformities and swelling. Observe the heel from the rear. The axis of the heel is normally a continuation of the long axis of the lower leg. If the heel is tilted toward the midline of the body, the heel is in varus, a deformity frequently associated with club foot and cavus foot. If the axis is tilted away from the midline, the heel is in valgus, a deformity associated with flat feet. A bony prominence just anterior to and below the medial malleolus suggests the presence of an accessory navicular bone, which may be associated with flat feet and which makes proper shoe-fitting difficult. A swelling over the medial aspect of the metatarsophalangeal joint of the great toe is termed a bunion and may be associated with a medial deviation of more than 15 degrees when compared with the lateral four metatarsals (metatarsus primus varus), or with lateral deviation of the great toe (hallux valgus). Inspect the toes for abnormalities such as clawtoe, in which the proximal phalanx is hyperextended and the proximal interphalangeal joint acutely flexed, and hammertoe, in which the proximal interphalangeal joint is acutely flexed while the metacarpophalangeal joint remains in neutral position. Inspect the sole of the foot for the presence of calluses over the metatarsal heads. Observe the state of nutrition of the tissues of the foot by palpating the skin, noting the presence or absence of hair on the dorsum of the toes, and palpating the dorsalis pedis and posterior tibial pulses.

Test inversion and eversion of the foot by rocking the heel medially and laterally while stabilizing the lower leg, rotating the foot by supinating and pronating the metatarsals as a unit while stabilizing the heel, and flexing and extending the toes. Test the individual muscles by having the patient perform the appropriate movements against the resistance of the examiner's hand.

Palpate for points of tenderness indicating inflammation of bursae or joints. Diffuse tenderness over the calcaneus itself is often associated with early rheumatoid arthritis or with Reiter's syndrome. Inflammation detected in any of the joints of the foot may be caused by septic arthritis, rheumatoid arthritis, osteoarthritis, or gout. Tenderness to palpation over the metatarsal heads on the ball of the foot indicates metatarsalgia; tenderness between the metatarsal heads on the dorsum of the foot suggests interdigital neuroma. The latter possibility is strengthened if a sensory abnormality can be demonstrated on the opposing surfaces of the contiguous toes supplied by the involved interdigital nerve, and if medial to lateral compression of the forefoot induces metatarsal pain which radiates into the involved toes.

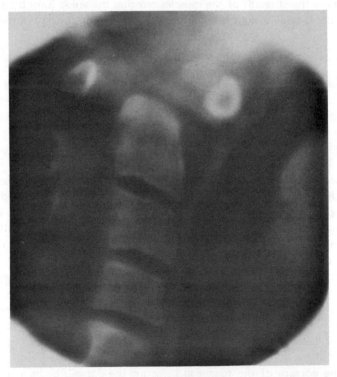

FIGURE 18-5. Lateral polytomogram of the upper cervical spine. Congenital absence of the midportion of the odontoid process and anterior displacement of C-1 on C-2 are demonstrated.

SPECIAL TECHNIQUES

X-ray At least two x-ray views taken at 90 degrees to each other are indispensable for the thorough evaluation of a bone or joint. In addition, special views such as oblique projections and stress and weight-bearing views may be necessary.

Tomography. Tomography (Figure 18-5) allows detailed study of regions of the skeleton that are difficult to study with plain x-ray views. By this technique details of bone structure and relationship may be accurately determined.

Discography. Discography (Figure 18-6) is a technique by which the interior architecture of a disc is visualized by injecting a radio-opaque substance into the nucleus pulposus of the disc. With rupture of the annulus fibrosus, extravasation of the radio-opaque material is noted on x-rays. The contrast medium is introduced into the disc percutaneously or during a surgical procedure.

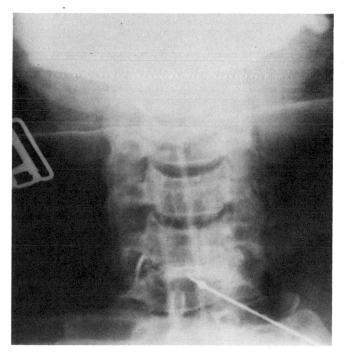

FIGURE 18-6. Anteroposterior view of a discogram performed at C-5-6 level. Extravasation of the dye through a rent in the annulus fibrosus of the disc is demonstrated.

Myelography. Myelography (Figure 18-7) is carried out by means of radio-opaque substance introduced into the dural sheath. X-rays then detect lesions which produce displacement of the contrast material, such as herniated discs, bony lesions of the spine, and spinal cord tumors.

Arthrography. Arthrography (Figure 18-8) is performed by introducing radio-opaque material into a joint, rendering the joint surfaces visible upon x-ray

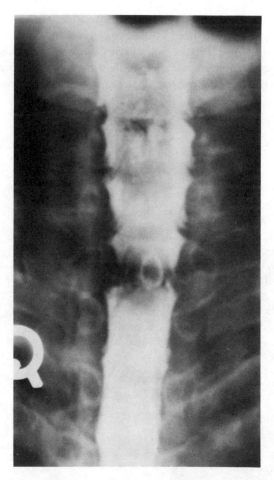

FIGURE 18-7. Posteroanterior view obtained during a myelogram of the cervical and thoracic region. A filling defect is present in the dye column at the cervical-thoracic junction.

examination. This technique demonstrates dislocations of a joint, irregularities of the joint surfaces, and loose fragments of cartilage within the joint which may not be visible on plain x-ray views.

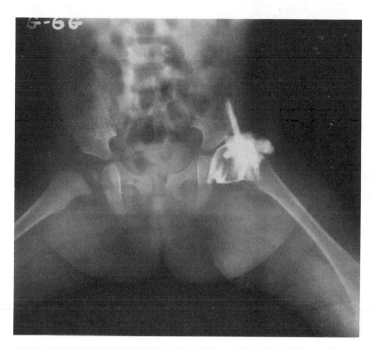

FIGURE 18-8. Arthrogram of the hip of an infant. The radio-opaque dye outlines the cartilaginous femoral head, which is normally shaped and in proper position within the acetabulum.

Arteriography (Figure 18-9). The vascular supply to a bone, joint, or bone lesion is well demonstrated by injecting radio-opaque material into the artery supplying the anatomic region to be studied, and taking a series of x-ray films following the injection. Some bone tumors are not readily apparent on plain x-ray views and the extent of the tumor can be well demonstrated only by this technique.

Cinefluorography. Cinefluorography is a technique whereby motion of the skeleton is observed by means of image-intensifier fluoroscopy and recorded on movie film to allow more detailed study at a later date. This technique reveals dynamic abnormalities, such as subluxation of vertebrae or a block to normal motion of a joint, which may not be visible on plain x-ray views.

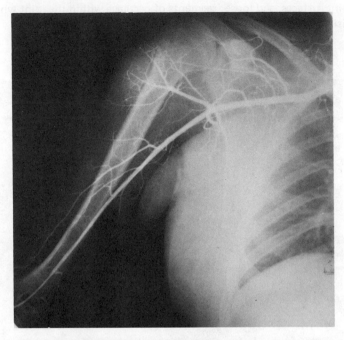

FIGURE 18-9. Arteriogram demonstrating the vascular supply to a tumor arising from the periosteum of the proximal humerus.

Synoviocentesis. Synoviocentesis is a technique whereby joint fluid is removed for study. It involves placing a needle into the joint cavity following aseptic preparation of the skin through which the needle is inserted, and aspirating fluid for study. Strict asepsis is observed throughout the procedure. Studies which may be performed upon joint fluid include observation of the color and consistency of the fluid, microscopic examination of the fluid for the number and type of cells that may be present, gross observation of protein content by means of the mucin clot test, microscopic examination and cultures for bacteria, and characterization of immune globulin which may be present.

Synoviocentesis provides an invaluable technique in the diagnosis of local and systemic conditions that may involve a joint.

REFERENCES

1. Aegerter, E., and Kirkpatrick, J. A., Jr. *Orthopedic Diseases.* Philadelphia: Saunders, 1958.
2. American Academy of Orthopaedic Surgeons. *Joint Motion: Method of Measuring and Recording.* Chicago, 1965.

3. Mercer, W., and Duthie, R. B. *Orthopaedic Surgery*. Baltimore: Williams & Wilkins, 1964.
4. Perkins, G. *Orthopaedics*. London: The Athlone Press (University of London), 1961.
5. Shands A. R., Jr., and Raney, R. B. *Handbook of Orthopaedic Surgery* (7th Ed.). St. Louis: Mosby, 1967.
6. Turek, S. L. *Orthopaedics: Principles and Their Application* (2nd Ed.). Philadelphia: Lippincott, 1967.

THE ACUTELY INJURED PATIENT   19

George D. Zuidema
Gerhard Schmeisser, Jr.

GENERAL PRINCIPLES

The aim of this textbook is to present the principles of examination and diagnosis. The first aim of the examining physician is the preservation of life. When dealing with acute trauma it is often impossible to separate diagnostic and therapeutic measures; indeed, it would be improper to completely dissociate these features. The care of the acutely injured patient imposes certain important restrictions upon the examiner. It may be impossible to obtain a detailed or even cursory history from the patient. The examiner is often forced to rely heavily on physical findings for diagnosis. The initial examination is as likely to be performed in the field or beside a highway as in a well-equipped hospital emergency room.

When confronted with an acutely injured patient you should ask yourself the following questions, in rapid order:

Is the airway patent?
Is there significant hemorrhage?
Is there serious or potential brain or spinal cord injury?
Is there a fracture?
Is there a chest injury?
Is there acute intra-abdominal injury?
Is there injury to the urinary tract?
Is there peripheral nerve injury?

The most urgent requirement is the evaluation of the patient's airway. Death may ensue in minutes if adequate ventilation is not possible. The problem may be compounded by unconsciousness, by aspiration of blood or vomitus, or by serious injuries to the chest wall or lung parenchyma. In evaluating the status of the patient's airway the patient should be turned on his side or face down with his head in the dependent position to minimize the danger of aspiration. It may be necessary to exert traction on the tongue to maintain an oral airway. If tracheal obstruction exists, an emergency tracheostomy may be indicated. If ventilation is inadequate, mouth-to-mouth resuscitation should be initiated without hesitation. Prompt recognition of this need may be life-saving. It may be necessary to combine these measures with closed-chest cardiac massage. It has been demonstrated that an adequate peripheral circulation can be maintained by this technique. Unless ideal circumstances exist in an operating-room environment with intratracheal intubation and anesthesia apparatus, open-chest cardiac massage will seldom be indicated.

The second important threat to life is massive hemorrhage. In the patient with an injured extremity, the steady oozing of blood is best controlled by constant pressure. This may have to be maintained manually. Elevation of the limb will help to control blood loss. Only in exceptional instances will arterial blood loss be a major problem. In these rare circumstances a tourniquet should be applied proximal to the wound. The limb should be observed carefully, occasionally loosening the tourniquet, to permit an attempt at reconstructive surgery. By the time the patient reaches operating-room facilities, continuous slow, steady oozing of blood may have seriously depleted the circulating blood volume in an insidious manner. It is easy to underestimate this type of blood loss unless one is alert to this possibility.

When adequate airway and ventilation are assured and hemorrhage is controlled it is time to perform a rapid physical examination to assess the extent of coexisting injuries. This should be initiated as soon as possible and, although brief, should be thorough and systematic. Failure to carry out this kind of survey will lead to errors in diagnosis and management which may have serious consequences or result in loss of life. The information obtained by this preliminary examination provides valuable base-line data by which to follow the patient's progress and also aids subsequent management.

The examination should rule out the possibility of spinal injury, since if such a patient is moved roughly or improperly, sudden paralysis may occur. Pain medications should be withheld until a clear indication for them exists and brain injury is ruled out. If narcotics are given before the examination is completed, diagnostic evaluation becomes clouded and neurologic signs are difficult to interpret. When given, the analgesic should be administered intravenously because of the uncertainties of absorption associated with hypotension which often accompanies massive trauma.

Examination of the Open Wound

There is a great tendency to probe, explore, and investigate open wounds. Our natural curiosity leads us to do this, but in most instances this should be avoided. In general, except to control hemorrhage, open wounds should be explored only in the operating room. Anesthesia should be adequate to permit thorough study; cleansing should be performed with copious amounts of sterile saline; and sterile technique should be strictly observed. Anything short of this fosters an incomplete and inadequate examination, is likely to lead to infection, and may result in failure to find foreign bodies. The ideal method of early management calls for application of a dry, sterile (or at least clean) dressing to prevent additional soilage. Definitive care should be given as soon as conditions permit within the controlled environment of the operating room.

THE RADIOLOGIC EXAMINATION

X-ray examination is a valuable adjunct to the physical examination in evaluating the extent of injuries. Selected studies often provide information that may be obtained in no other way. Numerous examples have been mentioned previously. In addition, a standard-size chest film is worth obtaining. In many ways physical examination of the heart and lungs is limited, and every physician realizes that the x-ray is much more accurate and effective in detecting subtle pulmonary, mediastinal, or cardiac changes.

A flat film of the abdomen is also important if abdominal trauma has occurred. In addition to the specific studies cited above, free air from a perforated viscus may be detected by an upright or lateral decubitus film.

X-ray examination of the skull is an important part of the evaluation of patients with head injury. It is important, however, to select the proper time for x-ray study. To obtain satisfactory films it is usually necessary to enlist the cooperation of the patient. This is frequently not possible in the early state of injury, and the manipulation of such a patient may be attended by considerable hazard. In most instances, such films may best be obtained when the patient's condition is stable. Furthermore, the types of intracranial hemorrhage requiring prompt surgical treatment will be detected by means of observation of vital signs and neurologic examination rather than on the basis of an x-ray study. The principal value of skull films lies in the recognition of skull fractures, which require specific treatment. There are also sound medicolegal reasons for obtaining such x-rays. As a general rule, radiologic studies should be completed as soon after head injury as the general condition of the patient permits.

Common sense should dictate when radiologic studies will contribute to successful management. Under certain urgent circumstances early operative intervention is more important than obtaining a complete set of films. In this type of situation only films that vitally affect decisions are indicated.

HEAD INJURIES

Cerebral Trauma

Patients with head injury will frequently have airway obstruction or ventilatory problems. This requires primary attention. The neurologic examination may then be performed. There are three aims of emergency neurologic evaluation: (1) to determine whether the patient is in need of emergency surgical intervention; (2) to establish the diagnosis of any existing head injury; and (3) to obtain base-line neurologic information for comparison purposes later.

A time-consuming, detailed, elaborate neurologic examination is not appropriate for the early care of patients with head injury. On the contrary, a few carefully selected studies may be obtained using simple equipment. The state of consciousness should be evaluated. The patient's response to painful stimulation should be recorded. Examples are response to supraorbital pressure, pinprick, or pressure on the sternum with the knuckles. The condition of the pupils, their relative size, equality, and response to light should be observed and recorded. *Small, contracted pupils* which do not respond may be associated with midbrain damage. This may be misleading, however, if the patient has been medicated, or has used alcohol. *Dilated fixed pupils* have a poor prognosis. Inequality of the pupils may reflect local brain damage. *Unilateral dilatation* of the pupils occurring under observation is strongly suggestive of intracranial hemorrhage.

Character of Respiration. Irregular or depressed respirations may accompany severe intracranial injury. If this situation exists in the presence of an adequate airway, the prognosis is grave.

Degree of Motor Activity. If the patient is conscious, this may be appraised by having the patient squeeze the examiner's hands or by testing his ability to resist passive motion of the extremities. In the comatose patient the degree of flaccidity may be evaluated by lifting the extremity and letting it drop. It is important to examine both sides of the patient. This permits comparison and provides base-line information should his condition deteriorate under observation. Extensor rigidity of all extremities carries bad prognostic significance. Although alcoholism or drug intoxication may confuse this observation, complete flaccidity and areflexia usually indicate severe central nervous system damage.

Body Temperature. When associated with evidence of intracranial injury, the development of hyperthermia to *temperatures of 103°F.* or above may be associated with bad prognostic significance.

Evaluation of Deep Tendon Reflexes. A detailed examination may be inappropriate at first. Study of the triceps, biceps, radioperiosteal, plantar, knee, and ankle reflexes together with the test for the presence of ankle clonus, should be adequate for initial evaluation.

Types of Head Injuries. A simple linear *skull fracture* may be of minor significance and should not in itself alter the overall program of management. The importance of this finding will depend on the central nervous system signs and symptoms detectable in the individual patient. *Depressed skull fracture* occurs from direct trauma with an instrument or missile. Here, too, the resulting brain damage is extremely variable, and no general statements are possible. When dealing with lacerated or contused wounds of the scalp, one should suspect a possible depressed fracture, and careful neurologic examination should be performed. Depressed skull fracture is usually an indication for prompt surgical treatment. Failure to provide it may be associated with the progression of neurologic signs or infection.

Penetrating wounds of the skull may be misleading, since extensive intracranial injury may be associated with a small wound of entrance. Early surgical exploration and debridement are indicated.

Cerebral concussion is associated with a loss of consciousness and memory regarding the accident. The patient may appear to be well when seen initially, but observation is important, since there is the possibility of delayed intracranial hemorrhage.

Extradural or *subdural hemorrhage* is associated with head trauma. Characteristically, a brief period of unconsciousness may be noted, followed by a "lucid interval." Confusion, drowsiness, and progressive coma then supervene. With extradural hemorrhage, the sequence of events tends to be fairly rapid, developing over a matter of hours. With subdural hemorrhage the course of the condition may be extended over several weeks or months. In either event the presence of lateralizing neurologic signs, asymmetric dilatation of the pupils, and alteration in the deep tendon reflexes and motor responses should be sufficient reason to prompt emergency neurosurgical intervention.

Examination for Facial Injuries

In accidents in which intracranial injury occurs, fractures of the facial bones are common. Fracture of the nose is often obvious because of deviation from the normal contour. Edema and discoloration of the skin may, however, render this diagnosis difficult. The patient should be examined from above as well as from the front if minor degrees of asymmetry are to be detected. Bimanual palpation and intranasal speculum examination may also be of help.

A blow on the cheek may produce a *fracture of the zygoma*. This fracture is frequently associated with considerable edema and subcutaneous hemorrhage, diplopia, subconjunctival hemorrhage, and anesthesia below the eye due to injury of the second division of the trigeminal nerve.

Fractures of the facial bones do not constitute surgical emergencies although they may be associated with some degree of airway obstruction. Operative treatment may be delayed for several days if necessary to permit adequate treatment of the patient's other injuries.

INJURIES OF THE SPINE

Fractures or fracture dislocations of the spine result from violent trauma. These may occur as a result of automobile accidents, football injuries, and diving accidents. It is essential to complete the examination without disturbing the patient. If a fracture of any portion of the spine is suspected, it is of utmost importance to complete the diagnosis and transport the patient for initiation of treatment without producing injury to the spinal cord. The cervical spine, because of its mobility, is particularly susceptible to injury. For this reason, in moving a patient gentle traction should be exerted on the head in the long axis of the spine. The patient should not be allowed to flex his neck. A general rule is to permit the patient as little motion as possible and to transport him in the prone or supine position, depending on the circumstances, without permitting rotation, flexion, or extension of the spine.

Injury of the spinal cord is evaluated by asking the patient to move his legs and toes. If he is able to do this he has escaped major cord damage. If, however, the legs are paralyzed but the patient can move his hands the cord lesion is located below the cervical region. If arm function is interfered with, cervical spine involvement is suggested. It is possible to confirm the location of cord injury by testing for loss of sensation to pinprick.

INJURIES OF THE CHEST WALL

Rib fractures are common injuries. They may or may not be associated with intrathoracic damage. To permit proper examination the patient should be stripped to the waist and asked to take a deep breath. In the presence of rib fracture there is usually limitation of motion associated with pain on the affected side. The palpation of each rib in order should be carried out to rule out the possibility of subcutaneous emphysema. Compression of the chest cage in an anteroposterior direction and laterally may elicit pain when rib fractures are present.

Percussion and auscultation of the chest should be performed in every instance to detect the presence of some type of intrathoracic injury. It is well to omit compression of the chest cage if physical signs of pleural effusion, pneumothorax, or mediastinal shift are present.

Fractures of the sternum are usually secondary to considerable violence and may occur in steering-wheel injuries. This fracture is usually associated with considerable pain and rapid and shallow respiration. The sternum may show a visible depression and subcutaneous hemorrhage may be prominent. Characteristically, the patient is seen to hold his head forward and rigid. This lesion is often associated with a contusion of the heart, hemopericardium, or injury to the intrathoracic aorta. Cardiac arrhythmias or murmurs may be present.

With severe crushing injuries of the ribs, resulting in multiple fractures, a portion of chest wall may become freely movable. This condition is known as *flail chest*. With inspiration the mobile portion of the chest wall is sucked inward, resulting in decreased ventilation. In effect, the involved side ceases to function in ventilatory exchange. The result is an increase in shunting of blood through the affected lung (physiologic shunt). Diagnosis is not difficult, since respiration is associated with exquisite pain, dyspnea, and cyanosis. The chest wall is seen to move paradoxically. It is imperative to achieve prompt stabilization of the chest wall. Intrapulmonary hemorrhage, or "wet lung," may also be present. Tracheostomy may be indicated to reduce dead space and permit adequate intratracheal toilet.

Nonpenetrating wounds of the chest wall should be distinguished from crush injuries. They show the characteristics of soft-tissue wounds in general. Penetrating closed wounds of the chest are usually associated with some degree of intrathoracic visceral damage. The possibilities include hemothorax, tension pneumothorax, subcutaneous or mediastinal emphysema, and cardiac tamponade.

With an *open wound* of the chest there is free communication between the pleural space and the outside. This is usually associated with great distress and signs of asphyxiation. Cyanosis and hypotension with a rapid and thready pulse are usually present. Inspiration is labored and accompanied by an audible sucking sound. Expiration is forced and often accompanied by frothy serum or blood issuing from the wound (Figure 19-1). Subcutaneous emphysema is common. A large sucking wound of the chest is a surgical emergency which demands immediate treatment. It can usually be closed by applying a clean dressing. This should be carried out immediately, using materials at hand without regard to sterility. Ventilation may then be improved by having the patient lie on his injured side.

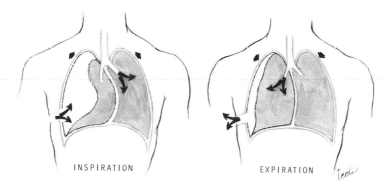

INSPIRATION   EXPIRATION

FIGURE 19-1. Intrathoracic dynamics in patient with sucking wound of the chest. Note shifts in lung and mediastinum with inspiration and expiration.

Hemothorax. The physical signs of hemothorax are those of pleural effusion; these are diminished breath sounds at the base posteriorly on the involved side and dullness to percussion. The mediastinum may be shifted, and this may be detected by percussion and palpation of the trachea for shift. All degrees of hemothorax may occur, depending on the site of hemorrhage. They may also be associated with pneumothorax if the parenchyma of the lung is involved. In this situation increased resonance and absent breath sounds will be present above the area of dullness. The physical signs of pneumothorax tend to obscure the signs of pleural effusion.

Pneumothorax.   Physical signs of pneumothorax depend on the amount of air present. Small amounts of pneumothorax are difficult to identify, but one of any significant size should be readily recognizable. Respiratory rate is increased and dyspnea is present. The chest wall on the affected side shows decreased movement, and cyanosis may be present. Percussion may indicate a shift of the heart and mediastinum. This may be confirmed by palpation of the trachea. The percussion note over the involved side is characteristically increased in resonance and is tympanitic.

Tension pneumothorax occurs when injury involves pulmonary parenchyma or bronchi. It may develop when an open sucking wound of the chest is closed by packing or strapping, or it may be spontaneous due to rupture of a pulmonary bleb. This injury is serious and demands prompt surgical treatment. With each inspiration air enters the pleural space on the involved side, increasing the collapse of the lung and pushing the mediastinum toward the uninvolved side. This further reduces the function of the good lung. Clinically, the situation is characterized by pronounced dyspnea and cyanosis. Vascular collapse with hypotension and a rapid thready pulse are due to decreased venous return to the heart. It is important to differentiate between circulatory collapse secondary to hemorrhage or shock elsewhere in the body and that which is secondary to tension pneumothorax. The trachea and heart will be shifted toward the uninvolved side, and the percussion note is usually hyperresonant and tympanitic. It is possible, however, for tension pneumothorax to exist with minimal signs of hyperresonance and tympani. The mediastinal shift toward the uninvolved side may also be detected by percussion. Breath sounds on the involved side are generally absent or muffled. Emergency treatment, consisting of aspiration of the trapped air, should be instituted promptly and may be of life-saving importance. Sufficient air should be aspirated to produce relief of symptoms. The patient should be followed carefully to prevent recurrence.

Subcutaneous Emphysema.   Subcutaneous emphysema occurs when air gains access to tissue planes around the wound or injury. Considerable subcutaneous spread is possible. This is characterized by local swelling, edema, and crepitation on compression. This is a common accompaniment of compression injuries to the chest involving rib fractures. It may also be seen in rupture of the parenchyma of the lung with dissection beneath the visceropleura into the mediastinum. It may then spread rapidly to produce swelling of the neck, face, and chest wall, and may even extend to the abdominal wall and scrotum. It may be associated with minimal respiratory distress. If, however, dyspnea and cyanosis are present, one should suspect a coexisting tension pneumothorax. With tension pneumothorax it is possible to develop sufficient mediastinal pressure to embarrass the venous return to the heart. On auscultation over the base of the heart the characteristic crackling "mediastinal crunch" may be detectable.

Blunt or Penetrating Chest Wounds. These wounds may produce an accumulation of blood within the pericardium. This results in progressive compression of the heart with obstruction of the great veins. Cardiac filling is impaired and reflected in a decreasing cardiac output. This may progress to death unless aspiration of the pericardium is carried out. *Cardiac tamponade* is associated with a high venous pressure. This may be recognized by distention of the neck veins and will be accompanied by dyspnea and cyanosis. If tamponade occurs rapidly, the area of cardiac dullness may not be increased, and the diagnosis may be missed on percussion or even on fluoroscopy. On auscultation the heart sounds are distant and rapid. Systemic blood pressure is low, with a narrow pulse pressure created as systolic pressure falls and diastolic pressure rises. A paradoxic pulse is present and may be demonstrated by maintaining the blood-pressure cuff pressure at the level at which systolic sounds are first heard. With each inspiration the systolic sounds disappear. Fluoroscopy will demonstrate decreased cardiac pulsations. Treatment consists of aspiration of blood from the pericardium.

With blunt injury of the chest it is sometimes possible to encounter extensive pulmonary damage in the presence of an intact chest wall. An *intrapulmonary hematoma* is one consequence and may be associated with hemoptysis or frothy sputum, and dyspnea. On percussion dullness may be noted, whereas on auscultation breath sounds will be diminished and may be associated with coarse, bubbling rales. X-ray examination is an important diagnostic measure, and a high temperature may be present in the postinjury state.

Pulmonary edema may occur secondary to reflex stimulation from the intrathoracic viscera or from fluid overload. This may produce "traumatic wet lung." The physical findings consist of cyanosis, dyspnea, and production of frothy white or blood-stained sputum. Coarse rhonchi may be palpable, and moist rales may be heard on auscultation.

Contusions of the heart may be associated with cardiac irregularities or syncope. The heart may have associated valvular damage or rupture.

Injuries of the aorta may accompany steering-wheel trauma. Delayed rupture of the aorta may occur in the postinjury period. The site of rupture is usually located in the descending aortic arch in the region of the left subclavian artery. Frequent x-ray examination of the chest should be employed to detect early signs of enlargement at this point. These injuries are difficult to manage and many have a fatal outcome.

ABDOMINAL INJURIES

As a general rule, penetrating wounds of the abdomen require surgical exploration. If treatment is to be successful, early operation is an absolute necessity. If more than eight hours elapse from time of injury to time of exploration, even simple wounds are associated with a high mortality rate.

Hypotension occurring early after injury is likely to be associated with blood loss, whereas hypotension developing after several hours' delay may indicate widespread infection or peritonitis.

A complete examination is extremely important. It must take into account the type of agent inflicting the wound and the position of the patient at the time of injury. When the physician is faced with multiple wounds it is easy to be misled and to overlook small wounds of entrance. The buttocks, perineum, and anal canal should be carefully inspected as a general routine. The appearance of the wound may provide information regarding the nature of the injury. For example, the presence of intestinal contents or bile may denote specific visceral injury. Under these circumstances prompt exploration is indicated, and little is to be gained from prolonged detailed physical examination. Where wounds of entrance are small, however, detailed examination is important. This is particularly true when the wounds are located in such a way that intra-abdominal damage is not definitely established. Careful physical examination will then be directed toward eliciting evidence of even minor degrees of peritoneal irritation. This should certainly include rectal examination and may include sigmoidoscopy without the use of air insufflation. Injection of stab wounds with water-soluble contrast medium (Hypaque) is often helpful in determining peritoneal or visceral penetration. X-rays are taken in lateral and oblique projections. Our policy is to explore patients with peritoneal penetration. This contrast injection technique should not be used on patients with gun-shot wounds or multiple stab wounds of the abdomen, or on patients with stab wounds of the chest. It should be remembered that spinal cord injuries may produce abdominal pain, rigidity, and hypotension. The neurologic examination in such instances is an important part of the evaluation process.

Intra-abdominal hemorrhage may be produced by laceration of the liver, spleen, mesenteric vessels, or retroperitoneum. The development of pallor, sweating, restlessness, and thirst within a few hours of the time of injury is significant. Hypotension ensues, and the pulse becomes rapid in rate and thready in quality. Dyspnea or "air hunger" may be present. Shifting dullness and rebound tenderness may be present. With massive hemorrhage, the abdomen becomes progressively swollen and full. When intra-abdominal hemorrhage is not massive, normal blood pressure may be maintained for several hours. Under these circumstances it becomes necessary to follow the pulse pressure and pulse rate with care. The course of these indices is more important than the actual initial value. With slow, continued bleeding, progressive abdominal tenderness and spasm may become evident; with sustained, slow blood loss, decompensation and hypotension may occur rather suddenly. A rising pulse rate may indicate impending decompensation.

With *perforation of a hollow viscus,* abdominal pain and vomiting may occur. These will rapidly become associated with a rigid, tender, silent abdomen. These features are most prominent if some time has elapsed following injury. The early signs may be overlooked in the presence of multiple injuries or if analgesics or sedatives have been administered. If shock develops after eight to twelve hours or more after injury, it may be caused by generalized peritonitis. It may be accompanied by tachypnea and characteristic anxious facies.

Several abdominal wounds are commonly associated with hypotension. Pain and syncope may contribute, but *blood loss* and *massive peritonitis* may rapidly contribute to *"irreversible shock."* Although the time period necessary for this to develop may vary, it is worth noting that therapy should not be withheld on the premise that the observed hypotension is irreversible. It is also true, however, that failure of blood pressure to rise after adequate replacement transfusion carries with it a poor prognosis. It is equally important to be certain that some remediable lesion is not contributing to the patient's poor clinical condition. For example, tension pneumothorax or cardiac tamponade may occur in association with intra-abdominal injury. The physician should be prepared to reevaluate the patient completely at frequent intervals to be certain that his working diagnosis is accurate.

Blunt Abdominal Injury

The liver, stomach, intestines, spleen, and pancreas are all subject to severe injury of a blunt or nonpenetrating nature. The injury may occur when the viscus is crushed against the vertebral column. Frequent examination of the abdomen is of considerable importance. If there is reasonable doubt about the possibility of intraperitoneal injury, exploratory celiotomy should be indicated. This is a matter of judgment, since the severity of other associated injuries must be weighed against the possibilities of a negative exploration.

Laceration of the liver results in intraperitoneal hemorrhage. This may vary in extent. With a sizable lesion, exsanguination and death may result. With minor degrees of laceration, bile peritonitis may occur. In either event the physical signs will be those of peritoneal irritation, possibly with shifting dullness, rebound tenderness, and generalized peritonitis.

Splenic rupture is a common injury and should be suspected following blows on the left flank or the left lower chest. The clinical picture is characterized by abdominal pain, pain in the left shoulder, and shock. On physical examination peritoneal irritation will usually be present, and will be most marked in the left upper quadrant. Pain in the left shoulder and dyspnea may result from diaphragmatic irritation. Diagnosis may be exceedingly difficult with minor lacerations of the spleen. Careful clinical evaluation is required.

Mild trauma may produce a *subcapsular hematoma of the spleen* which may rupture several days or weeks later with shock, intraperitoneal hemorrhage, and rebound tenderness. The problem may be difficult to diagnose. Lateral abdominal x-rays may be of help in differentiating this from retroperitoneal tumor masses. An upright film of the abdomen after the patient has drunk a carbonated beverage may reveal irregular hematoma in the gastrosplenic mesentery.

In some diseases such as malaria, leukemia, or infectious mononucleosis, splenomegaly is a prominent feature of the disease. In these patients the spleen may rupture spontaneously or following minor trauma.

Forcible compression of the small intestine may lead to laceration. The commonest location is just distal to the ligament of Treitz or in the terminal ileum, close to points of fixation. The duodenum in its position anterior to the spine is susceptible to rupture. When this occurs, however, bowel contents leak out and peritonitis develops posteriorly. The associated physical signs may be late to appear. Spasm and abdominal rigidity will be delayed.

Rupture of the large intestine is not common. However, lacerations of the large-bowel mesentery may occur, and necrosis and gangrene may result. The patient will present with considerable abdominal pain, but usually without signs of peritonitis. The diagnosis may be extraordinarily difficult to make.

INJURY OF THE URINARY TRACT

Injuries of the kidney are usually seen in association with injury of the abdominal viscera, either the blunt or the penetrating type. Flank pain and hematuria are fairly constant findings. Blood loss may be considerable but is rarely exsanguinating. Extravasation of urine may occur into the renal fossa and flank. The combination of hemorrhage and urinary extravasation may produce muscle spasm, tenderness, and flank fullness. A mass may be palpable and may even be noted on inspection. Other physical signs include ecchymosis in the flank, nonshifting dullness in the flank, and a positive psoas sign secondary to extravasation of blood and urine overlying the psoas muscle.

It may be difficult to distinguish between injury of the spleen or liver and a damaged kidney. If the patient's condition warrants and he is not in shock, an intravenous pyelogram is helpful. This examination provides information regarding the involved kidney as well as the functional state of the uninvolved side.

Bladder and urethral injuries are usually associated with fractures of the pelvis. In evaluating the possibility of this type of injury it is important to establish when the bladder was emptied prior to the accident. If the patient had not voided for some time and the bladder was known to be full at the time of accident, rupture of the bladder is much more likely than if the bladder had been empty. The passage of bloody urine following injury helps establish the diagnosis of bladder or urethral injury. If the patient successfully voids clear urine after the accident, it is safe to assume that no serious injury to the lower urinary tract has resulted. If there is evidence of injury of the bladder neck or membranous urethra, catheterization should be performed by a skilled urologist who is prepared to assume complete surgical management. Damage may be compounded by unskilled attempts to pass the catheter in presence of urethral damage.

Intraperitoneal rupture of the bladder occurs only if the bladder was full at the time of injury. Physical findings on examination consist of deep tenderness, muscle spasm, and peritoneal irritation. Rectal examination demonstrates tenderness and a normal prostate and membranous urethra. Upon completion of these initial diagnostic steps, catheterization may be performed. If bloody urine is obtained, urinary-tract damage should be suspected. A cystogram may be diagnostic.

Injury of the bladder neck or membranous urethra results in extravasation of urine into the tissues surrounding the bladder and lower abdominal wall. The extravasation extends laterally, and the area is markedly tender to gentle pressure. Rectal examination is important in localizing the area of injury. Damage to the prostatic urethra results in the presence of a boggy, tender mass which obscures the prostate gland. With laceration of the urogenital diaphragm, urine and blood extravasate into the perineum and perivesical space. These physical findings are indications for early surgical intervention.

Injuries to the lower urinary tract usually occur in conjunction with pelvic fractures. Lateral compression of the pelvis helps to make this diagnosis, although an x-ray will be helpful in determining the extent of the injury.

PERIPHERAL NERVE INJURIES

Wounds in the extremities including peripheral nerve injuries are encountered in military experience where extensive wounds of soft tissues and long bones occur. Peripheral nerve damage involves lower motor neuron axons, resulting in a flaccid type of paralysis. As a late result muscles distal to the lesion undergo atrophy, sensory loss, and autonomic changes. The skin distally becomes thinned, smooth, and pale or mottled. Sweating is absent. Fingernails and toenails become brittle. These latter signs are, however, late in appearance, and early diagnosis will depend on loss of voluntary muscle power or absence of perception of pinprick.

Some peripheral nerve injuries are commonly encountered in civilian medical practice. For example, the radial nerve may be injured with fractures of the shaft of the humerus; the ulnar nerve may be damaged in association with fractures about the elbow; common peroneal nerve involvement may be found in connection with fractures, soft-tissue wounds, or tight casts producing pressure about the knee. The sciatic nerve may be injured when dislocations or fractures of the hip occur. Lacerations about the wrist may produce damage to the median or ulnar nerves, and traction on the upper extremity may produce brachial plexus damage.

Partial damage to a peripheral nerve may be followed by a characteristic type of pain termed *causalgia*. This may develop rapidly after injury or may require several days to make its appearance. It is characterized by constant, intense, burning pain. It is made worse by moving, touching, minor trauma, excitement, and often by temperature change. The sciatic and median nerves are those most commonly involved. On neurologic examination the peripheral nerve injury is usually not complete. The skin of the involved member tends to be shiny and glossy, although not invariably so. Characteristically, anesthetic block of the related sympathetic pathways produces prompt relief of the pain. This observation is of value in diagnosis.

EXAMINATION OF THE MUSCULOSKELETAL SYSTEM FOLLOWING ACUTE TRAUMA

The circumstances under which the injuries sustained by an accident victim are first determined usually require a different examination sequence and technique from that used in diagnosing less urgent musculoskeletal problems in the outpatient clinic or private office. In a modern community, patients who may have severe injuries are usually not seen by a physician until after they have been rushed by ambulance to an emergency department. By the time the doctor sees them they have already been placed in a recumbent position on a stretcher. Following the initial evaluation only a limited degree of repositioning should be performed while x-rays or other special diagnostic studies are completed. Time may be very critical. Adequate functioning of vital organ systems must be established as soon as possible. Only then can attention be given to examination and management of other systems and regions. Frequently, no medical history can be obtained from either the patient or any other source; there is no useful information on the magnitude and direction of forces involved in the accident, and the patient may be totally unable to cooperate with the examiner.

Under these circumstances the examiner requires an examination plan which can be executed swiftly, almost instinctively. It should first identify malfunction of the most critical organ systems, then determine the existence of any significant injury, and finally clarify the details of any injury. It should involve minimum handling of the patient and allow for the concomitant initiation of urgent therapeutic measures. Such an examination sequence will be described in the last part of this chapter.

Fortunately, the symptoms and signs of acute bone or joint injury are few and easy to detect. Also, although bone and joint injuries involve adjacent soft tissues, the anatomy of each region is conducive to unique patterns of injury which are easily identified by the experienced examiner. For example, the shoulder may sustain either a fracture through the humeral neck, with associated soft-tissue injuries, or a scapulohumeral dislocation; similarly, the elbow may sustain either a supracondylar fracture or an ulnohumeral dislocation.

For quick determination of precisely which injury has occurred, routine x-ray studies are more reliable than the physical examination. Even the most experienced clinician cannot be certain of all details without x-ray studies. Therefore, attempts to distinguish injuries within a given region by lengthy clinical examination should be avoided, especially when the examination requires diagnostic manipulations which might further injure the soft tissues. In the initial survey, on the other hand, a good screening physical examination is more reliable than x-ray films in disclosing regions of injury. X-ray studies should not be performed, therefore, until all suspicious regions have been identified.

Frequently, the treatment of the injuries should be started during the course of the examination. As soon as a presumptive diagnosis of a fracture or dislocation is established, the involved region should be immobilized by splinting in order to minimize further damage to the soft tissues during further examination and x-ray procedures. These splints should not be disturbed either during the x-ray examination or afterward, until definitive treatment is started.

Positioning the Patient

For the initial examination, the accident victim should, if possible, be lying supine on a stretcher. Since moving an injured patient is hazardous, the stretcher frame should be detachable, so that the patient can be lifted to a table and x-rayed or operated upon while he is still on the stretcher (Figure 19-2). Although no single position allows complete inspection of the patient, full supination offers the greatest latitude for both examination and treatment. Turning an accident victim onto his back without regard for a possibly broken spine or other serious fracture is, of course, fraught with grave risks. On the other hand, if his condition requires immediate airway clearance or closed-chest cardiac massage, there is no choice but to turn him.

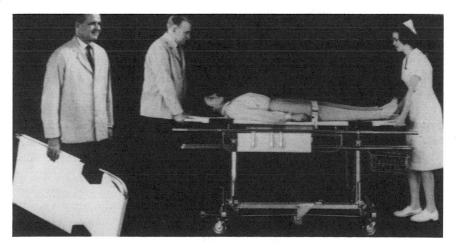

FIGURE 19-2. Stretcher with removable top to facilitate x-ray studies without disturbing fractures (by permission of Stryker Corporation). (Figures 19-2 through 19-10 from Schmeisser, G., Jr. The Initial Management of Fractures. In W. F. Ballinger II, R. B. Rutherford, and G. D. Zuidema [Eds.], *The Management of Trauma* [2nd Ed.]. W. B. Saunders Company, 1968.)

In less critical cases the patient should be questioned for location of pain and his limbs quickly inspected for evidence of deformities which might indicate fractures or dislocations. He should be asked to move his fingers and toes; the presence or absence of paraplegia or quadriplegia is thereby established. If the patient has pain in his neck or back, the presence of a vertebral injury should be presumed until confirmed or refuted by x-rays.

If the findings suggest a vertebral injury, it is possible to turn the patient, if necessary, by straightening his back or neck and carefully avoiding any bending or twisting. The patient should be rolled with someone holding his head and turning it in harmony with the rest of his body, with care to see that his arms are extended along the sides of his body. It is remotely possible that straightening an injured vertebral column may aggravate an injury, but this possibility must be accepted since completion of the examination and treatment ultimately require a supine position.

If a limb bone injury is suspected, the patient should be turned with an attendant supporting the injured limb and turning it in gentle harmony with the rest of the body. Although splinting a limb prior to turning has been advocated, splints are most easily applied with the patient supine and are of relatively little value in maintaining fracture alignment during turning. After turning has been accomplished, splinting will afford protection during further maneuvers.

General Types of Fractures

Fracture and *break* are synonymous terms, classically defined as "a dissolution of continuity of a solid structure," for example, a bone. If there are more than two fragments, the fracture is said to be *comminuted.* Self-explanatory terms such as *transverse, oblique, spiral, T, Y* are frequently used to describe the orientation of a fracture line. A fracture line may enter one side of the bone and then divide and extend across the remainder of the bone in two diverging branches, creating an extra piece of bone imaginatively denoted as a *butterfly* fragment.

When a long bone is subjected to bending forces, the portion which is under the greatest tension stresses usually breaks apart first. This type of break is similar to that of a broken piece of chalk. The fragments can move apart easily. In the body, the surrounding soft tissues restrain the fragments from displacement. Displacement and deformity are relative to the degree of soft-tissue disruption, particularly of the periosteum. In adults the periosteum is thin; therefore, widespread displacement and severe deformity are likely. In children the periosteum is thick and is frequently torn on only one side of the bone, thereby allowing only angular deformity with little or no displacement. In either age group, if the periosteal disruption is sufficiently severe, the ends of the fragments may displace. If the fracture line is oblique or spiral, or if sufficient lateral displacement occurs in a transverse fracture, some overriding of the fragments and consequent shortening of the limb are likely. One fragment may twist with respect to the other (*malrotation*). Any degree of angulation may occur. The ultimate position of the fragments and the external appearance of the limb are determined by the effects of neighboring muscles as well as by the effects of gravity and the direction and magnitude of the fracture force. Although shortening of a limb is commonly observed immediately after a fracture, lengthening almost never occurs.

Since children's bones are less brittle than adults', only one side of the bone may pull apart; the other side bends. This *incomplete fracture* is referred to as a *greenstick* fracture. It exhibits only an angular deformity. Occasionally, when a child's long bone is subjected to a bending or an axial compression force, it may buckle on the compressed side of the bone. This usually occurs in the flared (metaphyseal) area. The opposite cortex of the bone appears normal on an x-ray and the limb exhibits no deformity, merely localized pain and tenderness to palpation. This type of *incomplete* fracture is referred to as a *Torus* fracture.

Fractures near the ends of children's long bones have special significance. If a fracture either crushes or splits an epiphysis and its underlying epiphyseal (growth) plate, a growth disturbance may occur. More frequently, fractures in this area pass transversely along the metaphyseal side of the epiphyseal plate and then deviate into the metaphysis. The terminal fragment then consists of the epiphysis with the epiphyseal plate and a small piece of metaphysis. Although displacement of this composite fragment may be wide, little or no damage has been done to the epiphyseal plate, and an ultimate growth disturbance is unlikely.

If there is a wound extending from the skin surface to the fracture site, there is grave risk of infection of the injured bone regardless of whether this skin is pierced by a bone fragment from within or by a foreign body from without. Such fractures are termed *open* or *compound.* It is essential that any skin wound, however small, which might communicate with the fracture site be recognized and promptly treated. The blood in a fracture hematoma quickly darkens and contains visible globules of fat from the marrow. A skin wound in the neighborhood of a fracture which oozes fluid of this type is therefore one which communicates directly with the fracture. Any fracture, no matter how comminuted, which does not communicate with a skin wound is referred to as a *closed* or *simple* fracture. Because of frequent confusion in the designations *compound* and *simple,* the terms *open* and *closed* are now preferred in standard usage.

Small bones, such as those of the spine, ankle, and wrist, may be split or crushed by the forces of trauma. Displacement of the fragments is usually minor compared with that of fragments of long bones. Deformity is seldom evident if the bones are deep beneath the skin, as in the case of vertebral fractures; however, because of their proximity to the spinal cord and nerve roots, even minor displacements of vertebral fractures may produce signs of nerve damage. Fractures of small bones in the ankle and wrist seldom exhibit deformity unless displacement is severe, but they may alter the range of motion of associated joints, as indicated by pain on local palpation or attempted motion. Fractures extending into joints cause bleeding into the joints. Motion is painful and if the joint is a superficial one, such as the knee or elbow, swelling eventually occurs.

In summary, the cardinal symptom of a fracture is pain, which is usually most severe at the fracture site. Tenderness may be evoked by palpation directly over the fracture or by moving the bone fragments. The former technique is safe and reliable; the latter is hazardous. Motion of the fragments is also likely to cause audible or palpable crepitus. The patient may be able to report that he felt or heard his bone snap at the time of injury and has felt the grating of fragments when he has been moved. The examiner should not attempt to elicit crepitus. Instability is usually obvious, but occasionally a fracture may be quite stable, and the fact that a patient can use his injured limb does not rule out fracture. Visible or palpable deformity may or may not be present. Swelling, ecchymosis, and increased local heat occur to a variable extent with all fractures, but these are late signs.

General Types of Dislocations

If the articular surface of one bone is totally displaced from the articular surface of its partner, the joint is *dislocated.* A joint may be dislocated and spontaneously reduce itself, or the bones may be trapped by the surrounding structures and special maneuvers required for reduction. A *subluxation* is a partial dislocation; contact between the articular surfaces is less than normal. This term is best applied to certain malformed joints and is seldom applicable to traumatized joints. In trauma the joint either is or is not dislocated; there are no intermediate possibilities.

As with a fracture, the cardinal symptom is pain, and there is local tenderness to palpation or motion. The range of motion is altered and usually reduced; in fact, the joint seems locked in an abnormal position. The examiner should not attempt to elicit motion but should take the word of the patient on this point. Deformity of adjacent structures is evident in a superficial joint. The posture of the limb distal to the joint is usually abnormal. Dislocation of some joints, such as the shoulder, increases the overall length of the limb. Swelling, ecchymosis, and heat are late signs. Ideally, diagnosis should be established and the dislocation reduced before these have been allowed to develop.

X-rays should not be delayed in an effort to establish the details of a joint injury by physical examination, since films are far more reliable. Two points, however, should be quickly determined: the presence of a wound into the joint, since an *open* or *compound* dislocation is as vulnerable to infection as an open fracture, and the presence of vascular or neurologic insufficiency in the limb distal to the injury.

Ligaments, Tendons, and Muscles

The ligaments of interest in this chapter are those which span joints and thus stabilize and control the motion of the involved bones. In the joints of the arms and legs, motion is possible primarily within a single plane; strap ligaments on either side of the joint allow very limited lateral or medial motion (abduction and adduction). With excessive leverage improperly applied to the joint, a ligament may be ruptured, torn transversely, or separated from one of its points of attachment. In any case, the point of maximum pain and tenderness to digital palpation coincides with the point of injury. If the most tender spot lies over ligament rather than bone, a ligament injury is more probable than a fracture. If x-rays at this time indicate that no fracture is present, it is safe to test the ligament for partial or total disruption by carefully abducting or adducting the joint. If a ligament is totally divided, and if the patient is able to relax his muscles, excessive motion will be found in the joint. Such an examination is sometimes too painful for a patient to endure, and either an anesthetic must be administered or the joint allowed to "cool off" for a couple of weeks in some form of protective immobilization such as a cast. *Strain* and *sprain* are words frequently used imprecisely. *Strain* is correctly used to refer to a stretch or partial tear; *sprain* is correctly used to refer to a total tear or avulsion. It should be recognized that a dislocation cannot occur without serious tearing or avulsion of ligaments.

A tendon forms the connection between muscle and bone, converting the contraction of the muscle into motion of the bone. Tendons, musculotendinous junctions, and occasionally even muscles, can be torn apart if the bone is prevented from moving or is forced to move in opposition to a strong muscle contraction. In children, avulsion of a piece of bone at the point of tendon attachment is more common than a tear in the tendon or musculotendinous junction. The opposite is true in adults. Common examples of tendon tears in adults occur in the tendons of the long head of the biceps, the tendon of the supraspinatus, and the tendoachilles of the calf. These injuries are identified by correlating the site of pain with weakness or absence of function in a specific muscle. There may be a palpable soft-tissue defect at the point of pain and enlargement of the muscle indicating its uncontrolled recoil. X-rays which are taken for soft-tissue detail can sometimes confirm a clinical impression of tendon tear.

Arteries

Fractures or dislocations may impair distal arterial blood supply by compressing or lacerating the artery at the site of injury. This occurs most frequently at the knee and elbow but can occur with bone or joint injuries in many other regions. Therefore, the quality of the peripheral circulation in all four limbs should always be clearly determined and recorded during the initial examination, before any splint is applied or x-rays taken. Early signs of arterial insufficiency are lowered skin temperature, absence of wrist or foot pulses, and pallor. Later signs include hypesthesia and paralysis. At the site of arterial injury there may be a large and pulsatile swelling, indicating massive extravasation of arterial blood; however, this is a relatively infrequent complication.

Nerves

Wherever nerves lie close to bone, fractures can injure them. The region of greatest concern, of course, is the vertebral column. Elements of the lumbosacral plexus are occasionally damaged by fractures of the pelvis. A significant percentage of posterior dislocations of the hip damage the sciatic nerve and paralyze the dorsiflexors of the foot (foot drop). Severe knee injuries and fractures of the upper end of the fibula may damage the peroneal nerve, producing a foot drop. Fractures and dislocations of the shoulder joint are occasionally associated with injury to the axillary nerve resulting in weak abduction. Fractures of the shaft of the humerus, especially at the junction of the middle and distal thirds, may injure the radial, median, or ulnar nerves; radial palsy signified by inability to dorsiflex the wrist (wrist drop) is most common. Elbow injuries may be complicated by the same types of nerve injuries. At the wrist, compression of the median nerve in the carpal tunnel, with resulting numbness over the first two fingers, is common. To determine whether nerves have been injured, the motor and sensory function distal to any fracture must be evaluated and recorded on initial examination.

Specific Fractures and Dislocations

Vertebral Fractures and Dislocations. Paralysis and sensory loss signify spinal cord or nerve root damage with a vertebral injury. Analysis of the level of motor or sensory loss will reveal the level of vertebral injury. Charts of sensory dermatomes and muscle innervations are helpful but are not necessary if the examiner has memorized the nerve supply to a few key regions and muscles. If the patient is comatose or stuporous, the determination is difficult but not impossible. Usually such a patient will move those regions which are not paralyzed in response to painful stimuli. If coma is profound, normal muscle tone as well as response to painful stimuli may be absent. Usually, however, if one foot is raised and dropped directly over the other, the descending heel will not strike the other foot unless the falling limb is paralyzed. The upper limb of an unconscious patient can similarly be tested by dropping the patient's hand toward his face. When innervation is intact, the falling hand will usually either deviate to the side or decelerate just before impact.

Pain in the back or neck of an accident victim should be considered an indication for spinal x-ray. The film should include several vertebrae above and below the painful area, since pain is frequently referred below the level of injury and occasionally above. Palpation of the spinous processes for an area of maximum tenderness may facilitate precise localization of injury but should not be performed if it requires turning the patient.

Because of the deep position of the vertebral column in the body, external or palpable deformity is rare. Either pain or neurologic deficit is adequate evidence for diagnostic x-rays, and the examiner should not look for deformity or abnormal vertebral motion.

X-rays may show any of the following types of vertebral injuries occurring individually or in combination: subluxation, body fracture, unilateral or bilateral facet joint dislocation, or fracture of the arch, odontoid, spinous process, or transverse process (Figure 19-3).

Subluxation of an upper vertebra on a lower one is best seen in a lateral projection of the cervical spine in a neutral or forward flexed position. The interspinous ligament and sometimes the capsular ligaments of the posterior facet joints have been torn by the upper vertebra traveling too far forward on the lower, and sometimes the intervertebral disc space is narrowed. This condition represents a sprain, and is an example of the so-called whiplash injury.

A vertebral body fracture is usually caused by compression forces, and the consequent reduction in height of the involved vertebral body is most conspicuous on lateral x-rays. Frequently, pieces of the centrum are displaced and may impinge on the spinal cord. Compression of a thoracic vertebra into a wedge with the narrower portion anterior, and no resulting neurologic deficit, is a common finding in middle-aged and elderly persons with osteoporosis.

Dislocated posterior facet joints are rare except in the cervical spine. When unilateral dislocations occur, the vertebrae are locked in slight malrotation. For this reason the anteroposterior x-rays show a shift in alignment of the spinous processes, with those above the dislocation shifted about ¼ inch toward the side

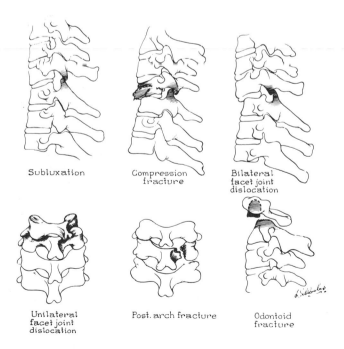

Subluxation   Compression fracture   Bilateral facet joint dislocation

Unilateral facet joint dislocation   Post. arch fracture   Odontoid fracture

FIGURE 19-3. Common types of vertebral injuries.

of the dislocated facet joint. No such rotation occurs with bilateral facet joint dislocations, but the extreme forward shift of the upper on the lower vertebra is likely to cause severe spinal-cord injury, in contrast to the minor neurologic deficit following a unilateral dislocation. The articular processes of dislocated facet joints can best be seen in lateral or oblique projections.

A posterior vertebral arch may sustain a fracture through the lamina, best visualized on an anteroposterior or oblique x-ray as a faint line near an articular facet. Usually when a ring of bone such as a vertebra is broken in one place, there is a second fracture through another part of the ring, which may be difficult to demonstrate on the x-ray. These fractures should be strongly suspected when there is a slight lateral shift of one spinous process. They may be associated with any degree of neurologic deficit. They may result from a hyperextension injury or a direct blow, and are likely to be associated with significant instability of the vertebral column. Hyperextension of the spine should be carefully avoided in handling these cases.

An odontoid fracture represents a perilous injury in view of potential instability of C-1 on C-2 and the limited space available to the spinal cord if displacement occurs. When cord injury does occur at this level, there may be immediate paralysis of all respiratory muscles and quick death. This fracture is usually best seen on an open-mouth anteroposterior view as a line across the waist or base of the odontoid. The overlying shadow of an upper incisor tooth frequently obscures the true outline of the odontoid in a manner which simulates a fracture. Congenital malformation of the odontoid with incomplete ossification may also confuse the findings. For a reliable diagnosis the x-ray findings must be correlated with the nature of the injury and physical symptoms.

The distal tip of a spinous process, frequently at C-7 or T-1, may be the site of fracture. These processes are easily palpated and are very tender if fractured. If there are no other vertebral fractures, the stability of the column is not compromised. These fractures are usually best seen on lateral x-ray projection.

Transverse processes on the lumbar vertebrae serve as points of attachment for the psoas muscles and may be fractured by avulsion. Retroperitoneal hemorrhage may occur, producing abdominal tenderness. The lateral edge of the psoas muscle may be obscured on the x-ray. The fractured transverse processes are best seen on an anteroposterior x-ray of the lumbar spine. Congenital malformation with incomplete ossification is common and may lead to a false diagnosis of fracture. Structural stability of the vertebral column is not impaired.

Pelvic Fracture. When evaluating the possibility of fracture, the examiner should think of the normal pelvis as a symmetric ring of bone with various projections serving as points of attachment for muscles. The projections may be sheared off by a direct blow or avulsed by strong muscle action. An example of a shearing injury is a fracture of the iliac crest caused by impingement of the handle-bar grip of a motorcycle against the wing of an iliac bone as the rider is hurled forward in a collision; local contusion and loss of normal pelvic contour are apparent. An example of avulsion is the fracture of an ischial tuberosity by a sprinter as he forcibly contracts his hamstrings while pushing away from the starting blocks at the beginning of the race. Pain and tenderness at the point of a specific muscle attachment and on contraction of the involved muscle are diagnostic signs (Figures 19-4 and 19-5).

Because of the elasticity of the symphysis pubis, the pelvic ring may be broken at only one point, but simultaneous fracture in two separate areas is more likely, and also more serious because of the resulting instability. Weak areas are present near the sacroiliac joints, at the ischiopubic junctions, and through the symphysis pubis. A fracture through one sacroiliac joint and the ischiopubic junctions on the same side (Malgaigne fracture) creates a large lateral fragment to which the entire lower limb is attached. This fragment is frequently displaced proximally by the trunk muscles. It may be tilted medially or laterally. Pelvic asymmetry and shortening of the lower limb are present and easily detected. Motion on gentle compression or distraction of the iliac crests is a conclusive finding of this injury. A fracture through the ischiopubic junctions bilaterally creates an anterior fragment which may be associated with bladder rupture. If the symphysis pubis can be moved by gentle pressure, the diagnosis of this fracture is confirmed. This injury is frequently caused by falling astride some large, rigid object. Contusion of the perineum may be present and the

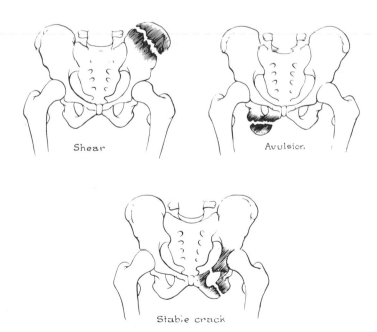

Shear

Avulsion

Stable crack

FIGURE 19-4. Benign fractures of the pelvis.

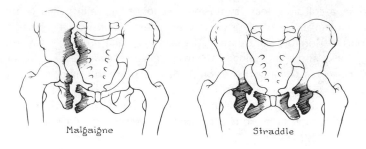

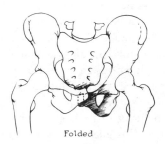

FIGURE 19-5. Unstable fractures of the pelvis.

scrotum becomes distended with blood. In any of these injuries the abdomen may be tender to palpation and exhibit signs of retroperitoneal hemorrhage. Various branches of the lumbosacral plexus may be injured, producing hypesthesia or muscle weakness.

Hip Joint and Upper Femur. Among the types of fractures sustained by the upper femur are those with stability adequate to permit ambulation initially but which ultimately separate with unpleasant consequences. Significant events in the history are a fall or misstep followed by pain in the hip. Objective physical findings may be entirely negative. Good quality x-rays are necessary to establish the diagnosis, which may be that of a crack from greater to lesser trochanter, or a fracture across the neck of the femur just beneath the impacted femoral head. Although a fracture through the upper femur (broken hip) may occur at any age, it is more common among elderly persons, especially women, and more especially those with osteoporosis. Typically, the limb lies in external rotation. The distance between knee and iliac crest is reduced and any motion of the hip joint is painful.

The fractures are classified according to their anatomic location as determined by x-rays. Proceeding from proximal to distal, commonly used terms are *subcapital, high cervical, midcervical, low cervical, intertrochanteric,* and *subtrochanteric.* Since the capsule of the hip joint is attached to the femur near the base of the neck, fractures proximal to this line are *intracapsular* and those distal to it are *extracapsular.* This is an important point, since most of the blood to the head of the femur is supplied by vessels which approach the femur in the capsule, travel along the surface of the neck, and finally enter the bone in the subcapital area. Blood supply to the femoral head may therefore be interrupted by displacement of an intracapsular fracture but not by an extracapsular one (Figure 19-6).

Fractures of the acetabulum are frequently caused by forces which drive the femoral head into it. The femoral head may or may not be displaced medially and/or cephalad with the acetabular fragments. This is termed a *central fracture dislocation* of the hip. Frequently, however, this injury can be distinguished from an intracapsular femoral fracture only by x-rays.

If a person is in a sitting position with his hip in flexion and adduction at the moment of impact of a force on his knee, the femoral head may be driven over or through the posterior rim of the acetabulum and come to rest against the sciatic nerve just behind the socket. This is known as a *posterior dislocation.*

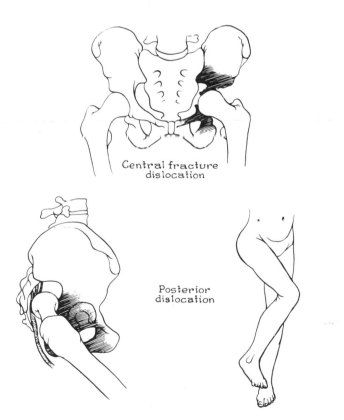

Central fracture
dislocation

Posterior
dislocation

FIGURE 19-6. Central fracture dislocation of hip and posterior dislocation of hip.

Paralysis of foot dorsiflexors (foot drop) from the sciatic nerve injury is common. The posture of a patient with this type of dislocation is unique. He lies with his involved limb in adduction, flexion, and internal rotation. He prefers to lie on his uninjured side and resists any attempt at normal positioning of his leg. Two other relatively uncommon dislocations of the femoral head occur, one into the *obturator* foramen, and the other into the region of the *inguinal* ligament. The former is an inferior dislocation and the latter, an anterior one. Both of these are caused by excessive abduction. A patient with an anterior dislocation lies in moderate abduction and external rotation and the head is palpable beneath the inguinal ligament. A patient with an inferior dislocation lies in extreme abduction and external rotation and the head is not palpable.

Shoulder Joint (Scapulohumeral) and Upper Humerus. The shoulder is designed to facilitate placement of the hand in an unlimited number of positions in space. This is accomplished by a very shallow socket (glenoid) in the head of the scapula, articulating with the head of the humerus. The clavicle has a sinusoidal curve and is commonly fractured by a compression force applied longitudinally. Such a force may be applied directly by a blow on the tip of the shoulder or indirectly by a fall on the outstretched hand. This injury is one of the commonest fractures in children and also in adults. The sternoclavicular joint can be disrupted by a similar force, with the result that the medial end of the clavicle dislocates medially and behind or in front of the sternum. An abrupt force downward on the tip of the shoulder can dislocate the acromioclavicular joint. Any of these injuries is easily detected by inspection and palpation, since the structures lie close beneath the skin. Dislocation of the medial end of the clavicle behind the sternum may create a sensation of airway obstruction due to pressure on the trachea. Clavicular fractures are frequently associated with a sizable subcutaneous hematoma and, rarely, with injury to elements of the brachial plexus and even the apex of the lung (Figure 19-7).

The common "dislocated shoulder" is an anterior or anteroinferior scapulohumeral dislocation. When the humerus is abducted to the limit of its normal range of motion, it impinges against the outer edge of the acromium. If abduction is forced beyond this limit, the head of the humerus is levered over the anteroinferior edge of the rim of the glenoid. The patient is then unable to bring his arm in against his side or to rotate it internally. This type of dislocation frequently recurs. The first episode is extremely painful, but each recurrence is less painful, and the patient may learn to reduce the dislocation himself. The patient's appearance is unique. The deltoid bulge is flattened and the acromium unusually prominent. The distance between the acromium and olecranon on the afflicted side is greater than on the opposite side. Sometimes it is possible to palpate the humeral head high in the axilla.

Fractures of the head or neck of the humerus can occur from a fall on the outstretched hand. They may be found in any age group but are more common in the elderly. Any active motion of the humerus is painful. In contrast with

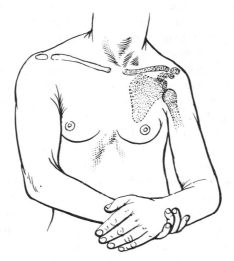

FIGURE 19-7. Anterior dislocation of scapulohumeral joint.

dislocations, the deltoid bulge is accentuated by displacement, or overriding of the fragments, and by accumulation of extravasated blood. The distance between acromium and olecranon is unchanged or decreased.

Although these injuries can be identified by physical examination, this should not be done if adequate x-ray facilities are available. The initial examination of a specific area should be limited to determining that a skeletal injury exists; x-rays of that area are then indicated as soon as all other areas have been checked.

Shaft of Femur and Humerus (Single-Bone Segments). Since the thigh segment of the lower limb and the arm segment of the upper limb each contain a single long bone, fracture of the shaft of that bone creates such conspicuous instability of the segment that diagnosis is usually obvious. Palpation of the bones may be difficult because of the thick muscles in these areas, but motion at the fracture site is revealed by any attempt to move the limb. The tone of the muscle tends to cause overriding of the fragments and therefore a decrease in length of the segment. If overriding is present or a hematoma has accumulated, the girth of the limb is enlarged. The ends of the sharp fragments may injure the neighboring soft tissues. In the femur, one or more of the deep veins may be torn, causing internal hemorrhage. The fractured humerus frequently damages the radial nerve, producing a wrist drop. Open fractures of these bones are especially serious injuries, and a careful circumferential check of the skin for any wound should always be made.

Knee and Elbow. In the knee and elbow, bone fragments may easily be displaced; therefore, no diagnostic manipulations should be performed on a deformed or painful joint until x-rays have been examined and found negative.

These joints link single-bone segments with double-bone segments; thus, each articulation involves three long bones. The motion of both joints is primarily within a single plane of flexion and extension. Each joint has a bony prominence which increases the mechanical advantage of the extensor muscle. For the knee, this prominence is the patella; for the elbow it is the olecranon. The knee, which is subjected to high compression forces in many positions, has two semilunar wedges of cartilage, the menisci, which function as lubrication wedges, shock absorbers, and shims. Neurovascular structures lie close to the flexor aspects of distal humerus and femur (Figures 19-8 and 19-9).

Fractures which jeopardize the arteries in these areas occur just above, between, or through the condyles, and are termed *supracondylar, intracondylar,* and *condylar,* respectively. Such fractures of the humerus are especially common in children. One of the fragments, usually the distal, may be tilted by the flexor muscles toward the artery. Deformity of the elbow may range from insignificant to severe. Sometimes it may simulate a "gun stock," with medial or lateral deviation, a fullness in the upper part of the antecubital space, and a depression posteriorly just above the olecranon. Supracondylar fractures of the femur are more common in adults. There is an increase in the circumference of the thigh just above the knee. The fracture causes hemorrhage into the knee joint with consequent swelling. The distal end of the shaft fragment may penetrate the skin near the patella. In view of the strong possibility of arterial injury, circulation to the limb beyond the fracture must always be verified.

Both the olecranon and the patella are vulnerable to fractures from direct blows and from avulsion by muscle action. The former fractures are more often comminuted.

The patella can dislocate to the lateral side of the knee, a condition which is easily diagnosed by inspection and palpation. Dislocations of the tibia on the femur, or vice versa, cause extensive injuries, usually accompanied by neurovascular injury. Deformity is severe and the diagnosis self-evident.

An unfractured olecranon can dislocate only in association with the rest of the ulna, which may be displaced posteriorly and sometimes laterally. The deformity seen with this posterior dislocation resembles that of a supracondylar fracture of the humerus.

If either the elbow or knee is subjected to excessive abduction or adduction stress, either the ligament fails on the side subjected to tension forces, or the

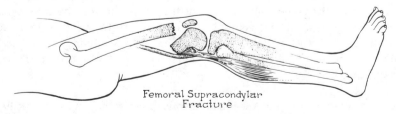

Femoral Supracondylar
Fracture

FIGURE 19-8. Injury to femoral artery by supracondylar fracture of femur.

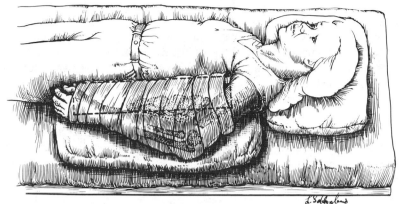

FIGURE 19-9. Supracondylar fracture of humerus.

bone crumbles on the side subjected to compression forces. Such fractures in the knee are called tibial plateau or tibial condyle fractures. In the elbow the counterpart of such an injury is a fracture of the radial head or neck. These fractures produce pain at the fracture site, pain on joint motion, and joint swelling. Ligament and meniscus injuries produce similar symptoms. Fractures should always be ruled out by x-rays before subjecting a joint to abduction or adduction stresses in order to ascertain ligament rupture.

If a meniscus is torn, the free part of it may become trapped between the femur and tibia and obstruct joint motion. This obstruction may be unyielding or it may be limited and variable. To detect the obstruction it may be necessary to flex and extend the joint with it twisted internally or externally and abducted or adducted. Such manipulation should never be performed if the possibility of fracture exists. A small area of acute tenderness to palpation is frequently present at the joint line close to a meniscus tear. Joint swelling is likely with this or any other knee injury and may, of itself, limit joint motion.

Monteggia's Fracture Dislocation. This injury consists of a fracture of the ulna and a dislocation of the proximal end of the radius. It is usually caused by a blow on the ulnar side of the forearm near the elbow. The force breaks the ulna and then acts on the radius. If the radius does not break, the proximal radioulnar articulation is pulled apart. The annular ligament is broken and the proximal end of the radius is displaced from its normal position of articulation against the lateral condyle (capitellum) of the humerus. The existence of this injury may be presumed if a fracture of the ulna is identified and there seems to be pain or limitation of motion of the elbow. Diagnosis is confirmed by x-rays of both the elbow and forearm.

Leg and Forearm (Double-Bone Segments). The thin soft-tissue covering of the bones in the leg and forearm facilitates detection of an undisplaced fracture by palpation for point tenderness directly over the bone. This technique is especially valuable in establishing the possibility of an incomplete fracture of the Torus or greenstick type. Only the radius cannot be palpated throughout its length. In the proximal one-third of the radius, where this bone is covered by the extensor muscles, a fracture should be presumed if tenderness is produced by pronation or supination.

Instability may not be conspicuous if only one of the bones is broken or if both bones are broken at different levels.

Fracture of both bones in a double-bone segment is more common than fracture of only one bone. Since these fractures may occur at a considerable distance from each other, x-rays should always be taken to include the entire length of the segment.

Colles', Barton's, and Smith's Fractures (Figure 19-10). Falling on a dorsiflexed and outstretched hand may fracture the radius just proximal to the wrist joint. The distal fragment is pushed proximally and may override or impact on the proximal one. The ulnar styloid process may be avulsed, but the rest of the ulna remains intact. The hand is forced into radial deviation. The distal radial fragment is usually tilted backward and the volar edge of the end of the shaft fragment crowds the flexor tendons or median nerve where they enter the carpal tunnel. The volar side of the wrist is abnormally prominent. Abraham Colles described the deformity associated with this particular fracture as resembling a

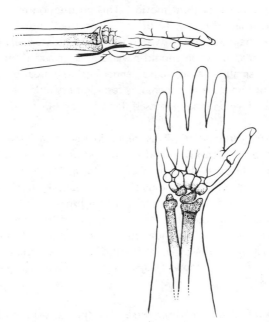

FIGURE 19-10. Colles' fracture.

dinner fork. Sometimes the main mass of the radius remains intact and only the prominent dorsal margin of the articular surface is broken off, permitting the carpus to dislocate posteriorly. This fracture-dislocation is called a Barton's fracture. The clinical deformity is similar to a Colles' fracture. A fracture which might be defined as a Colles' fracture, except that the distal radial fragment is tilted toward the volar rather than the dorsal side, is called a Smith's fracture.

Ankle Injuries (Pott's Fracture). As with the shafts of the tibia and fibula, the thin, soft-tissue covering of the malleoli facilitates detection of an undisplaced fracture by palpation for point tenderness directly over the bone. When maximum tenderness to palpation is over the medial or lateral ligaments or anterior capsule, the injury is more likely a sprain than a fracture. No other diagnostic manipulation should be performed until fractures are ruled out by x-rays. X-rays may confirm that one malleolus or both is broken. In association with these fractures, the talus may be displaced medially or laterally. Sometimes the posterior articular margin of the tibia is also broken and displaced proximally (trimalleolar fracture). If the posterior fragment includes more than one-quarter of the articular surface of the tibia, the talus also dislocates posteriorly. Pott's fracture is a generic term for ankle fractures in general.

Foot and Hand. Whereas displaced fractures or dislocations of the bones of the feet and hands are easy to recognize on inspection, x-rays are necessary to identify certain undisplaced fractures, especially those of the carpal and tarsal bones. Fractures of the calcaneus are usually caused by a blow on the bottom of the heel, as when an accident victim strikes the ground in a standing position after falling from a ladder or scaffold. Vertebral compression fractures are commonly associated with this injury and should not be overlooked.

Fist fights result in broken knuckles and Bennett's fractures. The involved knuckle is usually depressed. The fracture is through the metacarpal neck, and the metacarpal head is tilted down into the palm. If a first metacarpal is driven proximally with excessive force, a fracture dislocation of the metacarpo-multangular joint is sustained. This is called a Bennett's fracture. The thumb is shortened, and pain is greatest at its base. Thumb abduction and extension are painful and limited.

A blow sustained on the tip of an outstretched finger may forcibly flex the tip before the extensor tendon can be relaxed. The insertion of the extensor tendon is avulsed and the patient is unable to lift the tip of his finger.

Emergency Examination Sequence

C ardio
R espiratory
A bdomen
S pine
H ead

P elvis
L imbs
A rteries (peripheral)
N erves (peripheral)

The CRASH PLAN mnemonic facilitates examination of an accident victim with injuries of unknown extent in a sequence which follows a descending order of priority, is easily remembered, requires minimum manipulation of the patient, is sufficiently complete to detect any significant injury, and is capable of expansion or acceleration as individual circumstances may require.

Evaluation of cardiorespiratory function may be started as one walks toward the patient. If he is talking, he is doubtless circulating oxygen to vital centers, and consideration of immediate cardiac compression or forced pulmonary ventilation may be abandoned. Upon arrival at the patient's side, vital signs are checked and details of pulmonary exchange, cardiac function, and thoracic injuries noted. Attention is then given to examination of the abdomen in search for signs of a ruptured viscus or hemorrhage. By this point in the examination there will be some indication of appropriate therapeutic and diagnostic steps, such as tracheal intubation or tracheotomy, intravenous infusion, central venous pressure monitoring, electrocardiography, gastric intubation, and catheterization.

After evaluation of the abdomen and while life-supportive measures are being instituted, attention should be directed toward the central nervous system as signified in the mnemonic by reference to the spine and head. The presence or absence of pain along the spine and any gross motor or sensory loss is determined. The state of consciousness should be noted and the head inspected for evidence of fractures.

The possibility and extent of a pelvic fracture should be established by the technique described earlier in this chapter. The limbs should then be inspected for fractures, dislocations, or other serious injuries, and any lacerations noted for subsequent treatment.

It is particularly important when limb bone fractures might be present not to conclude the emergency examination without verifying the arterial pulses, or at least the adequacy of circulation in the hands and feet.

At least one of the functions of each of the major peripheral nerves should be checked and the result recorded. The following tests are easily applied and interpreted: The patient is asked in sequence to abduct his shoulder (axillary nerve), flex his elbows (musculocutaneous nerve), dorsiflex his wrists (radial nerve), lift his thumbs up away from his palms (median nerve). Since standard muscle tests for ulnar nerve function are easily misinterpreted, detection of light touch or pinprick over the ulnar aspect of the little fingers is a desirable alternative.

A similar test sequence for the function of the three major peripheral nerves of the lower limbs is performed by asking the patient to extend his knees (femoral nerve), dorsiflex his toes (peroneal nerve), and plantar flex his toes (tibial nerve).

A quick check for detection of light touch and pinprick over both upper and lower limbs indicates the extent of a suspected major nerve lesion.

REFERENCES

1. Cornell, W. P., Ebert, P. A., and Zuidema, G. D. Preliminary report: X-ray diagnosis of penetrating wounds of the abdomen. *J. Surg. Res.* 5:142, 1965.
2. Schmeisser, G., Jr. The Initial Management of Fractures. In Ballinger, W. F., II, Rutherford, R. B., and Zuidema, G. D. (Eds.). *The Management of Trauma* (2nd Ed.). Philadelphia: Saunders, 1973.

NERVOUS SYSTEM

Robert D. Currier

GLOSSARY

Anarthria Inability to articulate words (complete voicelessness) due to disease of the nerves or muscles of speech.

Aphasia Loss of the power of expression by speech, writing, or signs, or of comprehending spoken or written language, due to injury or disease of the higher cerebral centers concerned with expression and comprehension.

Ataxia Failure of muscular coordination; irregularity of muscular action.

Atrophy A wasting away of or diminution in the size of a muscle.

Bulbar palsy Paralysis and atrophy of the muscles of lips, tongue, mouth, pharynx, and larynx due to degeneration of the nerve nuclei of the floor of the fourth ventricle or of the motor nerves originating from them.

Dysarthria Imperfect articulation in speech.

Dysmetria Disturbance of the power to control the range of movement in muscular action; a sign of cerebellar dysfunction.

Dysphagia Difficulty in swallowing.

Dysphasia Incomplete degree of aphasia.

Dyssynergia Disturbance of muscular coordination; a sign of cerebellar dysfunction.

Extinction (suppression) Failure to perceive one of two identical bilateral simultaneous stimuli; if sensory perception is otherwise intact, extinction or suppression may indicate parietal cortex dysfunction.

Fasciculation Visible spontaneous contraction of a number of muscle fibers supplied by a single motor nerve filament.

Graphesthesia Sense by which figures or numbers written on the skin are recognized.

Homonymous hemianopsia Loss of vision in the nasal half of the visual field in one eye and the temporal half in the other.

Horner's syndrome Sympathetic paralysis causing ptosis of the upper lid, constriction of the pupil, and decreased sweating, all on one side of the head and face.

Lower motor neurons Peripheral neurons whose cell bodies lie in the ventral gray columns of the spinal cord and whose terminations are in the skeletal muscles.

Nystagmus Involuntary repetitive rapid movement of the eyeball.

Optic atrophy Pale appearance of the optic nerve head when the nerve has become demyelinated.

Optic neuritis Inflammation of the optic nerve which when acute produces a swollen appearance of the optic disc similar to that of papilledema.

Papilledema Edema of the optic disc ("choked disc") usually due to increased intracranial pressure.

Paresis (hemi-, para-, quadri-) Weakness or partial paralysis.

Plegia Complete paralysis. *Hemiplegia,* paralysis of one side of the body and the limbs on that side; *paraplegia,* paralysis of the legs and lower part of the body, usually caused by disease or injury of the spinal cord; *quadriplegia,* paralysis of all four limbs.

Proprioception Perception of movements and position of the body and joints.

Pseudobulbar palsy Weakness of the pharyngeal, laryngeal, and facial muscles, simulating bulbar paralysis but due to supranuclear (upper motor neuron) lesions.

Ptosis Drooping of the upper eyelid.

Quadrantanopsia Blindness in one of the quadrants of the visual field.

Scotoma Blind or partially blind area in the visual field.

Sensory "level" Level below which there is a decrease or loss of sensation corresponding to the level of dysfunction of the spinal cord.

Stereognosis Faculty of perceiving and understanding the form and nature of objects by the sense of touch.

Upper motor neuron Neuron which conducts impulses from the motor cortex to the motor nuclei of the cranial nerves or to the motor cells in the ventral gray columns of the spinal cord.

The examination of the nervous system has a reputation for being tedious, complicated, and difficult to interpret. This reputation is only partially deserved. The examination is time-consuming only when one is acquiring proficiency. With repetition, interest increases and the speed with which an adequate neurologic examination can be done also increases. A screening neurologic examination can be accomplished in fifteen minutes. This can be cut down further if various parts are assimilated into the general physical examination.

Some observers prefer to examine the nervous system by regions rather than by systems; that is, the legs are examined as to motor, sensory, reflex, and cerebellar responses, and then the arms are examined, and so forth. This approach has advantages, but probably should be left to those with considerable experience. It is not necessary to do the examination in the order listed here, however. It is often better to examine first the part or function complained of by the patient and then return to the established routine.

The examiner should have available a reflex hammer, tuning forks (128 and 256 cps), an ophthalmoscope, a small flashlight, tongue blades, sharp pins, and cotton or a camel's-hair brush. A stethoscope and otoscope are also often required.

TECHNIQUE OF EXAMINATION AND NORMAL FINDINGS

Cranial Nerves

I: Olfactory. Ask the patient to identify, with his eyes closed, any common, nonirritating odor, such as coffee, tobacco, vanilla, turpentine, or cloves. Each nostril is tested separately. If these scents are not readily available, items at the bedside, such as oranges, flowers, or cigarettes, may be used. Normally the patient should be able to approximately identify familiar odors. Care must be taken here to differentiate between inability to smell the substance and inability to identify it properly. Many patients smell the substance correctly but are unable to name it. For this reason very familiar odors are probably best. Loss of the ability to smell (anosmia) is abnormal *but does not usually indicate disease of the nervous system* and is therefore of little localizing value. Local nasal disorders are far more common as a cause of loss of smell.

II: Optic. Examination of the optic nerve is divided into three major subdivisions: (1) vision, (2) visual fields, and (3) funduscopic.

Visual acuity is best tested with standardized charts or with small, pocket-sized testing cards. If these are not available the examiner can use any printed material and compare his own visual ability (assuming he knows his own acuity) with the patient's. Since we are interested in lesions of the retina and optic pathways and not in errors of refraction, the patient should be tested with his glasses on. Test each eye separately. If the acuity is so diminished that the patient cannot read even large print, ask him to count extended fingers at various distances. If even this is impossible, test his ability to see moving objects or to distinguish light from dark. Visual acuity, with correction, should be 20/20. In youth, the visual acuity is usually better than this, and in fact an acuity of 20/20 can represent some loss of vision.

Check the *visual fields* by the confrontation method. Ask the patient to cover one eye and with his other eye look straight into your eyes. Slowly bring from behind the patient one constantly moving finger (wiggling), held 12 to 18 inches from his head. A small test object such as a cotton applicator may be used. As the test object is moved forward, the patient will let you know when he first sees it. All four quadrants should be tested at *45-degree angles from the horizontal and the vertical* (Figure 20-1). Each eye must be tested separately, but most field defects of central nervous system origin can be found with both of the patient's eyes open. Finer testing can be done with small test objects and screens made for the purpose. Many scotomas cannot be found without such special equipment. Extinction, or suppression, is tested by wiggling a finger first in one-half of the field, then repeating on the opposite side with a finger on the other hand. Then both fingers are moved simultaneously. The patient is asked the same question each time that one or both fingers are moved: "On which side do you see the finger wiggle?" Normally the visual fields are full, without any obvious blind spots, and there is no extinction; that is, both fingers are seen to wiggle simultaneously.

The *optic fundi* should be examined with the ophthalmoscope. The presence of a sharp disc outline and of pulsations of the veins in the discs is of particular interest from a neurologic standpoint, since these findings indicate a normal intracranial pressure. This and other techniques of optic examination are fully described in Chapter 6.

III, IV, and VI: Oculomotor, Trochlear, and Abducens. These nerves supply the muscles of eye movement and are tested as a unit. Each eye is tested separately and then both together. Ask the patient to follow an object such as a fingertip, keeping his head motionless. Move the object laterally from side to side, and then vertically up and down when lateral gaze is reached. Test elevation and depression with the eyes in the mid-position also. Pause for a moment at each end point and inspect for nystagmus and weakness of any eye muscle. The test object is then brought from a distance of 3 or 4 feet to within an inch of the patient's nose. In this way both convergence and the normal pupillary constriction to convergence are tested.

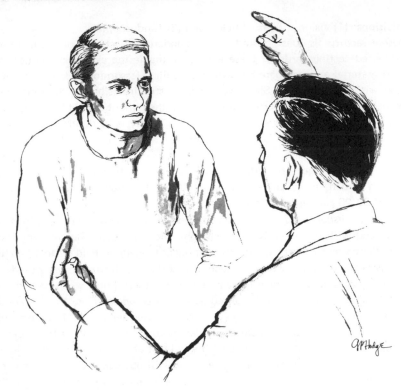

FIGURE 20-1. Confrontation method of testing visual fields.

The eyes are inspected for ptosis, and any difference between the widths of
the palpebral fissures is noted. Check the pupillary reaction to light. The light is
shone directly into each eye while the other eye is shielded from its rays. The
presence of constriction in the eye into which the light is shone (*direct light
reflex*) and in the other eye (*consensual light reflex*) can thus be observed. The
resting diameter of the pupils and their equality or inequality are next estimated.
If double vision (*diplopia*) is complained of, further testing to determine the
muscle or nerve weakness can be carried out as described in Chapter 6.

Normally the extraocular muscles move in parallel in all directions without
diplopia and without any obvious weakness of either eye in any direction. If the
eyes are moved to the limit of lateral gaze in any direction, there normally may
be minimal nystagmus. This is not present if the eyes are not forced to the
extreme of lateral gaze. There should be no nystagmus on elevation and
depression of the eyes, even at the extremes.

The pupils normally are equal, about 2 to 3 mm. in diameter, and react
quickly to light both directly and consensually, and also to convergence.
Inequality of the pupils (*anisocoria*) may be an important finding but also may
be congenital and of no particular significance.

V: Trigeminal. The areas supplied by the three divisions of the trigeminal nerve are tested for their sensitivity to light touch (cotton), pinprick, and temperature. The *corneal reflex* is important and can be tested by carefully placing a fine, elongated wisp of the cotton on the cornea while the patient is looking away from the approaching cotton (Figure 20-2). *Both eyes should blink quickly with this stimulation.* Each eye is tested separately. The masseter and temporalis muscles should be palpated with the patient clenching his jaws. Deviation of the jaw on opening the mouth should be assessed. Test also the strength of the jaw on lateral movement. Check the strength of jaw closure by asking the patient to grip a tongue blade with his teeth on each side while you try to extract the blade. Normally all these modalities are intact and muscle strength is good.

VII: Facial. Test the facial nerve by asking the patient to show his teeth and smile, by strong closure of the eyes against resistance, by elevating the eyebrows, and by contracting the platysma. Each of these maneuvers may be first demonstrated to the patient by movements of your own. Test taste on the anterior two-thirds of the tongue with sugar or salt. Place a few grains on half of the anterior two thirds of the protruded tongue and instruct the patient to keep his tongue out until he tastes the substance. At times it helps to have him point to the word "sweet" or "salt" written on a piece of paper rather than to

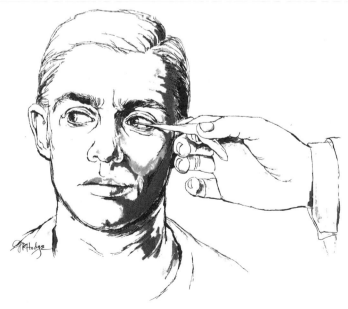

FIGURE 20-2. Method of eliciting corneal reflex.

withdraw his tongue and state what he has tasted. Each side of the tongue should be tested separately.

All facial movements should be equal bilaterally. Some individuals, however, habitually talk and smile more out of one side of their mouth than the other. These habit patterns must be kept in mind while looking for possible evidence of weakness.

VIII: Auditory. The two divisions of the auditory nerve, cochlear and vestibular, are tested separately. The patient's ability to hear a watch tick or a whispered voice at a definite distance from each ear may be all that is necessary to test. Hearing may also be tested with a tuning fork of 256 or more cycles per second. Normally the patient's hearing should be equal in both ears, and air conduction should be about twice the duration of bone conduction. Tests of hearing are described in detail in Chapter 8, as is caloric testing for vestibular function.

IX and X: Glossopharyngeal and Vagus. These two complex cranial nerves are tested together and, in fact, are rather simple to evaluate. Note the quality of the patient's voice and his ability to swallow. Ask him to open his mouth and say "ah." The position of the soft palate and uvula at rest and with phonation is observed. Check the gag reflex with a tongue blade by touching each side of the posterior pharyngeal wall separately. Taste on the posterior one-third of the tongue can be examined when an abnormality is suspected.

Normally the uvula and palate rise in the midline with phonation. It is important to test this function with the patient's head in the midline and not turned to either side.

XI: Spinal Accessory. Ask the patient to turn his head to one side against the examiner's hand on that side. Meanwhile palpate the opposite sternocleido-mastoid muscle (Figure 20-3). Palpate the trapezius muscle and test the strength of the patient's "shrug" while you press down on the shoulders with both hands. Make sure that there are no fasciculations present. Normally the strength of the sternocleidomastoid and trapezius muscles is equal bilaterally.

XII: Hypoglossal. The tongue has been described in detail in previous chapters. The patient is asked to protrude his tongue in the midline, and any deviation or atrophy is observed. If atrophy is suspected, the tongue may be palpated. Another test for weakness or deviation of the tongue is to ask the patient to stick his tongue into his cheek while the examiner presses against the bulging cheek.

Motor System

Inspection. Observe the patient's musculature for abnormal movements either large (capable of moving a joint) or small (within the belly of the muscle itself).

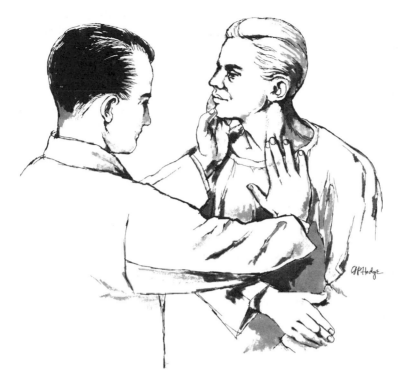

FIGURE 20-3. Method of testing for spinal accessory nerve function.

Observe for atrophy and hypertrophy. If atrophy is present or if a progressive disease that might cause atrophy is suspected, the circumference of the muscle should be measured bilaterally and the point at which the measurement is taken recorded. The limbs of the normal individual at rest do not demonstrate any movements either of the joints or in the body of the muscles themselves except for an occasional rare and random fasciculation. A concept of normal muscle bulk is gained only by experience. Great variation is normally present among persons of different ages, sexes, and occupations. Chronic illness in itself may produce generalized wasting of the muscles.

Strength. In testing the strength of all appropriate muscle groups, pay particular attention to the relative strength of the two sides and between proximal and distal groups. Special attention should be paid to any area mentioned by the patient as being weak. It is not necessary to do a complete detailed muscle examination on every patient, especially if he does not complain of weakness. A screening examination would logically include testing of strength of dorsiflexion and plantar flexion of the feet, extension and flexion of the knees and hip, strength of the grip, extension of the wrist, and flexion and extension of the forearm and shoulders. Strength can be graded from 0 to 5: 0 — no movement of the muscle; 1 (trace) — a trace of muscular contraction incapable of moving the joint; 2 (poor) — minimal movement with gravity eliminated; 3 (fair) — strength equal to moving the joint against gravity; 4 (good) — good power but not normal; 5 (normal) — normal strength. A system of grading by percentage of the normal may also be used.

There is a great variation of strength in different individuals, all of whom may be considered normal. Normal hand grip for a society matron would certainly not be considered normal for a farmer. The difference in strength between the nondominant and dominant sides of the body is *usually very minimal and should not be used as an explanation for unilateral weakness.*

Muscle Tone. Check muscle tone by both fast and slow flexion and extension at the elbows, wrists, shoulders, knees, and ankles. Ask the patient to relax as much as possible during this examination. In healthy cooperative adults the muscles are normally somewhat hypotonic. The normal patient should be able to relax his extremities sufficiently so that the examiner can easily test for tone. In older or senile patients or in patients with decreased intelligence, there is sometimes marked difficulty in producing the necessary relaxation of the extremities. The patient tries to cooperate and voluntarily moves the arm in the same manner as the examiner, thus defeating the examiner's purpose. In a few people it may be impossible to assess tone adequately for this reason.

Cerebellar Function

Coordination and the ability to perform skilled and rapid alternating movements in the upper and lower extremity are tests of cerebellar function.

Upper Extremities. The patient is asked to supinate and pronate his hand alternately as rapidly as possible on each side. Direct him to touch the index finger of his outstretched arm to the tip of his nose slowly, with his eyes closed (finger-to-nose test). The ability to perform rapid fine movements of the fingers, such as touching the tip of the index finger to the thumb, can also be assessed.

Lower Extremities. Ask the patient to tap his foot against the floor as rapidly as possible; then direct him to touch one heel to the opposite knee and run it slowly down his shin to the big toe (heel-to-knee-to-toe test). For other detailed tests of cerebellar function a textbook on the neurologic examination should be consulted.

Normally the patient is able to perform these maneuvers without any wavering, tremor, or ataxia. The heel-to-knee-to-toe test should show no falling off of the heel from the shin. The patient should be able to tap the floor quickly with either foot and should be able to perform rapid alternating movements nearly as well with his nondominant as with his dominant hand.

Reflexes

Always check the biceps, triceps, radial periosteal, patellar, and Achilles reflexes in all patients. The tendon or bone is struck directly and smartly with the reflex hammer. The limb to be tested should be relaxed and in a flexed or semiflexed position. The biceps reflex is best tested by tapping the examiner's finger or thumb which has been placed over the tendon. Detailed instructions on the exact method of obtaining each tendon reflex are not given here. Demonstration is superior to the written word in this respect.

In normal young adults the reflexes may be minimal or apparently absent and may be elicited only by asking the patient to "reinforce" by pulling with one hand against the other at the time the tendon is tapped. A reflex is not considered truly absent until it has been proved that it cannot be elicited by this maneuver. In certain patients the reflexes may be extremely active, but *clonus* (persistent involuntary flexion and extension of the joint under extension or flexion pressure) is never present in the absence of central nervous system disease.

It is good to develop the habit of charting all the reflexes completely. This is easily done by drawing the figure of a man, labeling the sides right and left, and placing the state of the reflex represented by 0 through 4+ at the appropriate joint. An abbreviated chart may also be used for this purpose. Reflexes may be represented as follows: 0, absent; 1+, decreased; 2+, normal; 3+, hyperactive; 4+, hyperactive with clonus.

Pathologic Reflexes. The "pathologic" reflexes are extremely important and should be carefully looked for. There are two basic pathologic reflexes, the Babinski and its variants, and the Hoffmann and its variants.

The *Babinski* reflex is tested by stroking the sole of the foot with a pointed instrument from the heel, along the lateral edge of the foot, and then across the ball of the foot medially. The *Chaddock* reflex is elicited by stroking the lateral edge of the foot from the heel to the toes just above the sole. The *Oppenheim* reflex is elicited by firmly pressing down on the shin with the knuckles from the knee to the ankle. The *Gordon* reflex is tested by squeezing the calf firmly. Normally, after infancy, there is no extensor response of the foot and toes to any of these maneuvers. Therefore, the presence of these reflexes is an unequivocal sign of disease of the pyramidal tract.

The *Hoffmann* reflex is tested by quickly extending or flexing the last joint of the middle finger. While suspending the limp hand by its middle finger, quickly flip the tip of the finger upward or downward. Normally there is very little if any response to this maneuver, but a flexion response of the thumb and fingers is not an unequivocal sign of pyramidal tract disease. It is significant only when markedly exaggerated or unilaterally present.

Clonus is tested by quickly flexing or extending a joint and maintaining the tension.

Superficial Skin Reflexes. The main superficial reflexes are the abdominal and cremasteric reflexes. The *abdominal reflex* is elicited by stroking with a firm or slightly irritating object over all four quadrants of the abdomen. The stroke can be directed either toward or away from or at right angles to the umbilicus. Normally the umbilicus moves toward the stimulus. The *cremasteric reflex* is elicited by stroking along the internal aspect of the upper thigh. A normal response results in an upward movement of the testicle on the side stimulated due to involuntary contraction of the cremaster muscle.

Both of these reflexes are normally present in the young, relaxed patient, but their absence is common and may be due to a variety of causes.

Sensory Examination

The sensory examination is difficult to perform well and interpret correctly. It is subject to great variation, depending on the experience and skill of the examiner and the cooperation and emotional balance of the patient. It is, however, of primary importance and can give clues to a diagnosis that are obtainable in no other way. Sensory testing is usually done with the patient's eyes closed.

Vibration, Motion, and Position. With a 128-cycle tuning fork test the duration of vibration perceived on a bony prominence in all four extremities. Use the lateral malleolus on each ankle and the first knuckle of each hand. The tuning fork should be struck with the same strength and held with the same degree of firmness against each prominence. In this way a fairly good quantitive estimate of vibratory sense can be obtained. Next, grasp one toe or finger of each of the four extremities, move it up and down, and ask the patient to identify the digit with his eyes closed. After moving the digit up and down repeatedly, leave it in the upward or downward position and ask him to identify its direction. Ideally, the digit should be grasped by the sides rather than the top and bottom but this is not always feasible. The vibratory sense is the best and most assessable posterior column function and therefore should be tested in all patients. Position and motion sense is usually not decreased until vibratory sense is markedly diminished or lost.

Pain and Temperature. With a sharp pin, test perception of superficial pain in all four extremities. A standard pattern might include going laterally to medially across the dorsum of the foot and the back of each hand. If any loss is found, it is important to map the extent of this loss, *going from the area of decreased sensation toward the area of normal sensation.* The pinprick should also be tested up the whole length of the body on each side. In certain cases it is important also to test the "saddle" or perianal area.

Temperature sensation can be tested with any cool or warm object. The tines of the examiner's tuning fork will often serve well enough. Often temperature testing serves better than pinprick to check the spinothalamic system since it does not produce withdrawal.

Test deep pain sensation by squeezing the Achilles tendon as near to the bone as possible or by squeezing the muscles at the junction of the thumb and first finger. In a normal person this should be quite painful and will cause him to rise slightly in protest. Other areas in which deep pain can be tested are the testicles, the trapezius muscle at the base of the neck, and the supraorbital notch above the eye.

Fine Sensory Modalities

These include touch, two-point tactile sensation, stereognosis, graphesthesia, and extinction. Test for *light touch* with a wisp of cotton on the hairy surfaces of the body. Include the trunk and all four extremities, and test one side against the other. Normally a very light wisp of cotton can be felt only on the hairy surfaces, and the movement of one hair can usually be perceived. However, very light touch that does not indent the nonhairy skin such as the palms or soles frequently cannot be perceived.

Two-point tactile sensation is tested with two dull points. One side is compared with the other, and the upper and lower extremities are tested. The threshold at which the patient feels definitely two points, or conversely at which he feels them as one, is recorded. Two-point sensation is usually tested on the tips of the fingers. The threshold on the tip of the index finger on a normal person is about 3 mm.

Stereognosis is tested in the hands. Ask the patient to close his eyes and identify objects placed in his hands by moving them around and feeling them with his fingers. Any convenient objects such as coins or keys can be used. A normal person can differentiate between a penny, nickel, quarter, and half-dollar.

Test *graphesthesia* by drawing numerals on various parts of the patient's skin and asking him to identify them. The size of the identifiable numeral will naturally vary with the area.

Extinction (suppression) can be tested by the method of double simultaneous stimulation. Have the patient close his eyes, and touch him, for example, on the back of the hand. Ask him where he was touched. Touch him then on the identical spot on the other side of his body and repeat the question. The third time, touch him in both places simultaneously and with equal pressure. Suppression is normally not present, and the patient feels both sides. If it is present it is usually significant.

Gait and Station

Ask the patient to walk as normally as possible for some distance. Notice his general posture, the size of his steps, the lateral distance between his feet as he places them on the ground, the amount of associated arm swinging present, and his balance while walking. Test his ability to start and stop on command.

Sometimes walking will be normal, but running impossible. If walking balance is to be further tested, the patient can be asked to walk a straight line, putting one heel directly in front of the toes of the other foot. This is called *tandem walking.* The patient's standing balance is then tested by asking him to stand with his feet together first with his eyes open and then with his eyes closed. The examiner should be ready to catch the patient as there is often a tendency to fall in this position *(Romberg's sign).* Finer tests of balance include standing on one foot and hopping or jumping on one foot.

A normal person should be able to walk in a fairly straight line for 10 or 20 feet with his eyes open or closed. He should not lose his balance with his feet together and his eyes closed. A normal person should be able to balance and hop on each foot with his eyes open and closed.

Meningeal Signs

The simplest and most easily tested of the meningeal signs is that for *nuchal rigidity* (stiff neck). Ask the patient to relax, and then with both hands flex his neck on his chest. If nuchal rigidity is present, there will be resistance and pain. *Brudzinski's sign* is flexion of the hips when the head is flexed in this manner. Test for *Kernig's sign* by flexing the hip on the trunk, allowing the knee to flex at the same time. When the hip is flexed to 90 degrees, extend the knee. In the presence of meningeal irritation there will be resistance, pain, and sometimes a tendency toward flexion of the neck.

Aphasia

In most cases it is not necessary to do detailed testing for aphasia. A good idea of the patient's ability to express himself and to understand what is said to him can be gained in normal contact. If aphasia is suspected, simple tests can be done. Ask the patient to name objects pointed out by the examiner, such as thumb, nose, ear, eye, watch, watch crystal, tie, and colors in the tie. Auditory receptive aphasia can be tested by asking the patient to perform at the command of the examiner, such as (without gestures), "Touch your finger to your nose," or "Touch your left first finger to your right ear." Testing can be done to determine the patient's ability to read by asking the patient to perform written instructions, such as, "Touch your finger to your nose." The patient's ability to recognize printed words and simple objects can be tested even in the presence of a marked expressive aphasia by printing the names of simple objects, such as key, pencil, coin, on a piece of paper and placing these objects next to the paper. The patient quickly understands that he is to match the printed word with the object. For further testing of aphasia, standard texts on the neurologic examination should be consulted.

General Examination

Palpate the common carotid arteries in the neck and the temporal arteries just in front of the ears. Palpate both radial arteries at the wrist, noting not only strength of impulse but also simultaneousness. Palpate both dorsalis pedis arteries.

With the stethoscope bell listen for a bruit over each eyeball by asking the patient to close both eyes. Place the bell over one eyeball firmly and ask the patient to open the other eye and look steadily at an object (Figure 20-4). Listen also for a bruit over both carotid arteries high in the neck at the angle of the jaw and in each supraclavicular space. If a bruit is heard in either of the latter two spaces, compare the blood pressures in the arms.

Palpate the ulnar nerves between the olecranon and medial epicondyle and the common peroneal nerves laterally just below the head of the fibula.

CARDINAL SYMPTOMS AND ABNORMAL FINDINGS

Numbness. Numbness usually indicates peripheral nerve or spinal cord (posterior column) involvement but may also be due to sensory pathway interruption in the brain stem, thalamus, or parietal cortex.

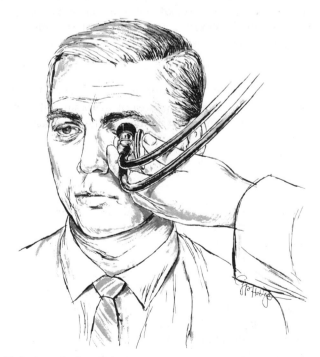

FIGURE 20-4. Auscultation of the eye.

Weakness. This may result from abnormalities of the muscle, myoneural junction, or nerve, or of the corticospinal tract in the spinal cord, brain stem, internal capsule, or motor cortex. Involvement of other central pathways connecting to this main motor system also produces weakness.

Fainting or Loss of Consciousness. This results from loss of function of the brain due to loss of blood supply, inadequate nutrition (oxygen, glucose), imbalance of nutrients, abnormal cerebral electrical discharge (epilepsy), or cerebral destruction due to any cause.

Dizziness. This usually means one of three things: faintness, vertigo, or imbalance. *Faintness* is discussed above. *Vertigo* is caused by dysfunction of the inner ear balance mechanisms or the vestibular portion of the acoustic nerve or its central connections. *Imbalance* may result from peripheral nerve, posterior column, cerebellar, or main motor pathway involvement.

Pain. This is due to distortion, transection, or irritation of pain endings and pain fibers of peripheral nerves or of the central pain pathways (spinothalamic tracts) in the spinal cord, brain stem, or thalamus.

Headache. This is caused by distortion of or traction on blood vessels or the meninges or other covering structures of the brain, or by pressure, distortion, traction, or displacement of almost any extracerebral structure in the head, including the skull, paranasal sinuses, scalp, and posterior suboccipital muscles of the neck.

Defective Memory or Thinking. Difficulty with these two cerebral functions may result from any lesion in almost any area of the brain, although a specific memory defect may result from small lesions in the hypothalamic-thalamic-temporal lobe structures.

Cranial Nerves

I: Olfactory. The most common neurologic causes of anosmia are head injuries and tumors, especially meningiomas of the olfactory groove. The ability to smell commonly decreases with advancing age.

II: Optic. The two most common funduscopic abnormalities which are of concern during neurologic testing are swelling or edema of the optic disc and optic atrophy. Papilledema first appears as a loss of distinctness of the disc margins which progresses until all evidence of margin disappears. It then may become raised above the surrounding retina, and the edema may spread into the retina. One should be careful not to attach significance to minimal indistinctness of the disc margins along the nasal or medial aspect of the disc, since this is often normal. Optic atrophy is usually manifested by a combination of pallor of the lateral aspect of the disc and a lack of normal vascularity in that area.

The most common gross visual-field defect of neurologic importance is that of a homonymous hemianopsia, in which, for instance, the right visual field is lost in the right and left eyes. This would indicate that the left optic tract or visual radiation is diseased somewhere between the optic chiasm and the left visual cortex. A homonymous quadrantanopic defect usually indicates disease involving the visual radiations as they pass through the parietal or temporal lobes. An upper quadrantanopic defect usually indicates involvement of the visual radiations in the opposite temporal lobe, and a lower quarter field defect indicates involvement of the opposite parietal lobe. Bitemporal field defects or involvement of the lateral half of the visual field in each eye indicates a lesion in the center of the optic chiasm interrupting the fibers as they cross from one optic nerve to the opposite tract. The most common lesion causing this defect is a tumor of the pituitary gland.

III, IV, and VI: Oculomotor, Trochlear, and Abducens. Anisocoria (unequal pupils) may be caused by a variety of abnormalities including a sympathetic paralysis (Horner's syndrome), oculomotor nerve weakness, Adie's syndrome, syphilis (Argyll Robertson pupil), diabetes, multiple sclerosis, and trauma or disease of the iris. If both pupils are markedly larger or smaller than normal, inquiry should be made into the use of drugs, especially stimulants such as d-amphetamine or eye drops for ocular disease such as glaucoma.

A unilateral dilated pupil often occurs with increased intracranial pressure, and in this case it is an early sign of third cranial nerve weakness. The dilated pupil is usually on the side of the lesion causing the increased pressure. The *Argyll Robertson pupil* is usually smaller than normal and reacts to convergence but not to light. *Adie's pupillary abnormality* is usually found in young women with decreased or absent tendon reflexes; the pupil is larger than normal and reacts only slowly to light.

A short summary of the common extraocular muscle paralyses will be given here, but for further details Chapter 6 should be consulted. Paralysis of the third cranial nerve results in a dilated pupil, external deviation of the eyeball, and ptosis of the upper lid. The fourth and sixth cranial nerves carry no parasympathetic fibers and their paralysis results only in weakness of the appropriate muscles. The eye is deviated inward with paralysis of the sixth cranial nerve (lateral rectus muscle weakness). Sixth-nerve weakness has only moderate localizing value because it may result from increased intracranial pressure regardless of cause, especially in children. Paralysis of the fourth nerve is not quite so easy to detect and is rare. There is weakness of internal rotation of the eyeball and of gaze downward and inward. The patient may tilt his head so that the eye with the paralyzed superior oblique muscle is elevated somewhat above the plane of the normal eye. This is done to bring the horizontal axes parallel and thus prevent diplopia.

Nystagmus often gives a definite clue as to the location and even the type of neurologic disease present. Of special importance is the *dissociated nystagmus* with lesions of the medial longitudinal fasciculus. In this type of nystagmus the eye on the side to which the gaze is directed participates strongly with a horizontal nystagmus, whereas the opposite eye will show less nystagmus but will show some weakness of internal rotation. Multiple sclerosis is the most common cause of this type of nystagmus.

Nystagmus is common in acute and chronic conditions of the cerebellum and brain stem, in which case it is usually more pronounced on looking toward the side of the lesion. There also may be a minimal to moderate nystagmus with strictly cerebral lesions, in which case it is usually more evident in looking away from the side of the lesion.

V: Trigeminal The corneal reflex may be diminished unilaterally from either a peripheral or a central lesion. A decreased corneal reflex may be an early clue to a peripheral lesion of the fifth cranial nerve. At times the corneal reflex may be consciously suppressed and thus may be decreased or absent in stoical persons or in conversion reactions. It is not difficult to differentiate the absence of a corneal reflex due to a lesion of a fifth nerve from that caused by a seventh nerve paralysis (weakness of the face).

Where there is a suspicion of a medullary or upper cervical cord lesion it is important to test pain, temperature, and touch on the face; touch may be intact with absence of pain and temperature with a lesion involving the descending tract of the trigeminal nerve. In such lesions the loss of pain and temperature sensation extends exactly to the midline. It may, however, involve only the first, second, or third divisions of the trigeminal nerve because of the arrangement of these divisions in the descending tract. For this reason it is important to test all three divisions.

With the usual type of cerebral infarct there may be some decrease in sensation on the opposite side of the face, but this usually does not extend exactly to the midline. There is usually no appreciable weakness of the masseter and pterygoid muscles on the side opposite a "central" (cerebral) lesion.

VII: Facial Nerve. In a "central" (upper motor neuron) weakness of the face, such as is typically seen with the usual cerebral infarct, there is moderate to marked weakness of the opposite *lower facial muscles,* including the platysma, the orbicularis oris, and sometimes the lower part of the orbicularis oculi. With a small lesion there may be only minimal drooping of the opposite corner of the mouth at rest and some weakness of that corner of the mouth on voluntary movement. In addition, the palpebral fissure may be somewhat wider than on the normal side. The patient will, however, be able to close his eye and wrinkle his forehead on the side opposite to the cerebral lesion fairly strongly. With weakness of the face due to a lesion of the seventh nerve nucleus or of the nerve itself (peripheral facial paralysis, lower motor neuron disease, *Bell's palsy*) there is weakness of all parts of the face on the side of the lesion. The patient will be unable to wrinkle his forehead, close his eye, or show his teeth.

Weakness of one side of the mouth, even though minimal, may be an important clue in determining whether the lesion responsible for the accompanying weakness of an arm, or an arm and a leg, on the same side is in the spinal cord, the brain stem, or the cerebrum. If the lesion is in the spinal cord, there should be no weakness of the face. If it is in the pons, the facial weakness will be on the side opposite to that of the affected arm and the leg. If the lesion is above the brain stem, weakness of the face, arm, and leg will be all on the same side.

Testing for taste on the anterior two-thirds of the tongue is of use when determining the site of the interruption in a peripheral seventh-nerve (Bell's) paralysis. Absence of taste on the anterior two-thirds of the tongue on the side of the facial weakness means that the lesion responsible is in the facial canal or central to it.

VIII: Auditory.   A lessening of the time of air conduction with preservation of bone conduction indicates that there is *conductive* (middle ear or external ear) deafness. If both air conduction and bone conduction are reduced, the deafness is *perceptive* (nerve) or mixed. With unilateral nerve deafness the patient will hear the tuning fork when it is placed on the midline of the skull (Weber's test) in his good ear, whereas if the defect is in the middle ear or external ear, the patient will hear it in his deaf ear (oftentimes much to his surprise). Lesions central to the cochlear nuclei do not cause unilateral deafness; therefore, a unilateral perceptive deafness indicates a lesion of the cochlear nuclei (rare), the eighth nerve, or the end organ (cochlea). To differentiate between these it is sometimes helpful to test the vestibular portion of the eighth nerve. The vestibular function will be preserved if deafness is caused by a lesion of the cochlear nuclei, but may be lost in lesions of the nerve or end organ (see Chapter 8).

IX and X: Glossopharyngeal and Vagus.   If there is unilateral paralysis of the ninth and tenth cranial nerves, the palate will deviate to the unparalyzed side during phonation. The distance between the soft palate and the posterior pharyngeal wall will be less on the paralyzed side. The arch of the palate at rest on the paralyzed side will tend to droop lower than on the normal side. The sensory component of the gag reflex is mediated by the ninth cranial nerve and the motor by the tenth. However, paralysis of the ninth cranial nerve alone is difficult to detect, since sometimes there may be no apparent sensory loss in the pharynx. If there is any doubt about the presence of a vagus paralysis, the vocal cords should be visualized to see whether or not there is a unilateral or a bilateral weakness.

Marked palatal and vocal-cord paralysis on the same side indicates a lower motor neuron lesion. However, some patients will show moderate and temporary palatal weakness of the side *opposite* a large upper motor neuron lesion such as caused by a large cerebral infarct. Marked unilateral paralysis of the palate, pharynx, and vocal cords without evidence of long sensory or motor pathway involvement nearly always indicates that the lesion is not within the brain stem (affecting the nucleus ambiguus) but involves the ninth and tenth nerves after they have left the medulla. Bilateral weakness can result from lesions of the upper motor neurons (bilaterally), giving a pseudobulbar palsy, or from involvement of the nuclei in the medulla, the nerves or the muscles, giving a bulbar palsy.

XI: Spinal Accessory. If unilateral weakness and atrophy of the trapezius and sternocleidomastoid muscles occur, one can be certain that the lesion responsible is outside the brain stem, since the nerve has its origin in the upper cervical cord. Marked bilateral weakness and atrophy of these muscles are often found in primary muscle disease such as muscular dystrophy.

XII: Hypoglossal. Unilateral tongue weakness is manifested by a deviation of the protruded tongue toward the weak side and often by atrophy and fasciculations on the affected side. *When the tongue is lying in the mouth, however, it will be pulled toward the strong side.* The reason for this is that different sets of tongue muscles are in action when the tongue is protruded than when it is resting in the mouth.

It is important to check the tongue carefully for fasciculations. They may be the earliest sign of lower motor neuron disease affecting the hypoglossal nerve nuclei. Normally a few small tremulous movements are seen in the tongue. These should not be confused with the continuous, "wormy" movements (fasciculations) over the entire tongue seen in amyotrophic lateral sclerosis. Occasionally, as with the palate and pharynx, there may be temporary unilateral weakness of the tongue on the side opposite a large acute upper motor neuron lesion. Associated with this there also may be a transient dysarthria. Both of these usually clear within a few days. With a bilateral upper motor neuron lesion there may be inability to protrude the tongue beyond the lips and anarthria.

Motor System

Inspection. Fasciculations may be emphasized by strong contraction of the suspected muscle before examination and by tapping the muscle with the reflex hammer. Since the presence of fasciculations is important in the diagnosis of diseases of the anterior horn cells, inspection should be careful and thorough. Atrophy and hypertrophy may give important clues in the diagnosis of muscle disease.

Abnormal Movements. Abnormal movements should be carefully observed and described. The abnormal movement of parkinsonism is a 4 to 5 per second *resting* (nonintention) *tremor* usually more evident in the hands but also found in the arms, head, face, tongue, and legs. It may be associated with rigidity or increased tone of the muscles and thus produce the typical *cogwheel effect,* which is a rhythmical increase and decrease in tone superimposed on passive movements produced by the examiner. The *essential* (familial or senile) *tremor* is not present at rest but is more pronounced with sustained posture or with movement. It is not associated with any alteration in tone and is more likely to involve the head than is the tremor of parkinsonism. The *intention tremor* characteristic of such diseases as multiple sclerosis is usually combined with some degree of ataxia and is brought out with purposeful movement such as the finger-to-nose test.

Choreic movements are irregular, involuntary, spontaneous movements usually involving more than one joint. They are purposeless but may not appear so at first glance. They are most easily characterized as being similar to the "fidgets" or normal restlessness seen in children. They involve any extremity, the face, the mouth, and the tongue. They are not associated with an increase in tone. *Athetoid movements* are characteristically slow, writhing movements involving mainly the proximal parts of the extremities and also the trunk and face. They are often associated with other neurologic abnormalities found in cerebral palsy.

Strength. It is important to test both distal and proximal muscle groups in both the upper and lower extremities. In some muscle diseases such as muscular dystrophy there may be marked weakness of the shoulder and hip girdle musculature and surprising strength of the hands and feet. If weakness is present, detailed tests of all muscle groups are indicated to determine the extent, distribution, and degree of weakness. Much diagnostic information can be obtained in this way.

It may be necessary to do muscle tests in the morning and evening to determine whether there is any change with time of day. The muscular weakness of myasthenia gravis is characteristically worse late in the day. Strength testing before and after a test dose of edrophonium is the most certain method of identifying this disease.

Feigned muscle weakness may fool the most sophisticated observer. It is said that a tendency to "give way" in a jerky fashion is characteristic of nonorganic muscular weakness, but some tendency toward giving way may be found in certain definitely organic diseases.

Muscle Tone. Muscle tone is normally decreased in diseases of the anterior horn cells and peripheral nerves and in uncomplicated diseases of the cerebellum. An increase in tone is found in upper motor neuron diseases (pyramidal tract), in which case it is characterized as *spasticity.* Spasticity is the most common abnormality of tone and is classically of the "clasp knife" type. The tone varies from normal as one rapidly extends or flexes the joint to a marked increase when full flexion or extension is approached. With continued pressure the muscle then gradually gives way.

On the other hand, *rigidity,* which characterizes the extrapyramidal group of diseases, such as parkinsonism, is characterized by an increase of tone that is present throughout the full range of motion of the joint, producing steady resistance to the examiner's superimposed movement. There may be in addition a "cogwheeling," as mentioned above.

A tendency to slow relaxation of the muscles is characteristic of such diseases as hypothyroidism, myotonia, and myotonic dystrophy. It is most evident in patients with myotonia. It may be demonstrated by percussion of affected muscles such as the quadriceps, the forearm groups, or the thenar muscles with the reflex hammer. The percussed muscle will contract and then relax at a rate several times slower than normal.

Cerebellar Function

Many things may interfere with cerebellar testing or may give false cerebellar signs. Among these are sensory loss, especially loss of proprioception. In addition, pyramidal-tract disease may give slowness in cerebellar testing, as also may marked weakness of lower motor neuron or peripheral nerve origin. A combination of these defects may produce cerebellar-like signs which make it difficult to determine whether or not the ataxia is really of cerebellar origin. Certainly one should be careful to do adequate sensory testing on any patient with apparent cerebellar disease to determine whether or not proprioception is intact.

Disease of one cerebellar hemisphere produces signs on the same side of the body. This is important to remember when attempting to localize a lesion, in view of the fact that thalamic and occasionally frontal lobe lesions of the opposite side may produce ataxia similar to that from involvement of the cerebellum itself. Since some diseases are characterized by the production of cerebellar signs that are more prominent in the distal or the proximal parts of the extremities, it is important to state the exact nature and location of the tremor or ataxia. In addition, since ataxia may be drug-induced and therefore transitory, it may be necessary to test the patient more than once.

Reflexes

The reflex examination is one of the most important parts of the neurologic examination because abnormalities in this sphere are difficult to feign and thus represent an objective assessment of the patient's neurologic status. Hypoactive reflexes and absent reflexes are found in a variety of diseases. Anything which interferes with the anterior horn cell, the sensory or the motor part of the reflex arc, the motor end plate, or the muscle may eliminate or decrease the reflex. On the other hand, a pathologic increase in reflexes almost invariably represents disease of the upper motor neuron or pyramidal tract. The lesion, of course, can be anywhere from the cerebral cortex to just above the appropriate anterior horn cell. Any isolated reflex change, either increase or decrease, requires special consideration by the examiner.

It is important to memorize a few reflex arc levels which are helpful in determining the level of spinal cord lesions:

Biceps	C-5, 6
Triceps	C-6, 7, 8
Radial periosteal	C-5, 6
Patellar	L-2, 3, 4
Achilles	S-1, 2

Pathologic Reflexes. The presence of an extensor plantar response (Babinski sign) is an unequivocal sign of disease of the pyramidal tract. The classic Babinski is a dorsiflexion (extension) of the large toe with fanning of the smaller toes. With extensive pyramidal tract disease there may be some withdrawal of the entire leg with flexion at the ankle, knee, and hip.

There are several sources of difficulty in interpreting the Babinski response. At times the response may be equivocal with perhaps only fleeting dorsiflexion of the large toe and no fanning of the other toes. There may be no response to plantar stimulation at all; when this occurs, it is usually significant if there is a difference in the responses of the two feet. Another source of difficulty is the tendency of some patients voluntarily to withdraw their lower extremity when the sole of the foot is stimulated. Often this can be overcome by simply asking them not to withdraw. Some Babinski responses are strongly abnormal, with withdrawal of the entire lower extremity, and yet are stated to be equivocal because it is thought incorrectly by the examiner that the withdrawal is voluntary and not part of the reflex. It is only with experience that the examiner learns to differentiate between these two types of withdrawals.

The Babinski response is one of the most reliable signs in neurology and should be tested for in all patients. The Hoffmann response, on the other hand, is not an unequivocal sign of pyramidal tract disease. It should be equated more with the tendon reflexes and is only significant when markedly exaggerated or unilaterally present.

Superficial Skin Reflexes. Diseases of either the upper or the lower motor neurons will eliminate the superficial reflexes. In addition, the superficial reflexes may be difficult to elicit in the tense, obese, pregnant, or uncooperative patient. Although they have been said to be of special significance when absent, they are not always of great diagnostic importance. Unilateral absence or absence of either the upper or lower abdominals may be of significance and may be helpful in ascertaining the "sidedness" of a lesion or the level of a lesion of the spinal cord. In this regard it is important to remember that the upper abdominals are innervated by segments T-7 through 9 and the lower by T-9 through 11.

Sensory Responses

Posterior Column Sense. In addition to giving an index to the posterior column function, vibratory sense may also be used to determine the level of a spinal cord lesion. Simply "walk" up the body with a tuning fork, going from bony prominence to bony prominence until it is felt by the patient. The level of sensory perception in the bones corresponds well to that of the skin over them. If vibratory sense is intact it is usually not necessary to test motion and position sense. If vibratory sense is decreased or absent, motion and position sense should be tested. If motion and position sense is absent, for instance, in the toes, it should then be tested at progressively higher levels. Some diseases are characterized by posterior column sensory loss (posterolateral sclerosis) and will show marked loss of vibratory and motion and position senses.

Pain and Temperature Sensation. Pain and temperature sensation, since they travel in the same pathway, need not routinely both be tested. However, if pain is absent, then temperature sensation should be tested. This gives the examiner a check on his own findings and may be helpful in some cases of nonorganic sensory loss. Pinprick sensation is tricky to test and one may easily obtain abnormal responses in a normal person. This is partly because there are natural variations in the pressure put on the pin and of the sensitivity of different areas of the skin. Some patients, especially the neurotic and some highly intelligent, careful individuals, may report pinprick decrease that is not significant. In this case the examiner is wise to go to some other part of the examination and then return later so as not to emphasize the "abnormality" that he has found. Another source of error in examining with a pin occurs when testing one side of the body against the other. The second or last side tested will often apparently be more sensitive to the pinprick. In this case the examiner should reverse the order of the sides tested.

Pinprick testing is the commonest way of determining the sensory "level" caused by a spinal cord lesion. There is usually an area of increased response to the pin at the level of the lesion, normal response above the level, and decreased response below.

If an area of sensory loss is found for any modality, it should be mapped out by testing from the area of loss to the normal exterior zone in all directions. This can then be transferred to a sensory chart and placed in the patient's permanent record. Such sensory charts are helpful in determining whether an area of loss is in the distribution of nerve roots or of a peripheral nerve. In any patient with bowel or bladder difficulty it is very important to check the "saddle" area (the area of the buttocks and genitalia) for any sensory loss that might indicate a lesion of the conus medullaris or cauda equina.

Fine Sensory Modalities. Since motion and position and vibratory senses have already been checked to test the function of the posterior columns, most of the finer modalities do not need to be tested in the usual patient. However, they are useful in evaluating parietal cortex function and should be tested when disease is suspected in this region. If a patient demonstrates extinction or suppression, he will feel the stimulus only on his "good" side, which will be opposite his normal parietal lobe.

Since these finer sensory modalities travel in the posterior columns of the spinal cord, they also may be useful in the analysis of hysterical (conversion) sensory loss. Some patients with hysterical sensory loss have loss of vibration but good stereognosis. The reverse of this may also occasionally be true.

One should be careful to determine, before stating that a patient has astereognosis, that there is no more obvious sensory loss in the extremity. If there is a definite sensory loss in the extremity and stereognosis is also decreased, the loss would better be called stereoanesthesia and does not necessarily indicate parietal cortex dysfunction.

Gait and Station

Romberg's sign is positive only when the patient falls with his eyes closed and his feet together *and does not fall with his eyes open.* This is said to be diagnostic of posterior column disease but may also be found in unilateral or bilateral weakness, cerebellar disease, or disease of the peripheral nerves.

Abnormalities in gait should be described as closely as possible. A difference in length of stride between the feet should be mentioned. Any tendency toward dragging of the toes and high lifting of the knees due to weakness of the dorsiflexors of the feet (*"foot drop" gait*) should also be commented upon. The hemiplegic gait with circumduction of the affected leg, weakness of dorsiflexion of the foot, and some tendency toward flexion at the knee on the affected side is classic. Similarly, the wide-based *ataxic gait* (drunken gait) is the classic gait of cerebellar dysfunction. Also diagnostic is the typical *gait of parkinsonism,* with small shuffling steps, lack of normal arm swing, flexion of the trunk, and a tendency to increase the speed and fall forward. The dancing *gait of advanced Huntington's chorea* is distinctive, as are the wormlike, athetoid movements of the limbs and trunk in athetosis. Congenital spasticity (Little's disease) produces a *scissors gait,* with a tendency toward internal rotation of both legs and scraping together of the semiflexed knees as the patient drags them forward. The typical *gait of tabes dorsalis* is characterized by ataxia, foot slapping, and a tendency for the patient to watch his feet (since he does not know where they are) when he walks.

Further description of these classic gaits would be unrewarding, since once they are seen in the typical form they are not easily forgotten.

Meningeal Signs

Although normal patients do not demonstrate any stiffness of the neck or signs of meningeal irritation, there are many causes of false positive and false negative meningeal signs. Patients with tetanus, extensor rigidity, parkinsonism, bony fixation of the neck, and extensor muscle spasm of the neck from any cause may have neck stiffness and decreased ability to flex the neck painlessly. Similarly, abnormalities of the muscles and joints of the lower extremity may produce an abnormal response on straight-leg raising. Conversely, patients with definite meningitis or subarachnoid hemorrhage occasionally have no signs of meningeal irritation at all, especially if they are unconscious or moribund.

Aphasia and Speech Defects

A normal patient will of course demonstrate no dysarthria or aphasia. *It is important to know the handedness of the patient, although if this cannot be ascertained, it is usually safe to assume that the left cerebral hemisphere is dominant.* (About 98 percent of all people, including at least half of those who are left-handed, have left-hemisphere dominance.) Involvement of the dominant hemisphere by any type of disease often produces aphasia or dysphasia (partial aphasia).

In the common cerebral infarct in either hemisphere there will be some short-lived dysarthric slurring of speech. It should not be confused with aphasia. The ataxia of speech of cerebellar disturbances and multiple sclerosis is also to be differentiated from the two above-mentioned interferences with communication. Lesions of the brain stem or bilateral cerebral lesions produce a bulbar or pseudobulbar speech which is similar to the dysarthria produced by a lesion of one hemisphere, but is more marked and more likely to persist. The significance of abnormalities found on testing for aphasia is a confused and complicated subject and will be discussed only briefly. If the patient's aphasia is mainly expressive, his lesion is likely to be in the front half of the dominant hemisphere; if the aphasia is receptive, the lesion is in the back half. If the aphasia is mixed, as it usually is, the lesion may include both frontal and parietal lobes or be deep in the hemisphere.

Conversion Reactions

The diagnosis of hysteria or conversion reaction on the basis of findings on neurologic examination is dangerous. It is true that a stocking or glove anesthesia, a sensory defect extending exactly to the midline, a sensory defect with changing borders, paralysis of the legs or of half the body, "giving way" on strength testing, and other "hysterical" neurologic findings can be and sometimes are nonorganic, but they are also found in bona fide neurologic disease. A stocking or glove sensory loss is also found in peripheral neuritis, a hemisensory defect to the midline is found in several types of lesions of the brain stem and spinal cord, and normally there is some variability in the borders of most true sensory defects. The organic causes of paraplegia are many, and anyone with such a defect, if it is recent, will naturally demonstrate some anxiety and perhaps appear hysterical. The same is true of a hemiparesis. Giving way on strength testing may be found when testing patients with true weaknesses. One must beware of making the easy diagnosis of hysteria. Many a spinal-cord tumor has gotten in its fatal blow while the physician was investigating the psyche. This is a greater error than treating a patient for organic disease when there is none.

GENERAL EXAMINATION

The carotid pulses are usually equal and normal, and when absent or markedly decreased may suggest an occlusion. However, complete or partial occlusion can take place in the presence of what is apparently a normally palpable carotid artery. If the common carotid artery in the neck gives a full pulse, but the pulse in the temporal artery on the same side is decreased or absent, there is presumptive evidence of occlusion of the external carotid. Retinal artery pressure readings (ophthalmodynamometry) give a more reliable clue as to the patency of the internal and common carotid systems than does simple palpation. The ophthalmodynamometer may also be used to determine the blood pressure in the temporal artery. The subclavian, brachial, and radial pulses and also the dorsalis pedis, posterior tibial, and femoral pulses, if absent or definitely decreased, may give hints of occlusion of the main branches of the aorta, and this may be helpful in neurologic diagnosis.

Auscultation of the skull may reveal a bruit on the side of increased or decreased blood flow through a carotid or vertebral artery. Increased flow may be due to occlusion on the opposite side or to an arteriovenous anomaly or tumor on the same side. A bruit due to decreased flow occurs with partial or complete occlusion on the same or opposite sides, or even in the absence of any vascular disease.

A bruit in the supraclavicular space may indicate partial or complete occlusion of the subclavian artery with possible production of the "subclavian steal" syndrome. In this syndrome there may be also a decrease in the blood pressure in the affected arm and a delay in the radial pulse on the same side.

REFERENCES

1. DeJong, R. N. *The Neurologic Examination* (3rd ed.). New York: Hoeber Med. Div., Harper & Row, 1967.
2. Mayo Clinic, Sections of Neurology and Section of Physiology. *Clinical Examinations in Neurology* (3rd ed.). Philadelphia: Saunders, 1971.

3. Steegman, A. T. *Examination of the Nervous System* (3rd Ed.). Chicago: Year Book, 1970.
4. Van Allen, M. W. *Pictorial Manual of Neurologic Tests.* Chicago: Year Book, 1969.

PEDIATRIC EXAMINATION   21

George H. Lowrey

For several reasons this chapter could be conveniently divided into three parts — the infant, the young child, and the older child and adolescent. The physician approaching the infant must rely totally upon the parents for the history of illness, and in his physical examination he must depend almost entirely upon objective methods. Although observation of behavior is an essential part of this examination, he cannot obtain the verbal responses that are of such value in an older patient. In the preschool and younger school child some history may be obtained directly, but much reliance must still be placed on the parent. Fear of the doctor and what he may do will influence responses during the history taking as well as during the examination. In the older child and adolescent, considerable faith can be put in the responses obtained, though anxiety may still modify these to a considerable extent. In dealing with children the physician has an advantage in that he is usually confronted with facts unchanged by theories or imaginary ills. Symptoms and physical signs are true and dependable when elicited from the child. Nevertheless, the examiner must be even more alert to note the wince that indicates tenderness, the facial expression associated with nausea, the position or posture of greatest comfort, the presence or absence of a smile to an appropriate stimulus, to list just a few examples.

In interviewing children, the physician should carefully select his words to be understood. His method of questioning should be slow and deliberate, and his attitude should convey real interest in what the patient says. In very young children, a few minutes of play of a casual type may be most helpful in "breaking the ice."

The outline of the history given below will vary in order and in detail depending upon what is desired in terms of completeness or upon the nature of the patient's illness. The history of a two-year-old who appears to be retarded would require considerable detail in data relating to the course of pregnancy, method of delivery, birth weight, early feeding problems, times of accomplishing the developmental milestones, and similar elements. The family history relative to genetic factors would be important. Conversely, the history of a twelve-year-old who has had no previous problems but now has a fever and a sore throat would require relatively little information about his birth and neonatal history.

The history of a sick child should indicate the background of the child and his family and answer to some degree the question, "In what sort of a child did this sickness develop?" As with the adult, the informant may be guided by the examiner, but he or she should be given the freedom to give a complete record in his or her own words.

HISTORY FORM

Name _____ Informant _____
Birth date _____ Reliability _____

Chief Complaint: Usually a single symptom in the informant's own words, duration.

Present Illness: Initial symptoms, date of onset, subsequent symptoms chronologically, pertinent negative data by direct questions.

Past History:

1. Birth and neonatal: prenatal care, illnesses during pregnancy; gestation time; labor, delivery (position, instruments, etc.); birth weight; immediate cry, cyanosis (duration, therapy); jaundice (duration, therapy); days hospitalized; early feeding (breast, bottle, difficulties); early weight gain.

2. Developmental milestones (average time accomplished given in parentheses): smiles (2 months); head held steady (2 months); lifts head when prone (2 months); eyes follow (2 months); sits unsupported (7 months); grasps toys (7 months); creeps (8 months); pulls to feet at rail (10 months); one to two meaningful words (11 months); walks alone (14 months); toilet training; grade in school, progress and problems.

3. Feeding history (mainly for young children): breast feeding (duration, etc.); formula (ingredients, changes in formula and why); schedule; duration and quantity per feeding (apparent cause of prolonged feeding time); weight at various ages; solids (when started and how received); vomiting (relation to feeding, character of material, projectile, etc.); stools (frequency, quantity, color, consistency).

4. Immunizations (reactions): pertussis, polio, tetanus, diphtheria, smallpox, measles, others (include boosters).

5. Illnesses (frequency, severity, complications, operations, fractures, accidents, allergies, etc.).

6. Habits (sleep, naps, bowel and bladder, nail biting, tics, behavior with other children, etc.).

Family History: Many complaints about a child may result from problems within the family. Examples are parental conflicts, chronic or recurrent absence of one or both parents, intense sibling rivalry, rigidity of discipline or disagreement of parents in its methods or use, unrealistic expectations of the child's performance. Since many of these factors may not be recognized or may be suppressed by either parent or child, a direct question concerning them is often unrewarding. Clues may be obtained from remarks about changes in behavior, withdrawal from friends, poor schoolwork, etc. In addition to these aspects the usual family history as previously outlined should be recorded.

Social History: Because of its influence upon the child this may be very important — type of home, own room for child or shared, number of people in home, income, interfamily relations, etc.

Systems Review: Similar to that for the adult.

THE PHYSICAL EXAMINATION

General Rules of Procedure

Observation of the patient during the interview often reveals evidence of mental retardation, parent-child conflicts, parental attitudes concerning discipline, posture related to pain or weakness, facial expression related to specific questions in the older child. The mother's handling of the infant while dressing or undressing him and while feeding is often revealing of her level of understanding and emotional reaction to the infant; it may also indicate errors in feeding techniques. Ambulation, relative to the age of the child, is an important observation too often neglected in the usual examination. Giving the child objects with which he can play will also help in the examination of general dexterity and his developmental level.

Respect the child's sense of modesty and level of understanding. In the young child undressing may be interpreted as a loss of personal identity and is best done in stages. A sheet should be provided for the older child and adolescent, and the undressing should be accomplished in the physician's absence. The examiner must proceed nonchalantly and in a manner of confidence. Especially in children from about 1 to 6 years, some resistance is to be expected. A friendly attitude with conversation and casual play at the child's level, but never condescending nor allowing the child to take command of the situation, is helpful. The physician must never convey feelings of either frustration or anger. An explanation of what is to be done and showing the child the instruments to be used beforehand may contribute to his cooperation. Often, part or all of the examination may best be accomplished with the patient in the mother's lap. In the hospitalized patient, several attempts with utmost patience may be necessary to perform abdominal palpation or some other portion of the examination. Skill in this respect comes with experience, and the student should not be discouraged by initial failures.

Order of procedure should delay the most objectionable parts until last. It is sometimes desirable to examine that part of the body first from which the chief complaint arises. This is because cooperation is often best early in the examination before fatigue or discomfort is experienced. Listening to the chest or palpating the abdomen before the child frets or cries may be important. Examination of the throat and ears is frequently disagreeable to the infant and young child and may be delayed to the end. The fact that restraint is often necessary in this portion of the examination accounts in part for the patient's objection.

Warm and clean hands and instruments are appreciated by both patient and parent.

Too much emphasis cannot be placed upon these preliminary considerations. They may mean the difference between a satisfactory and unsatisfactory physical evaluation. Both physician and patient can enjoy the examination, and this attitude should prevail. It might also be emphasized that the short time the examiner spends with the child should not be used as an opportunity to try to correct faults in disciplinary training.

With the very ill child or the smaller premature infant, the physical examination may be carried out in brief stages to permit periods of rest. In the premature this may also be necessary in order to conserve body temperature and to maintain adequate humidity or oxygen administration in an incubator.

The eyes are examined as in the adult, but examination often requires more patience and time in young subjects. Participation in a game or following a toy moved about by the examiner will be valuable in ascertaining the ocular movements and pupillary responses.

The ear, nose, and throat examination is best delayed to the end in the young child, as he often requires restraint. Figures 21-1 and 21-2 illustrate the technique. The electrical otoscope is as much a badge of the pediatrician as is the stethoscope. The anatomy of the ear in children differs only slightly from that of the adult, but the frequency of abnormalities here, especially otitis media, is so great that it must be stated that omission of a detailed discussion here does not indicate lack of importance.

The abdominal examination, because of crying, may be somewhat more difficult than in the adult. If the child is frightened, repeated attempts may be necessary. Distraction away from the examining hand can often be accomplished by conversation of interest to the child or by attracting his attention with a toy.

The rectal examination is extremely important in the child. Digital examination must be done with adequate lubrication, slow and steady pressure with the finger until it passes the sphincter, and the use of the little (fifth) finger in infants and young children. A nasal speculum may be used for examining the anal region in infants.

The neurologic examination will often evaluate behavior and responses to be expected to certain stimuli as well as the more formal elicitation of reflexes, muscle tone, and sensation. For example, what the nine-month-old infant does

FIGURE 21-1. Technique of ear examination in the young child. Note that the examiner's hand holding the otoscope rests on the patient's head, so that any sudden movement will be transferred to both the hand and instrument and prevent trauma to the ear canal.

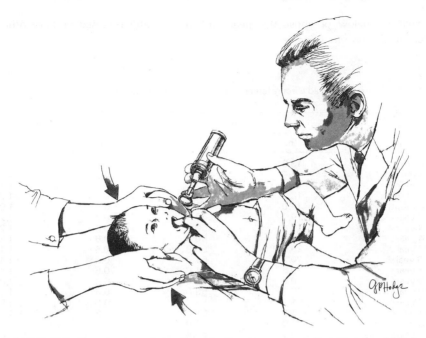

FIGURE 21-2. Examination of the throat in the young patient. Note that the child's head is held still by exerting pressure on his arms against the sides of the head. In this manner one person can immobilize both the head and arms.

with three cubes or blocks may be as important as his Achilles reflex. Does he grasp firmly, reach accurately, transfer from hand to hand, use fingers as well as palm in grasping?

NORMAL PHYSICAL CHARACTERISTICS

The child is not a miniature adult. The characteristics and measurements to be outlined constitute important references with regard to both history and the physical examination and serve as guides in evaluating normality or abnormality. In contrast to the nearly static adult, children are everchanging organisms.

Head size is relatively larger in children than in adults; the younger the child, the more this is evident. *Head circumference measurements* have a relatively narrow range for any age and are directly related to intracranial volume; they therefore permit an estimation of brain growth. Rate of growth is of vital importance in patients with suspected hydrocephalus or microcephalus. The measurement is obtained by passing a tape measure over the occipital protuberance and just above the supraorbital ridges (Table 21-1). The importance of this measurement can be appreciated when it is realized that more than 85 percent of children with a circumference of more than two standard deviations below or above the mean will be mentally retarded or have other neurologic abnormalities.

Table 21-1. Average Head Circumference of American Children

Age	Mean		Standard Deviation	
	In.	Cm.	In.	Cm.
Birth	13.8	35.0	0.5	1.2
3 months	15.9	40.4	0.5	1.2
6 months	17.0	43.4	0.4	1.1
12 months	18.3	46.5	0.5	1.2
18 months	19.0	48.4	0.5	1.2
2 years	19.2	49.0	0.5	1.2
3 years	19.6	50.0	0.5	1.2
4 years	19.8	50.5	0.5	1.2
5 years	20.0	50.8	0.6	1.4
6 years	20.2	51.2	0.6	1.4
7 years	20.5	51.6	0.6	1.4
8 years	20.6	52.0	0.8	1.8
10 years	20.9	53.0	0.6	1.4
12 years	21.0	53.2	0.8	1.8
14 years	21.5	54.0	0.8	1.8
16 years	21.9	55.0	0.8	1.8
18 years	22.1	55.4	0.8	1.8
20 years	22.2	55.6	0.8	1.8

Note: Tables 21-1 through 21-3 are from G. H. Lowrey, *Growth and Development of Children* (6th ed.), Chicago: Year Book Medical Publishers, 1973. They are reproduced by permission of the author and the publisher.

Height and weight measurements are always a routine part of the examination. Together, they have great value in estimating the state of nutrition, general health, and some aspects of endocrine balances and maturation. Often the first recognized sign of disease is failure to gain normal increments in weight or stature or actual loss of weight (Tables 21-2 and 21-3). In the final analysis of the growth of a child, the expected rate of gain is of greater value than any single measurement.

The extremities are comparatively short during the first few years of life. (The span of the outstretched arms is less than the standing height until approximately age 10 years in boys and 14 years in girls.) These proportions plus the abundant subcutaneous fat give the infant a rotund appearance. Going into the second and third years, the child normally becomes more linear and lean. Recognition of these changes is important in counseling worried parents and avoiding feeding problems precipitated by parents who desire to maintain the plumpness as a mistaken indication of healthfulness.

The spine is more flexible in the infant and child than in the adult. Some degree of lordosis and "pot belly" is natural until midchildhood.

Table 21-2. Weight and Height Percentile Table: Girls (Birth to 18 Years)

Weight in Pounds			Weight in Kilograms			AGE	Height in Inches			Height in Centimeters		
10%	50%	90%	10%	50%	90%		10%	50%	90%	10%	50%	90%
6.2	7.4	8.6	2.81	3.36	3.9	Birth	18.8	19.8	20.4	47.8	50.2	51.0
8.0	9.7	11.0	3.3	4.2	5.0	1 mo.	20.2	21.0	22.0	50.4	52.8	55.0
9.5	11.0	12.5	4.1	5.0	5.8	2 mo.	21.5	22.2	23.2	53.7	55.5	59.6
10.7	12.4	14.0	4.85	5.62	6.35	3 mo.	22.4	23.4	24.3	56.9	59.5	61.7
12.0	13.7	15.5	5.3	6.2	7.2	4 mo.	23.2	24.2	25.2	59.6	61.0	64.8
13.0	14.7	17.0	5.9	6.8	7.7	5 mo.	24.0	25.0	26.0	60.7	64.2	67.0
14.1	16.0	18.6	6.4	7.26	8.44	6 mo.	24.6	25.7	26.7	62.5	65.2	67.8
16.6	19.2	22.4	7.53	8.71	10.16	9 mo.	26.4	27.6	28.7	67.0	70.1	72.9
18.4	21.5	24.8	8.35	9.75	11.25	12 mo.	27.8	29.2	30.3	70.6	74.2	77.1
21.2	24.5	28.3	9.62	11.11	12.84	18 mo.	30.2	31.8	33.3	76.8	80.9	84.5
23.5	27.1	31.7	10.66	12.29	14.38	2 yr.	32.3	34.1	35.8	82.0	86.6	91.0
25.5	29.6	34.6	11.57	13.43	15.69	2½ yr.	34.0	36.0	37.9	86.3	91.4	96.4
27.6	31.8	37.4	12.52	14.42	16.96	3 yr.	35.6	37.7	39.8	90.5	95.7	101.1
29.5	33.9	40.4	13.38	15.38	18.33	3½ yr.	37.1	39.2	41.5	94.2	99.5	105.4
31.2	36.2	43.5	14.15	16.42	19.73	4 yr.	38.4	40.6	43.1	97.6	103.2	109.6
32.9	38.5	46.7	14.92	17.46	21.18	4½ yr.	39.7	42.0	44.7	100.9	106.8	113.5
34.8	40.5	49.2	15.79	18.37	22.32	5 yr.	40.5	42.9	45.4	103.0	109.1	115.4
38.0	44.0	51.2	17.24	19.96	23.22	5½ yr.	42.4	44.4	46.8	107.8	112.8	118.9
39.6	46.5	54.2	17.96	21.09	24.58	6 yr.	43.5	45.6	48.1	110.6	115.9	122.3
42.2	49.4	57.7	19.14	22.41	26.17	6½ yr.	44.8	46.9	49.4	113.7	119.1	125.6
44.5	52.2	61.2	20.19	23.68	27.76	7 yr.	46.0	48.1	50.7	116.8	122.3	128.9
46.6	55.2	65.6	21.14	25.04	29.76	7½ yr.	47.0	49.3	51.9	119.5	125.2	131.8
48.6	58.1	69.9	22.04	26.35	31.71	8 yr.	48.1	50.4	53.0	122.1	128.0	134.6
50.6	61.0	74.5	22.95	27.67	33.79	8½ yr.	49.0	51.4	54.1	124.6	130.5	137.5
52.6	63.8	79.1	23.86	28.94	35.88	9 yr.	50.0	52.3	55.3	127.0	132.9	140.4
54.9	67.1	84.4	24.9	30.44	38.28	9½ yr.	50.9	53.5	56.4	129.4	135.8	143.2
57.1	70.3	89.7	25.9	31.89	40.69	10 yr.	51.8	54.6	57.5	131.7	138.6	146.0
59.9	74.6	95.1	27.17	33.79	43.14	10½ yr.	52.9	55.8	58.9	134.4	141.7	149.7
62.6	78.8	100.4	28.4	35.74	45.54	11 yr.	53.9	57.0	60.4	137.0	144.7	153.4
66.1	83.2	106.0	29.98	37.74	48.08	11½ yr.	55.0	58.3	61.8	139.8	148.1	157.0
69.5	87.6	111.5	31.52	39.74	50.58	12 yr.	56.1	59.8	63.2	142.6	151.9	160.6
74.7	93.4	118.0	33.88	42.37	53.52	12½ yr.	57.4	60.7	64.0	145.9	154.3	162.7
79.9	99.1	124.5	36.24	44.95	56.47	13 yr.	58.7	61.8	64.9	149.1	157.1	164.8
85.5	103.7	128.9	38.78	47.04	58.47	13½ yr.	59.5	62.4	65.3	151.1	158.4	165.9
91.0	108.4	133.3	41.28	49.17	60.46	14 yr.	60.2	62.8	65.7	153.0	159.6	167.0
94.2	111.0	135.7	42.73	50.35	61.55	14½ yr.	60.7	63.1	66.0	154.1	160.4	167.6
97.4	113.5	138.1	44.18	51.48	62.64	15 yr.	61.1	63.4	66.2	155.2	161.1	168.1
99.2	115.3	139.6	45.0	52.3	63.32	15½ yr.	61.3	63.7	66.4	155.7	161.7	168.6
100.9	117.0	141.1	45.77	53.07	64.0	16 yr.	61.5	63.9	66.5	156.1	162.4	169.0
101.9	118.1	142.2	46.22	53.57	64.5	16½ yr.	61.5	63.9	66.6	156.2	162.5	169.2
102.8	119.1	143.3	46.63	54.02	65.0	17 yr.	61.5	64.0	66.7	156.3	162.5	169.4
103.2	119.5	143.9	46.81	54.2	65.27	17½ yr.	61.5	64.0	66.7	156.3	162.5	169.4
103.5	119.9	144.5	46.95	54.39	65.54	18 yr.	61.5	64.0	66.7	156.3	162.5	169.4

Table 21-3. Weight and Height Percentile Table: Boys (Birth to 18 Years)

Weight in Pounds			Weight in Kilograms			AGE	Height in Inches			Height in Centimeters		
10%	50%	90%	10%	50%	90%		10%	50%	90%	10%	50%	90%
6.3	7.5	9.1	2.86	3.4	4.13	Birth	18.9	19.9	21.0	48.1	50.6	53.3
8.5	10.0	11.5	3.8	4.6	5.2	1 mo.	20.2	21.2	22.2	50.4	53.0	55.5
10.0	11.5	13.2	4.6	5.2	6.0	2 mo.	21.5	22.5	23.5	53.7	56.0	60.0
11.1	12.6	14.5	5.03	5.72	6.58	3 mo.	22.8	23.8	24.7	57.8	60.4	62.8
12.5	14.0	16.2	5.6	6.3	7.3	4 mo.	23.7	24.7	25.7	60.5	62.0	65.2
13.7	15.0	17.7	6.2	7.0	8.0	5 mo.	24.5	25.5	26.5	61.8	65.0	67.3
14.8	16.7	19.2	6.71	7.58	8.71	6 mo.	25.2	26.1	27.3	63.9	66.4	69.3
17.8	20.0	22.9	8.07	9.07	10.39	9 mo.	27.0	28.0	29.2	68.6	71.2	74.2
19.6	22.2	25.4	8.89	10.7	11.52	12 mo.	28.5	29.6	30.7	72.4	75.2	78.1
22.3	25.2	29.0	10.12	11.43	13.15	18 mo.	31.0	32.2	33.5	78.8	81.3	85.0
24.7	27.7	31.9	11.2	12.56	14.47	2 yr.	33.1	34.4	35.9	84.2	87.5	91.1
26.6	30.0	34.5	12.07	13.61	15.65	2½ yr.	34.8	36.3	37.9	88.5	92.1	96.2
28.7	32.2	36.8	13.02	14.61	16.69	3 yr.	36.3	37.9	39.6	92.3	96.2	100.5
30.4	34.3	39.1	13.79	15.56	17.74	3½ yr.	37.8	39.3	41.1	96.0	99.8	104.5
32.1	36.4	41.4	14.56	16.51	18.78	4 yr.	39.1	40.7	42.7	99.3	103.4	108.5
33.8	38.4	43.9	15.33	17.42	19.91	4½ yr.	40.3	42.0	44.2	102.4	106.7	112.3
35.5	40.5	46.7	16.1	18.37	21.18	5 yr.	40.8	42.8	45.2	103.7	108.7	114.7
38.8	45.6	53.1	17.6	20.68	24.09	5½ yr.	42.6	45.0	47.3	108.3	114.4	120.1
40.9	48.3	56.4	18.55	21.91	25.58	6 yr.	43.8	46.3	48.6	111.2	117.5	123.5
43.4	51.2	60.4	19.69	23.22	27.4	6½ yr.	44.9	47.6	50.0	114.1	120.8	127.0
45.8	54.1	64.4	20.77	24.54	29.21	7 yr.	46.0	48.9	51.4	116.9	124.1	130.5
48.5	57.1	68.7	22.0	25.9	31.16	7½ yr.	47.2	50.0	52.7	120.0	127.1	133.9
51.2	60.1	73.0	23.22	27.26	33.11	8 yr.	48.5	51.2	54.0	123.1	130.0	137.3
53.8	63.1	77.0	24.4	28.62	34.93	8½ yr.	49.5	52.3	55.1	125.7	132.8	140.0
56.3	66.0	81.0	25.54	29.94	36.74	9 yr.	50.5	53.3	56.1	128.3	135.5	142.6
58.7	69.0	85.5	26.63	31.3	38.78	9½ yr.	51.4	54.3	57.1	130.6	137.9	145.1
61.1	71.9	89.9	27.71	32.61	40.78	10 yr.	52.3	55.2	58.1	132.8	140.3	147.5
63.7	74.8	94.6	28.89	33.93	42.91	10½ yr.	53.2	56.0	58.9	135.1	142.3	149.7
66.3	77.6	99.3	30.07	35.2	45.04	11 yr.	54.0	56.8	59.8	137.3	144.2	151.8
69.2	81.0	104.5	31.39	36.74	47.4	11½ yr.	55.0	57.8	60.9	139.8	146.9	154.8
72.0	84.4	109.6	32.66	38.28	49.71	12 yr.	56.1	58.9	62.2	142.4	149.6	157.9
74.6	88.7	116.4	33.84	40.23	52.8	12½ yr.	56.9	60.0	63.6	144.5	152.3	161.6
77.1	93.0	123.2	34.97	42.18	55.88	13 yr.	57.7	61.0	65.1	146.6	155.0	165.3
82.2	100.3	130.1	37.29	45.5	59.01	13½ yr.	58.8	62.6	66.5	149.4	158.9	168.9
87.2	107.6	136.9	39.55	48.81	62.1	14 yr.	59.9	64.0	67.9	152.1	162.7	172.4
93.3	113.9	142.4	42.32	51.66	64.59	14½ yr.	61.0	65.1	68.7	155.0	165.3	174.6
99.4	120.1	147.8	45.09	54.48	67.04	15 yr.	62.1	66.1	69.6	157.8	167.8	176.7
105.2	124.9	152.6	47.72	56.65	69.22	15½ yr.	63.1	66.8	70.2	160.3	169.7	178.2
111.0	129.7	157.3	50.35	58.83	71.35	16 yr.	64.1	67.8	70.7	162.8	171.6	179.7
114.3	133.0	161.0	51.85	60.33	73.03	16½ yr.	64.6	68.0	71.1	164.2	172.7	180.7
117.5	136.2	164.6	53.3	61.78	74.66	17 yr.	65.2	68.4	71.5	165.5	173.7	181.6
118.3	137.6	166.8	53.89	62.41	75.66	17½ yr.	65.3	68.5	71.6	165.9	174.1	182.0
120.0	139.0	169.0	54.43	63.05	76.66	18 yr.	65.5	68.7	71.8	166.3	174.5	182.4

Secondary sexual development shows great individual variation at time of onset. The following list gives averages of the range for normal American children.

Female fat deposition about pelvis	8-10 years
Initial breast hypertrophy	9-11 years
Mature breast development	14-18 years
Female pubic hair	9-12 years
Female axillary hair	10-13 years
Enlargement of penis and testes	10-12 years
Male pubic hair	10-13 years
Male axillary hair	11-15 years
Male facial hair	12-15 years

Temperature is usually obtained rectally until the age of 3 or more. A normal temperature in infancy and early childhood may be a degree or more above the adult average (98.6°F. or 37°C.).

Pulse and respiratory rates are more rapid in younger children and change with age. These should be obtained when the child is quiet. Both of these are fairly sensitive measures of fever, increasing about 15 to 20 percent with each degree of rise in temperature. The respiratory rate and depth of breathing may increase with either respiratory or metabolic acidosis, and such changes are often the initial physical findings of the underlying abnormality. Cardiac and pulmonary disease are also reflected by deviations from the normal.

Variations in Respiratory Rates
(Quiet Breathing)

Age	Rate per minute
Premature	40-90
Newborn	30-80
1 year	20-40
2 years	20-35
4 years	20-35
10 years	18-20
Adults	15-18

Average Heart Rate at Rest

Age	Rate per minute
Birth	130-150
1-6 months	120-140
6-12 months	110-130
1-2 years	110-120
2-4 years	90-110
6-10 years	90-100
10-14 years	80-90

Blood pressure also changes with age. A cuff of proper width is necessary for accurate readings. It should cover approximately one-third to one-half of the upper arm. Most children need to be reassured that the procedure will not be very uncomfortable, and the readings should be considered true values only when the subject is quiet and emotionally undisturbed.

Average Blood Pressure (mm. Hg)		
Age	Systolic	Diastolic
1 year	60	40
3 years	85	65
5 years	90	70
8 years	95	70
10 years	95	70
15 years	100-110	70-80

The skin and subcutaneous tissues reflect the general state of hydration and nutrition. The status of tissue turgor in the infant and child is of particular importance, and it is best demonstrated by picking up a fold of abdominal skin between the thumb and index finger. Normally upon release the skin rapidly returns to its former position. In states of dehydration or undernutrition the skin remains creased and raised for a varying period of time. In the premature and newborn infant the skin appears thin and almost transparent. It is red and wrinkled under normal conditions. Small red patches (nevus vasculosus) which are not raised and which blanch with pressure may be present over the occiput, forehead, and upper eyelids. These are commonly seen in the newborn. The soft, moist, white or clay-colored material covering all newborn infants is the vernix caseosa. Some flaky desquamation occurs shortly after birth and varies in degree with individuals. For the first few weeks of life very small white to yellow, raised lesions that are discrete are present normally in groups, especially over the face. They are caused by plugging of the as yet poorly functioning sebaceous glands, and collectively they are known as milia. Miliaria is the red "prickly heat" rash noted during the summer or in overly dressed infants. A blotchy blue appearance of the hands and feet (acrocyanosis) is normal in early infancy but is not a constant finding. Bluish, irregularly shaped areas that are not raised and vary greatly in size are sometimes present over the sacral and buttock areas of the darker-complexioned infants and are called Mongolian spots. These disappear with increasing age. Physiologic jaundice is present to a mild degree in many infants starting after the first day of life and usually disappearing by the eighth or tenth day. Jaundice that appears during the first 24 hours of life usually indicates excessive hemolysis, *hemolytic disease of the newborn,* due to the presence of maternal antibodies against the infant's red cells. If jaundice persists and gradually becomes more intense over the first few weeks of life, congenital anomalies of the biliary tree with obstruction should be suspected.

The lymph nodes have a distribution in children similar to that in adults but are more prominent up to the time of puberty. They are easily palpable as shotty, small, bean-sized nodules, and usually undergo considerable hypertrophy in response to infections throughout all of childhood.

The head of the newborn often undergoes some distortion in shape as it passes through the birth canal. There may be overlapping of the large flat bones, which are easily palpable. Depending upon the degree of molding, as this process is called, from a few days to a few weeks may pass before normal anatomic relationships are reestablished. Soft-tissue molding of the scalp at the time of birth results in *caput succedaneum*. This is a soft, poorly outlined swelling that pits on pressure from the edema present. It may overlie the suture lines. Asymmetry of the head may result from premature closure of some of the sutures. It may also be seen in the normal infant who always lies in the same position since the bones at this time are very soft. Flattening of a portion of the cranium may occur in normal infants but is more often associated with certain pediatric diseases, such as torticollis or mental retardation, due to the tendency for such children to maintain a constant position. *Cephalhematoma* is a swelling resulting from bleeding beneath the periosteum of the cranium and is therefore limited to a single cranial bone. Palpation usually reveals a small firm elevated margin of the lesion which is becoming organized into a clot and later may be calcified. This and the superficially similar-appearing caput succedaneum are limited to the newborn period. Careful and frequent measurements of head circumference constitute an important method of appraisal when compared with tables for normal growth rates. The posterior fontanel is closed to palpation in a few months. The anterior fontanel varies greatly in size throughout early infancy but is usually palpable only as a slight depression by 12 to 18 months. Normally, until they close, some arterial pulsation is transmitted through the fontanels.

Head control and movement of the head are important in evaluation of neuromuscular development. By 2 months, the head is held relatively steady when the baby is supported in an erect position, and he can raise it from a prone position. By 4 months, head control is good, with no unsteadiness. Complete coordination of eye movements may not develop until near the end of the first year; however, if there is any question of strabismus, this should be carefully evaluated to prevent amblyopia. Visual testing with a Snellen E or other appropriate chart can be accomplished by age 2½ to 3 years.

The *tonsils and adenoids,* as with all other lymphoid tissue, are relatively large and cryptic in many children. The presence of enlarged tonsils does not necessarily indicate chronic infection.

The chest in the infant and young child has a relatively greater anterior-posterior diameter than in the adult. The chest wall is so thin that disease of underlying structures may be more easily discovered by auscultation and percussion than in the adult. Small nodular breast hypertrophy is found in most newborn infants and may be associated with small amounts of milklike secretion for a few days. Some breast hypertrophy, not always symmetrical, is usually present transiently in adolescent boys. Obese children often have apparent breast hypertrophy; however, this is due to adipose tissue and not to glandular hypertrophy.

Because of the thinness of the chest wall in young children, auscultation reveals breath sounds that normally are loud, harsh, and somewhat bronchial in character as compared with the adult. Pathology is actually more readily apparent than in older subjects, once experience has been gained by listening to the normal chest. Percussion note over the lung fields is more resonant in the child and even approaches being tympanic in quality. In the infant respiration is largely under control of the diaphragm, with little or no intercostal movement. This leads to the so-called abdominal type of respiration until about 6 years of life. Examination of the chest in a crying infant or child has considerable value and should not be considered as meaningless. Deep respiratory sounds are actually enhanced. Even slight changes in position of the infant, such as turning the head, may influence the relative positions of the intrathoracic structures and therefore the intensity of breath sounds or the degree of resonance.

The heart in early life fills relatively more of the thoracic cavity than in later life, and the apex is one or two intercostal spaces above that which would be considered normal in the adult. Sinus arrhythmia is a physiologic phenomenon prominent throughout infancy and childhood. This finding is so constant that its absence should be looked upon as suggestive of cardiac abnormality. The heart sounds during childhood are of a higher pitch and shorter duration with greater intensity than during later life. Until adolescence, the pulmonary second sound is regularly louder than the aortic. Functional murmurs are the rule during childhood. They are less common in the newborn than later. Between the ages of 6 and 9, over half the children have murmurs that are obvious to the examiner. The commonest areas of maximum intensity in the order of frequency are the third to fourth intercostal area at the left border of the sternum, the pulmonic area, and the apex. Parasternal murmurs become less frequent as adolescence is approached, while pulmonic ones become more prevalent. These murmurs usually are of grade 3 intensity or less, well localized, and either blowing or vibratory in character. Change in intensity or complete disappearance may follow a change in position. A venous hum is also common in childhood. It is a continuous purring sound that is best heard either above or below the clavicles. It is accentuated in the upright position. It should not be confused with the murmur of patent ductus arteriosus.

The liver edge is often palpable in infancy and childhood and the spleen is normally palpable on deep inspiration in some children.

The size of *the genitalia* must be evaluated in relationship to age and not necessarily to body size. The penis and scrotum often appear disproportionately small in obese boys.

The changing pattern of *reflex behavior* in early infancy is important in estimating neurologic integrity. At birth tonicity and activity are equal bilaterally. The premature newborn has decreased tonicity and activity as compared to the normal infant. The premature infant lies in a flaccid position with hands open. The full-term infant assumes a flexed position with hands fisted and efforts to straighten out the extremities meet considerable resistance. When the baby is supported by one hand under the abdomen in a horizontal position, he raises his head and legs toward the plane of his body. This *Landau reflex,* normally easily elicited in the newborn, may be absent in the premature infant. The full-term infant firmly grasps an object, such as a finger, placed in its palm and can be lifted up so that most or all of its weight is supported. The greater the degree of prematurity, the less strong this *grasp reflex* becomes. Under 36 weeks of gestation the response may be absent or very weak. Sucking is vigorous on a finger placed in the mouth. When the cheek is lightly stroked the infant turns his head toward the stimulated side and the lips may protrude in preparation for sucking. This reaction is termed the *rooting reflex*. The newborn responds to sudden change in position, jarring, or loud noises by the *Moro reflex*. This is characterized by a tensing of muscles, a wide embracing motion of the upper extremities, and some extension of the legs. Normally this reflex disappears by 2 months of age, and its persistence beyond that time indicates neurologic abnormality.

Blinking, sneezing, gagging, and coughing to appropriate stimuli are easily elicited in the full-term infant and in all but the smallest of those born prematurely. Although a typical *Babinski reflex* is seldom demonstrable in the newborn, plantar flexion of the great toe to the usual stimulus is present in most and may persist throughout much of the first year. Due to the often relaxed Achilles tendon from fetal positioning, the ankle jerk may not exist, but all other *deep tendon reflexes* are present at birth. All the *superficial reflexes* — abdominal, anal, and cremasteric — are present at birth, although they may be somewhat difficult to elicit.

CARDINAL SYMPTOMS AND ABNORMAL FINDINGS

Posture, muscle tone, and coordination may be observed during the history-taking. Pain in the abdomen can result in flexion of the thighs upon the abdomen in infants and younger children. *Opisthotonus* is indicative of meningeal irritation. A "position of protection" is often assumed in the presence of pain or tenderness. Lack or limitation of motion may indicate paralysis, fracture or dislocation, joint inflammation, or an intracranial lesion. Spasticity, scissors gait, and poor coordination are found as a result of cerebral injury, often present since birth. Many of the muscular dystrophies are first manifested in an abnormal gait and either hypotonicity or hypertonicity. The general lack of tone of all muscle groups characterizes *Down's syndrome* (mongolism).

Examination of the Newborn Infant

Because no other period of life carries as great a risk of morbidity and mortality as the first weeks of life, it is appropriate to devote some special emphasis to the examination of the newborn infant. The 1-minute and 5-minute Apgar scores are a general indication of the viability of the infant and effects of labor (Table 21-4). Scores of 6 or less indicate actual or potential problems.

A limited examination including a search for major defects should be done immediately. If no problems are apparent, a more complete evaluation can be delayed for several hours. Auscultation of heart and lungs and palpation of the

Table 21-4. Apgar Score for the Newborn

	Score		
Signs	0	1	2
1. Heart rate	Absent	< 100	> 100
2. Respiratory effort	Absent	Weak	Good
3. Muscle tone	Limp	Some flexion	Well flexed extremities
4. Response to stimulus of feet	None	Some motion	Motion and crying
5. Color	Pale or blue	Acrocyanosis	Completely pink

The score is computed at 1 and 5 minutes following delivery by assigning 0, 1, or 2 to each item. A total score of 10 indicates optimum. A total score of 0 indicates a moribund infant. (From Apgar [1].)

abdomen can best be accomplished while the infant is quiet or asleep. At birth or soon thereafter the umbilicus should be examined for the presence of a single artery. Normally there are two arteries and one vein. A single artery is often associated with anomalies of the heart, central nervous system, or gut.

Not all infants of low birth weight, under 2500 grams, are premature. A significant number have had fullterm gestation but suffered intrauterine malnutrition from maternal disease or poor placental function. An attempt should be made to correlate the baby's gestational age and birth weight. The premature infant is vulnerable to *sepsis* and *hyaline membrane disease* (respiratory distress syndrome). He will also show organ immaturity, e.g., *hyperbilirubinemia* with jaundice. The malnourished baby is particularly susceptible to *hypoglycemia*. Unusually large babies (over 3800 grams) are often born to mothers with diabetes mellitus or to prediabetics.

Since menstrual histories are often inaccurate, more objective means of estimating gestational age are required. These include the following: before 36 weeks, only one or two transverse creases are present on the sole of the foot, the breast nodule is less than 3 mm. in diameter, no cartilage is present in the earlobe, and testes are seldom in the scrotum, which has few or no rugae. By 40 weeks, many creases are present on the sole, the breast nodule exceeds 4 mm., cartilage is present in the earlobe, and the testes have descended into the scrotum, which is covered with rugae. Increasing muscle tone with assumption of a posture of predominantly flexed extremities is another sign of increased maturity.

The premature infant often has brief periods of apnea lasting up to 20 seconds. Respiratory distress is indicated by an increased rate, grunting, retraction of intercostal and subcostal spaces and suprasternal notch, see-saw sinking of the chest with rising abdomen in contrast to the normal synchronous motions, and flaring of the nostrils.

Airway patency can be assured if a soft catheter will pass through the nose, pharynx, and esophagus into the stomach. The resting newborn is an obligatory nose breather, so obstruction, as *atresia of choanae, syphilis,* or *reserpine* therapy in the mother, may result in serious respiratory difficulty. Passing the tube into the stomach can eliminate the possibility of esophageal obstruction. Excessive collection of mucus in the nose and mouth characterizes atresia of the esophagus.

Visual inspection as well as palpation with a gloved finger of the oral cavity will rule out such defects as a cleft palate or an aberrant thyroid at the base of the tongue. *Thrush* (moniliasis) is an infection of the mucous membranes with slightly raised dull white patches and can be easily distinguished from *Epstein's pearls,* which are pearly white nodules limited to the palate.

It is important to remember that respiratory distress, with or without cyanosis, is not limited to intrinsic lesions of the lungs at this period. Intracranial lesions including anomalies, hemorrhage, or damage due to anoxia may be responsible. Congenital heart disease and diaphragmatic hernia are further possibilities to be ruled out and often require a chest x-ray for definitive diagnosis.

Some breast hypertrophy at birth, occasionally with small secretions of "milk," is not uncommon and is transient.

The genitalia should be carefully examined for anomalies such as hypospadias, hydrocele, hernia, and ambiguous development indicating possible abnormal sexual development secondary to endocrine influences. Failure to pass meconium or urine within 24 to 48 hours requires investigation as to cause.

Evaluation of the central nervous system depends upon observations of spontaneous alertness and activity, strength and character of the cry, response to stimuli, vigor of sucking and feeding, and postural tone. Head measurements and examination have been described previously, and some of the reflexes have been mentioned.

General Pediatric Examination

Fever. This is probably the most common symptom experienced in childhood. Throughout the early part of a child's life his febrile response is usually considerably higher than that of the adult to a similar cause. Most often the fever is due to an infection in the respiratory tract. Infections of other systems or a generalized infection as in septicemia may cause fever. The premature and newborn infant may have little or no fever even with very severe infections, and such reactions as an irregular temperature course, poor appetite, vomiting, and irritability may be the only symptoms. Prolonged or recurrent fever with no apparent cause may result from neoplasms, leukemia, rheumatoid arthritis (initially there may be no joint involvement), hypersensitive reactions, and diseases of the central nervous system. Chills, delirium, and convulsions often accompany high fever in children.

Abdominal Pain. This is often difficult to evaluate in children, as some degree of periumbilical discomfort or pain is associated with many illnesses not directly involving abdominal contents. In young children, all abdominal pain tends to localize in the umbilical area. In infants one should suspect such pain in persistent screaming and crying often associated with flexion of the thighs on the abdomen, grunting respiration, and vomiting. *Colic*, a condition seen in the first few months of life, is characterized by crying and some degree of gaseous abdominal distention, and is often relieved by feeding. It tends to recur at the same time of day or night. One has to consider many possible causes of the symptom of abdominal pain in children: appendicitis, intestinal obstruction due to intussusception or volvulus, pancreatitis, peptic ulcer, and urologic diseases, especially infection. Such pain is also observed at the onset of many of the acute infectious diseases. Discovering and analyzing associated symptoms, such as vomiting and nature of the vomitus, diarrhea or constipation and nature of the stools, is important.

Vomiting.   Like abdominal pain, vomiting often accompanies disturbances unrelated to the intestinal tract or central nervous system, the two areas most frequently involved in serious disease. In young children vomiting is frequently associated with acute infections, indiscretion in diet, fear or severe anxiety, and pain. *Regurgitation* is a nonforceful vomiting of small quantities, often seen in early infancy. Occasionally this kind of vomiting may persist, as in the ruminating child. Esophageal atresia is manifested by vomiting shortly after birth and by the presence of large amounts of mucus in the baby's mouth. Choking and cyanosis indicate aspiration. In the newborn period, vomiting of bile-containing material always indicates bowel obstruction until proved otherwise. Vomiting caused by pyloric stenosis in infants is associated with visible peristaltic waves in the upper abdomen and becomes increasingly projectile, but since there is no nausea, refeeding is easily accomplished. In vomiting secondary to lesions of the central nervous system, nausea is often present. In the very young subject this may be possible to detect only by the facial expression or the preceding "stomach cough." Excessive dosage of many drugs, most commonly salicylates, will produce nausea and vomiting. Many metabolic disturbances may cause vomiting: diabetes mellitus with acidosis, galatocemia, adrenogenital syndrome with salt loss, excessive hydration resulting in cerebral edema, and dehydration with its opposite effect upon the brain.

Failure to Gain Weight and Loss of Weight These are important symptoms, since infants and young children normally show a progressive though somewhat variable weight gain. Even the older child and adolescent, except when purposely dieting, will show only brief periods when weight is not gained. Obviously, any of the causes discussed under vomiting will result in failure to gain, if the condition persists. Malnutrition, with or without economic privation, will produce the symptoms. Defects in assimilation of food, as in cystic fibrosis of the pancreas and the various malabsorption syndromes, leads to a failure to gain. Most chronic disease will eventually result in failure to gain, because of loss of appetite as well as other less obvious factors, including fever, pain, infection, and impairment of organ function, such as heart failure. Failure to progress in normal statural growth will frequently accompany poor weight gain, as in *hypothyroidism, hypopituitarism, achondroplasia,* and *hereditary dwarfism.* The abused ("battered child") or emotionally neglected infant or child may show profound weight loss or failure to grow. Observation of the mother's handling and feeding of her child may be most helpful in determining the proper cause for a failure to thrive. Obviously a careful investigation of the kind and quantities of food ingested is important and, where appropriate, a detailed analysis of formula preparation.

Stridor. This is a harsh, high-pitched crowing noise most distinct during inspiration. In contrast to wheezing, it originates high in the respiratory tract, usually in the trachea or larynx. It indicates obstruction of the airway and may be combined with cough, dyspnea, hoarseness, retractions of the chest wall with respiration, and tachypnea. The small size of the infant airway is conducive to increased frequency and severity of obstruction. Slight stridor with crying is normal in some babies. In the newborn period, congenital structural abnormalities are the commonest cause. These include flaccidity of the epiglottis, laryngeal web, cysts, and defects in the tracheal cartilaginous rings. In the older child, acute spasmodic laryngitis (croup) is the most common cause and typically has its onset suddenly and at night with little or no fever. Stridor may also be caused by laryngeal edema due to serum sickness, irritation due to smoke or chemicals, and obstruction by a foreign body. Extrinsic factors such as a neoplasm or abscess in surrounding tissue can result in obstruction.

Slow Development (Mental Retardation). This may be suspected by parents at any age, the most severe forms at an early age, by comparison of their children to siblings or other children. The presence of some physical stigmata (mongolism, microcephaly, hydrocephaly, and some of the chromosomal defects) may be important clues. One cannot outline the developmental diagnosis for each age in a brief space, but it can be emphasized that delay in appearance of normal achievements in several areas of behavior is almost always significant. The areas of behavior are divided into motor, language, adaptive (reaction to environment and manipulation of it), and personal-social. These overlap to a considerable extent. Mental retardation is a symptom with many causes. Any physician who deals with children should become adept at recognizing the child with mental retardation; the degree of impairment may then be determined by a trained psychologist.

Dyspnea. Labored respiration, or dyspnea, is a symptom that must be discussed in relation to the age of the subject. We have noted the changes in respiratory rate with age. In the premature infant a periodic pattern of breathing is normally encountered with short periods of apnea. Gradually this pattern disappears. In the newborn period, dyspnea may be associated with atelectasis, the respiratory distress syndrome (most common in premature babies and those born to diabetic mothers). Aspiration of amniotic fluid and congenital anomalies such as lung cysts and diaphragmatic hernia are often associated with dyspnea. Labored breathing may also be seen with congenital heart disease, with or without failure. Later in life, dyspnea is more apt to be caused by pulmonary infection and asthma. *Hyperventilation,* seen in diabetic acidosis, fever of any cause, aspirin poisoning, and occasionally with intracranial lesions, must be distinguished from dyspnea.

Convulsions. These form another symptom complex that varies in causation with age. In the newborn infant, intracranial damage or congenital defects of the brain are nearly always combined with definite neurologic abnormalities. *Hypocalcemic tetany* most commonly is seen in the first two months of life and is often accompanied by carpopedal spasms and laryngeal stridor. The convulsive seizures of this metabolic abnormality are not easily distinguished from any of the other seizures regardless of cause. *Epilepsy* is characterized by seizures of great variety from grand mal to petit mal, and may or may not be associated with other neurologic symptoms or signs between attacks. Their common characteristic is that they tend to recur over a long period of time. Throughout childhood the most common cause of convulsive seizures is high fever, regardless of whether the cause of the fever is an infection of the respiratory tract, meninges, urinary tract, or gastrointestinal tract. The threshold for "febrile convulsions" appears to rise with age. Convulsions are occasionally associated with metabolic abnormalities such as severe electrolyte imbalances, hypoglycemia, and drug intoxication or poisoning.

A systematic review of the physical manifestations of disease in the infant and child would lead only to a tedious reassertion of material already presented. However, a few exceptional findings, unique to the pediatric age group, deserve special emphasis.

Speech and cry are of great value. Hoarseness is often present with laryngitis, hypothyroidism, and tetany. A high-pitched, piercing cry in the infant may indicate increased intracranial pressure. Pharyngeal paralysis due to poliomyelitis or diphtheria will influence speech, producing a nasal quality. A monotone type of verbalization may indicate hearing loss.

The precocious appearance of sexual hair may be caused by adrenal lesions (in early infancy by congenital *adrenal hyperplasia*), brain lesions, gonadal tumors, and a few other rare conditions. Delayed appearance or absence of sexual hair may be found in pituitary, thyroid, or gonadal insufficiency and with certain chronic illnesses. Some degree of retarded growth is usually an accompaniment. Tufts of hair over the spine may indicate an underlying *spina bifida.*

Cradle cap or *seborrheic dermatitis* in the newborn is characterized by a greasy yellowish scale over the scalp which sometimes involves other areas of the head, especially behind the ears. Another common dermatitis of childhood is *tinea capitis,* or ringworm of the scalp. Hair on the involved area is broken off close to the scalp. Edema, reddening, and crusting are usually present.

Separation of the sutures which have previously been approximated and bulging with tenseness of the anterior fontanel are indicative of increased intracranial pressure regardless of cause. Prominence of the veins over the head may also be present in such cases. Microcephaly, a head circumference more than two standard deviations below the normal for a given age, may indicate premature synosteosis but is more commonly an associated finding in mental retardation with an underlying brain defect. Transillumination of the head is a valuable method of examination in infants. It is done in a dark room with a bright flashlight fitted with a soft rubber collar to insure a lightproof fit against the scalp. In severe *hydrocephalus* and *anencephaly,* nearly the whole skull will transmit light. In *hygroma,* a localized subdural collection of fluid, a sharply delineated area of transillumination is obtained on the side of the lesion.

The face is examined for shape and symmetry. Paralysis may be elicited only by making the child smile or cry. Thickening and puffiness of the features may be present with edema or hypothyroidism. *Chorea* is associated with un-controlled grimacing and must be differentiated from tics and habit spasms. Epileptic seizures may be localized to the face or begin in this area in children. A lack of expression is characteristic of severe mental retardation.

Dilation of the pupils is mandatory for funduscopic examination in infants. Many neurologic diseases in the child have retinal manifestations. A prominent epicanthal fold is sometimes familial but is a usual characteristic in mongolism.

With *acute pharyngitis* in children, the tonsils are nearly always involved. They may have small areas of whitish membrane that is usually not difficult to distinguish from the large, confluent, gray membrane of *diphtheria.*

Asymmetry of the chest with bulging over the heart may be present in children with prolonged cardiac enlargement. Softening of the rib cage with retraction of the lower ribs from the pull of the diaphragm results in Harrison's groove, a finding in rickets. Rhonchi transmitted from the trachea or large bronchi often confuse the student who is listening to the chest of an infant. Their character, position, and differentiation may be facilitated by holding the stethoscope an inch or two from the infant's mouth or nose and comparing these sounds to those heard over the chest.

Inspection of the abdomen of the crying child may best demonstrate hernia, diastasis recti, or localized bulging due to regional paralysis. Palpation is sometimes done with the child in a prone position, as he relaxes best if he is not looking directly at the examiner. In the infant, relaxation may be obtained during bottle feeding. This procedure may also be used to demonstrate peristaltic waves and for the palpation of a tumor such as that found in hypertrophic pyloric stenosis. *Umbilical hernia* is particularly common in infants, especially in association with hypothyroidism and mongolism.

Actual enlargement of the penis or clitoris, often accompanied by the appearance of pubic hair, is seen as a result of virilizing adrenal lesions or other causes of precocious development. Partial fusion of the labia minora is common in prepubertal girls. Because of the patency of the inguinal canal and the sensitive cremaster reflex, several examinations should be carried out before a diagnosis of undescended testicle is made.

The muscular and skeletal systems are examined as in the adult, taking into account the fact that ambulation is not present in the early months of life. Lack of motion, weakness, and distortion of normal relationships must be looked for even more carefully than in older subjects. Congenital dislocation of the hip, brachial plexus injuries, osteochondrosis, rickets, scurvy, amyotonia congenita, and other muscular dystrophies are particularly important in the childhood age group. In early infancy, there may be insufficient development of the nervous system to give reliable neurologic signs. Meningitis in this age group may not be associated with obvious nuchal rigidity, Brudzinski or Kernig signs. Lethargy, anorexia, vomiting, and other symptoms and signs of seemingly less specific significance may be the only findings to indicate meningeal irritation.

In this chapter an attempt has been made to emphasize some of the important differences between the child and the adult. The fact that the child is a changing organism is important to remember and extremely helpful in evaluating the history and the physical findings. Failure to grow in stature or weight is always significant. Delay in both physical and mental maturation as well as in growth may indicate endocrine or deficiency disorders or chronic infectious disease. These findings are lacking as important signposts in the adult. Fortunately, the immature organism is more labile in terms of fever, emotional response, fatigue, general behavior, intestinal upsets, and the like, than the adult, and these, therefore, are important indicators of disease. Special attention to behavior and facial expression before and during the actual examination may be of the utmost importance. If these facts are kept in mind, the examination of the infant or child can be an exciting and rewarding experience for the physician.

REFERENCES

1. Apgar, V., Holaday, D. A., James, L. S., Weisbrot, I. M., and Berrien, C. Evaluation of the newborn infant: Second report. *J.A.M.A.* 168:1985, 1958.
2. Farmer, T. W. *Pediatric Neurology.* New York: Hoeber Med. Div., Harper & Row, 1964.
3. Gesell, A. L., and Amatruda, C. S. *Developmental Diagnosis* (2nd ed.). New York: Harper & Row, 1949.
4. Green, M., and Richmond, J. *Pediatric Diagnosis* (2nd ed.). Philadelphia: Saunders, 1962.
5. Lowrey, G. H. *Growth and Development of Children* (6th ed.). Chicago: Year Book, 1973.
6. Nelson, W. *Textbook of Pediatrics* (9th ed.). Philadelphia: Saunders, 1969.
7. Schaffer, A. J., and Avery, M. A. *Diseases of the Newborn.* Philadelphia: Saunders, 1972.

THE PROBLEM-ORIENTED
MEDICAL RECORD

On the following pages are examples of the various components of a Problem-Oriented Medical Information System. This is not an actual record and all names are fictitious. This example is taken from an exhibit "The Problem-Oriented Medical Information System" prepared for the American Heart Association and the American College of Cardiology by J. Willis Hurst, M.D., Robert C. Schlant, M.D., W. Dallas Hall, M.D., and H. Kenneth Walker, M.D., all of the Department of Medicine, Emory University School of Medicine, Atlanta, Georgia.

Smith, Mr. John 000-001		COMPLETE PROBLEM LIST Permanent Part of Medical Record
Date problem entered	**Active**	**Inactive**
11-13-72 *1	Coronary atherosclerotic heart disease	
	a. Acute anteroseptal myocardial	
	infarction	
	b. Left ventricular dysfunction	
	c. Mitral regurgitation _11−14_ pap.	
	muscle dysf.	
11-13-72 2.	Adult onset diabetes mellitus (1952)	
	a. Peripheral atherosclerosis	
11-13-72 3.		Herniorrhaphy, R. Ing. (1954)
11-13-72 4.		Depression, situational (1970)
11-13-72 5.		Left BK amputation (1969)
11-15-72 6.	Diabetic retinopathy, as _____	
*Reason for admission		

Patient Identification___*Smith, Mr. John 000-001*_____   Date ___*11-13-72*____

CHIEF COMPLAINTS   *"Indigestion"*

HISTORY OF PRESENT ILLNESS

57 year old male who noted epigastric-lower sternal burning type of indigestion while climbing stairs at work 2 weeks ago. Took Tums and pain was lessened after 5—10 minutes. Does not recall any sweating, weakness or palpitations. Pain did not radiate above mid-sternum or to arms, neck or shoulder. It did not recur until —

Today while mowing the lawn he developed severe lower substernal "indigestion" discomfort that radiated to both elbows and was associated with nausea, weakness and a cold sweat. The pain has persisted over the past 3 hours and was not relieved by 2 Tums and 2 Alka-Seltzer tablets. He has noted some progressive shortness of breath since the onset of pain.

Instructions: Circle positive responses and comment appropriately. Underline negative
responses. Leave unaltered if information not available.

PAST MEDICAL HISTORY

(a) Pediatric and adult illnesses: (mumps,) (measles,) (chickenpox,) rheumatic fever,
arthritis, rheumatism, chorea, scarlet fever, pneumonia, tuberculosis, diabetes
mellitus, heart disease, renal disease, hypertension, jaundice.

(b) Immunizations: *All intact*

(c) Hospitalizations: *None*

(d) Trauma: *None*

(e) Transfusions: *None*

(f) Current medications: *Only as in present illness*

(g) Allergies: *None*

(h) Habits (drugs, alcohol, tobacco):

Drinks one six-pack beer/weekend
Smokes 2 packs cigarettes daily for 18 years (36 pack-years)

FAMILY HISTORY (Diagram pedigree if indicated)

(Diabetes mellitus) tuberculosis, cancer, stroke, hypertension, renal disease, deafness, gout/arthritis, anemia, (heart disease.)

—*Mother (65) controlled by diet*

—*Brother died of myocardial infarction age 44*

SYSTEMS REVIEW

(a) General: weakness, fatigue, change in weight _+ 2 lbs._ , appetite, sleeping habits, chills, fever, night sweats.

(b) Integument: color changes, pruritus, nevus, infections, tumor (benign/malignant), dermatosis, hair changes, nail changes.

(c) Hematopoietic: anemia, abnormal bleeding, adenopathy, excessive bruising.

(d) Central nervous system: headache, syncope, seizures, vertigo, amaurosis, diplopia, paralysis/paresis, muscle weakness, tremor, ataxia, dysesthesia.

(e) Eyes: vision, glasses/contact lens, date of last eye exam _none_ . scotomata, pain, excessive tearing.

(f) Ears: tinnitus, deafness, other.

(g) Nose, throat and sinuses: epistaxis, discharge, sinusitis, hoarseness, thyromegaly.

(h) Dentition: caries, pyorrhea, dentures.

(i) Breasts: masses, discharge, pain.

(j) Respiratory: (cough) (productive/non-productive), change in cough, amount and characteristics of sputum, duration of sputum production _1 year_ _36_ pack-years of tobacco usage, wheezing, hemoptysis, recurrent respiratory tract infections, positive tuberculin test.

— *Has smoked 2 packs cigarettes daily for the past 18 years; in the past year has noted a frequent cough, productive of small amounts of whitish sputum in the mornings.*

Note: positive responses are circled and elaborated upon. *Negative* responses are underlined. No alteration is made when information is not obtained — e.g., when the patient is comatose or when data must come from other sources; then "incomplete data base" becomes a "problem" requiring written plans.

SYSTEMS REVIEW (Continued)

(k) Cardiovascular: chest pain, typical angina pectoris, dyspnea on exertion, orthopnea, paroxysmal nocturnal dyspnea, peripheral edema, murmur, palpitation, varicosities, thrombophlebitis, claudication, Raynaud's phenomenon, syncope, near-syncope.

———See description in present illness; unable to recall any chest symptoms prior to two weeks ago.

(l) Gastrointestinal: nausea, vomiting, diarrhea, constipation, melena, hematemesis, rectal bleeding, change in bowel habits, hemorrhoids, dysphagia, food intolerances, excessive gas or indigestion, abdominal pain, jaundice, use of antacids, use of laxatives.

See present illness ———

(m) Urinary tract: dysuria, hematuria, frequency, polyuria, urgency, hesitancy, incontinence, renal calculi, nocturia, infections.

(n) Genitoreproductive system:

Male: penile discharge, lesion, history of venereal disease, serology, testicular pain, testicular mass, infertility, impotence, libido.

Female: *(Not applicable)*

A. Gynecologic history:
Age of menarche_____ , last menstrual period _____ , age at menopause _____ , post menopausal bleeding, abnormal menses, amount of bleeding, intermenstrual bleeding, postcoital bleeding, leucorrhea, pruritus, history of venereal disease, serology, last PAPs _____ , results_____.

B. Obstetric history:
Full-term deliveries_____
Pregnancies_____
Abortions_____
Living children_____
Complications of pregnancies, infertility, libido.

C. Methods of contraception:

SYSTEMS REVIEW (Continued)

(o) Musculoskeletal:

   (i) Joints: <u>pain</u>, <u>edema</u>, <u>heat</u>, <u>rubor</u>, <u>stiffness</u>, <u>deformity</u>.

   (ii) Muscles: <u>myalgias</u>.

(p) Endocrine: <u>goiter</u>, <u>heat intolerance</u>, <u>cold intolerance</u>, <u>change in voice</u>, <u>polyuria</u>,
              <u>polydipsia</u>, <u>polyphagia</u>.

(q) Psychiatric: <u>hyperventilation</u>, <u>nervousness</u>, <u>depression</u>, <u>insomnia</u>, <u>nightmares</u>,
             <u>memory loss</u>.

(r) Additional historical data·

PHYSICAL EXAMINATION

Vital signs:

   Pulse __*96*__ reg/irreg   Respirations __*22*__   Temp. __*99°*__ oral/rectal

   Blood pressure

      supine R arm __*130/80*__   L arm __*128/80*__   Leg __*145/90*__

      sitting R arm _____

      standing R arm _____

   Weight __*195*__ (scales used)

   Height __*71"*__

General:

Integument: <u>turgor</u>, <u>texture</u>, <u>pigmentation</u>, <u>cyanosis</u>, <u>telangiectasia</u>, <u>petechiae</u>, <u>purpura</u>,
            <u>ecchymosis</u>, <u>infection</u>, <u>lesions</u>, <u>hair</u>, <u>nails</u>, <u>mucous membranes</u>.

Lymph nodes: <u>cervical</u>, <u>postauricular</u>, <u>supraclavicular</u>, <u>axillary</u>, <u>ulnar</u>, <u>inguinal</u>.

Skull: <u>trauma</u>, <u>bruits</u>, other.

PHYSICAL EXAMINATION (Continued)

Eyes: lacrimal glands, cornea, lids, sclerae, conjunctivae, exophthalmos, lid-lag.

Fundi: discs, arteries, veins, hemorrhages, exudates, microaneurysms.

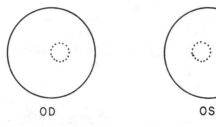

OD   OS

Ears: tophi, tympanic membranes, external canal, hearing,
air conduction ___>___ bone conduction, lateralization __*no*__ .

Mouth, nose, and throat: definition, gingiva, tongue, tonsils, pharynx, nasal mucosa,
nasal septum, sinuses.

Neck: mobility, scars, masses, thyroid, salivary glands, tracheal shift, bruits.

Breasts: masses, discharge, nipples, asymmetry, gynecomastia.

Chest:

Respiratory rate __*22*__ /min.   Amplitude: Shallow
                                                Deep
                                                (Normal)

Respiratory rhythm: (regular)
                    irregular
                    periodical
                    inspiration/expiration ratio

Chest wall: deformities
            motion
            lateral motion: (good,) fair, absent
            use of accessory muscles: yes, no

PHYSICAL EXAMINATION, Chest (Continued)

Auscultation: rales, wheezes, rhonchi.
breath sounds: increased, decreased, (normal.)
other:

(diagram location of abnormal breath sounds, transmitted voice, or abnormal percussion).

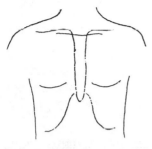

Cardiovascular System:

External jugular veins are distended to __4__ cms. above the angle of Louis at __45°__ degrees of truncal elevation from supine.

Point of maximum impulse is in the __4th__ I.C.S. _5 cm. lat to LLSB Plapable S4 and S3; late systolic bulge (see below)_ .

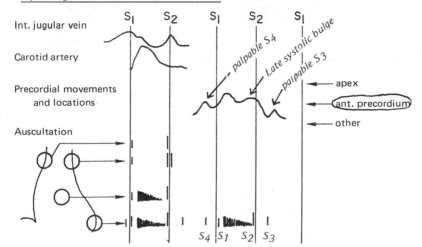

S_1 *Single; slightly diminished in loudness*

S_2 *Normal intensity and splitting (not paradoxical)*

Gallops *S_4 (atrial) and S_3 (ventricular) at LLSB and apex*

Systolic murmur *Blowing mid-late decrescendo systolic grade 3/6 murmur at apex with radiation toward LSB; no change with inspiration*

Diastolic murmur *None*

Other *No palpable thrill*

PHYSICAL EXAMINATION, Cardiovascular System (Continued)

Peripheral pulses

	Carotid	Brachial	Radial	Aorta	Femoral	Popliteal	D.P.	P.T.
R.	3+	3+	3+	2+	2+	2+	1+	1+
L.	3+	3+	3+		2+	1+	0	0

0 = absent   1+ = thready   2+ = decreased   3+ = normal   4+ = hyperactive

Extremities: <u>edema</u>, <u>cyanosis</u>, <u>stasis</u>, <u>ulceration</u>, (hair distribution,) <u>clubbing</u>.

Left lower leg amputation   ╱—*Male pattern baldness; Absent hair
on toes of right foot*

Abdomen: <u>obesity</u>, <u>contour</u>, <u>scars</u>, <u>tenderness</u>, <u>CVA tenderness</u>, <u>masses</u>, <u>rebound</u>,
<u>rigidity</u>, <u>fluid wave</u>, <u>shifting dullness</u>, <u>frank ascites</u>, <u>bruits</u>, <u>hernia</u>,
<u>venous collaterals</u>.

Bowel sounds: (normal,) absent, hyperactive, hypoactive, obstructive.
Organomegaly: <u>liver</u>, <u>spleen</u>, <u>kidneys</u>, <u>bladder</u>, <u>gall bladder</u>.
Liver size___9___cms. (total dullness)
Liver tenderness: (absent,) increased
Liver edge: smooth, irregular, nodular.

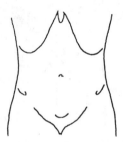

Male: genitalia: <u>penis</u>, <u>scrotum</u>, <u>testes</u>, <u>epididymis</u>, <u>masses</u>, <u>other</u>.

rectal: ⎫ perineum, hemorrhoids, sphincter tone, prostate, bleeding, masses.
Stool ⎭ ___*deferred*___.

Female: external genitalia: labia, clitoris, introitus, urethra, perineum, other.

internal genitalia: vagina, cervix, adnexa, cul-de-sac, discharge.

Paps: done, omitted.

rectal: hemorrhoids, sphincter tone, bleeding, masses.

Stool_____.

PHYSICAL EXAMINATION (Continued)

Joints: deformity, rubor, calor, tenderness, edema.

       range of motion: fingers, wrists, elbow, shoulder, hips, knees, ankles.

       spine: deformity (kyphosis, lordosis, scoliosis), thoracic excursion.

Neurological:

   Cerebral function: (alert wakefulness,) lethargic, obtunded, stuporous, semicomatose,
                  comatose.

   Mental status: *Intact; oriented to time, person and place*

   Cranial nerves:

      I. (list test materials) *Not tested*

     II. discs, papilledema, venous pulses, optic atrophy, visual fields, acuity.

    III, IV, VI. ptosis, palpebral fissure.
               Pupils: R __*4*__ mm. L __*4*__ mm. Shape __*R*__

               Reaction to light: R = L
               Consensual reaction: R to L √   L to R √
               Reaction to near vision: R √   L √

                   Extra-ocular movements: (full) abnormal, dolls-eyes, cold calorics,
                                   gaze preference, nystagmus, optico-kinetic
                                   nystagmus.

     V. Sensory: 1st division   2nd division   3rd division

                R corneal   L corneal

       Motor: masseters, pterygoids, temporalis.

    VII. (intact,) R.-L. central,   R.-L. peripheral.

    VIII. (intact)

    IX, X. dysarthria, gag, phonation, uvula, soft-palate, swallowing, (intact)

    XI. Sternocleidomastoids, trapezii.

   XII. tongue in midline, deviation to R.-L., atrophy, fasiculations.

PHYSICAL EXAMINATION (Continued)

Gait and station: *(Not tested)*

   walking: normal, abnormal, heel walking, toe walking, tandem walking.

   truncal ataxia:

   Romberg: present, absent, R.-L.

   Involuntary movements:

Cerebellum: rapid alternating movements, finger-nose, finger-finger, heel-shin,
              past-pointing, rebound, posturing.

Sensory: pain, temperature, light-touch, joint-position, vibratory,
           two-point discrimination, stereognosis.

Associative functions: speech, writing, reading, apraxia, agnosia, other.

Motor: tone, mass, fasiculations, tremor

      __*0*__ hemiparesis   __*0*__ hemiplegia.

Reflexes

	Bi	Tri	Br	F	K	A	Plantar	Abdomen	Snout	Grasp	Jaw Suck
R.	*2+*	*2+*	*2+*	*1+*	*2+*	*2+*	↓	*+*	*+*	*0*	*0*	
L.	*2+*	*2+*	*2+*	*1+*	*2+*	*2+*	↓	*+*	*+*	*0*	*0*	*0*

0 = absent c̄ facilitation   tr = trace   1+ = decreased   2+ = normal
3+ = hyperactive   4+ = sustained clonus

LABORATORY DATA (This defined Data Base requires that the following
                  examinations be done on every patient)

Hematology:

   WBC *13,200 mm³*   Differential *79 segs, 20 lymphs, 1 band*
   Hct. *48 vol %*   Platelet estimation: *Normal*

LABORATORY DATA (Continued)

Chemistry:

Na$^+$ *135 mEq./L.*   Blood sugar   *182 mgm. %*

K$^+$ *4.5*

CO_2 *20.0*   BUN   *18 mgm %*

Cl   *100.0*

Urinalysis: *2+ glucose, negative acetone; otherwise normal*

Chest X-ray (diagram if appropriate); ~~routine,~~ (portable, A-P.)

   Mild cardiomegaly;
                        marked pulmonary edema

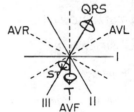

Electrocardiogram:

   rate *96*

   rhythm *Sinus*

   P-R___*.18*___   QRS___*.07*___   QT___*.28*___

Interpretation: *Marked S-T elevation and loss of initial forces in V$_{1-3}$ compatible*
                  with acute anteroseptal myocardial infarction; possible old inferior
                  infarction.

OTHER LABORATORY DATA AVAILABLE (not included in defined Data Base)

pH = 7.46

pCO$_2$ = 30 mm Hg

pO$_2$ = 62 mm Hg (room air)

PATHOPHYSIOLOGIC CLASSIFICATION

I. *HEART DISEASE*

CLASSIFICATION

Etiologic: *Coronary atherosclerosis*

Anatomic: *Coronary artery narrowing, sclerosis, myocardial infarction*

Physiologic: *Normal sinus rhythm, acute infarction, left ventricular dysfunction, papillary muscle dysfunction, mitral regurgitation*

Functional: *Class IV*

Therapeutic: *Class E*

II. *DIABETES MELLITUS*

CLASSIFICATION

Etiologic: *Congenital predisposition; obesity*

Anatomic: *Body obesity; pancreas normal to light microscopy*

Physiologic: *Inadequate beta cell response to stress of obesity*

Functional: *Not applicable*

Therapeutic: *Caloric restriction until ideal body weight attained*

Note: This page and the next page (SEQUENCE OF EVENTS) strictly speaking are not part of the POMR. However, these two items are useful in ensuring that the clinician does not jump too quickly from the raw Data Base to the formulation of a Problem List. In effect, these two features are added as stoplights hoping to encourage the clinician to manipulate and interpret the Data Base with great care.

SEQUENCE OF EVENTS (Diagrammatic display of pathophysiology.):

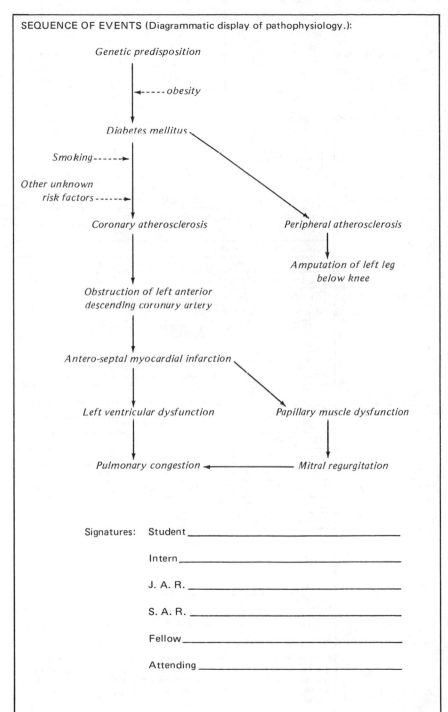

Genetic predisposition

◄----- obesity

Diabetes mellitus

Smoking ----- ►

Other unknown
risk factors ------►

Coronary atherosclerosis   *Peripheral atherosclerosis*

Amputation of left leg
below knee

Obstruction of left anterior
descending coronary artery

Antero-septal myocardial infarction

Left ventricular dysfunction   *Papillary muscle dysfunction*

Pulmonary congestion ◄—————— Mitral regurgitation

Signatures: Student _____

Intern _____

J. A. R. _____

S. A. R. _____

Fellow _____

Attending _____

Smith, Mr. John *000-001* Patient Identification	INITIAL PLANS

	Initial Plans
Problem #1	*Coronary atherosclerotic heart disease*
	a. Acute anteroseptal myocardial infarction
	b. LV dysfunction
	c. Mitral regurgitation
Diagnostic plans:	*Established by history and admission EKG; no other tests*
	required currently; later lipid profile in follow-up period
Therapeutic plans:	*Strict bed rest except bedside commode privileges*
	Diet, Na restricted and 1400 calories split for hs feeding
	CCU monitoring with prn rhythm strips; rpt EKG and portable
	chest X-Ray in am
	Slow IV drip of D_5W to keep vein open in case of emergency
	Colace to help prevent constipation and valsalva
	Nasal prong O_2 at 4 L/min flow
	Measure urine output
	Diuretic agents for pulmonary congestion
	Digitalize cautiously (see orders)
	Slow IV morphine prn pain; watch for bradycardia
	Elastic stockings to prevent peripheral thrombophlebitis
Educational plans:	*Pt. and family told of "heart attack" and that the monitor wires,*
	etc., are routine and for early detection of any problems.
	Explained to him that we will later go into rehabilitation, etc.

(Initial Plans are listed for each problem identified on admission, according to Problem Title and Number. Each plan contains three elements: Diagnostic Plans, Therapeutic Plans, Patient Education Plans.)

Smith, Mr. John 000-001 Patient Identification	INITIAL PLANS

	Initial Plans
Problem #2	*Diabetes mellitus*
Diagnostic plans:	*Established by GTT (1952)*
Therapeutic plans:	*10 u NPH in am; with supplementary regular insulin*
	See consultation and 10 pm progress note
	Fractional urines at 7-11-4-9
	Fasting plasma glucose in am
Educational plans:	*Explained to pt. why he may get regular insulin in addition to*
	NPH tomorrow and why we will need to draw frequent blood
	sugars for the first few days.

(Initial Plans are listed for each problem identified on admission, according to Problem Title and Number. Each plan contains three elements: Diagnostic Plans, Therapeutic Plans, Patient Education Plans.)

		TREATMENT ORDERS		
Hospital Policy: 1. The signature of a physician must accompany all orders. 2. Narcotics, hypnotics, sedatives, anticoagulants, and amphetamines require new orders every 48 hours. 3. Antibiotics require new orders every 7 days.			*Smith, Mr. John* *000-001* (Reserved Space)	

Ordered				Discontinued
Date & Hour	Doctor	Nurse		Date & Hour
11-13-72			*Prob. #1a. Acute Anteroseptal Myocardial*	
7:30 PM			*Infarction*	
			— Bedrest with bedside commode	
			— CCU monitoring with rhythm strip q1h	
			and prn	
			— Urine output q 4 h	
			— Elastic stockings	
			— Keep IV open with slow drip of D_5W	
			— Colace, 50 mg po bid	
			— Repeat EKG and portable chest X-Ray	
			in am	
			— Serum enzymes in am: SGOT, LDH, CPK	
			— Morphine, 5 mg slowly I.V. q 2 h prn	
			pain; call me if required more than twice	
			Prob. #1b. LV dysfunction	
			— Digoxin, 0.5 mg po stat	
			— Diuril, 0.5 gm po stat	
			— O_2 via nasal prongs, 4 L/min.	
11-13-72			*Prob. #2 Diabetes*	
9 PM			*— 1485 cal. ADA diet divided 3/10, 3/10,*	
			3/10, 1/10	
			— Urine fractional at 7-11-4-9; have pt.	
			empty bladder 30—45 min. before ob-	
			taining sample.	
			— Plasma glucose tomorrow at 7 am and	
			4 pm	
			— NPH insulin, 10 u SC in am	
			— 10 u regular insulin IV stat	

	PROGRESS NOTES	Smith, Mr. John
		000-001
Doctor		(Reserved Space)

Date	Notes		
11-13-72	# 1	_Anterior myocardial infarction_	
8:06 PM	S:	Some continued substernal aching pain despite 5 mg morphine	
		an hour ago	
	O:	BP fall to 106/80 in past hour	
		Monitor strip (7:50) shows coupled PVC's	
		S_3 and S_4 remain prominent; murmur unchanged	
	A:	Hypotension from worsening LV failure; ventricular irritability	
	P:	Watch monitor closely	
		Draw up lidocaine	
		Page Dr. Wilson stat	
11-13-72	# 1	_Ant. MI_	
8:17 PM	S:	As above; also has a "smothering" sensation	
	O:	New rhythm strip shows occ. runs of 4 PVC's as well as couples	
		BP now 112/80	
	A:	As above; intermittent ventr. tachy	
	P:	Begin lidocaine	
		Repeat pO_2 at 4 liter flow	
		Monitor hourly urine output	
11-13-72	# 2	_Diabetes_	
9 PM	S:	None	
	O:	Repeat plasma glucose 360 mg% at 8:30 PM 1 hr. after starting	
		D_5W; Urine fractional 3+ — neg.	
	A:	Worsening hyperglycemia due to acute stress situation plus IV	
		drip to keep vein open	
	P:	10 u IV regular insulin; repeat plasma glucose at midnight	

CONSULTATION

To _Endocrinology_____ Date Submitted _11-14-72_____ Hospital No. _000-001____

Ward _600 A_____ Patient _Smith, Mr. John_____

PROBLEM NUMBER AND TITLE   SPECIFIC QUESTIONS

 # 2 Diabetes mellitus   _Please advise re management of patient's_
diabetes. Should we continue NPH or
switch to regular insulin?

                                          Consultation requested by:

CONSULTANT'S REPORT (date and hour answered: _11-15-72 10 am___)
For each numbered problem categorize and state:
    Conclusions and recommendations
    Discussion (display the data used to formulate the conclusions and recommendations)
    New problems (with subjective and objective findings)

Conclusions and recommendations:

— _Continue NPH insulin in decreased dosage, 10 u q am. Supplement with regular insulin,_
  5–10 u as needed based on plasma glucose levels bid (7 am and 4 pm).
— _Continue diet as ordered, 1485 calories with 90 gm P, 180 gm C and 45 gm F divided_
  3/10; 3/10; 3/10; 1/10 until reaches ideal weight of 172 lbs. when diet should be
  revised to maintain weight.
— _Watch insulin requirements closely a) after acute stress of infarct subsides and_
  b) as weight declines.

Discussion:

    The combination of obesity and diabetes have undoubtedly contributed to his
accelerated atherosclerosis. Although we cannot reverse the large blood vessel damage
that has already occurred, hopefully we can slow the progression by returning his
metabolic status to a more normal state. The most important treatment from this
standpoint is to reduce body weight. He is currently 71" tall with a medium frame and
weight of 195; predicted ideal body weight is 172. After being on the diet for about 3
months, he should have a repeat lipid profile. Insulin requirements will probably decrease
as weight declines and physical activity increases.

New Problem # 6 Diabetic retinopathy, os →

S: No visual complaint
O: 3 small punctate hemorrhages characteristic of background diabetic retinopathy in
   the posterior fundus in the macular region os. No proliferative changes seen but
   pupils somewhat constricted (morphine) and fundus exam not entirely satisfactory.

When the back of the page is needed be certain to rearrange carbon paper so that the
duplicate copy will be complete. Send copy of consultation to Chairman of your
Department.

HOSPITAL RECORD AUDIT

CARDIOVASCULAR CONDITION AUDITED: _Acute Myocardial Infarction (AMI)_

PATIENT NAME: _Smith, John_   HOSPITAL: _GMH_

HOSPITAL NUMBER: _000-001_   DATE OF AUDIT: _11-27-72_

PATIENT TEAM: _600 A CCU, Team B_   AUDITED BY: _____

Physicians and Nurses

I.   DATA BASE
     A. Does the history include data regarding the presence
        or absence of the following: YES   NO
        1. Present Illness
           a) Pain compatible with AMI   x   ____
           b) Radiation of pain   x   ____
           c) Duration of pain   x   ____
           d) Time at onset of pain   ____   x
           e) Activity at onset of pain   x   ____
           f) Dyspnea   x   ____
           g) Sweating   x   ____
           h) Palpitation   x   ____
           i) Marked weakness   x   ____
           j) Severe apprehension   ____   x
           k) Nausea or vomiting   x   ____
           l) Syncope   ____   x
           m) Time from onset to reaching hospital   ____   x
           n) Time from reaching hospital to CCU   x   ____
           o) Prodromata within preceding 3 weeks   x   ____
     B.
     C.
II. . . V.

VI. PLANS FOR CORRECTION

    _The 73% score on present illness is acceptable; however, this is the 2nd con-
    secutive time the same deficiencies have been noted in this team's Data Base. I
    have notified Dr. Randall (CCU Cardiac Resident) who has planned to discuss the
    relevance of syncope and importance of timing in the first 24 hours in caring for
    patients with acute myocardial infarction._

Note: This audit contains only the Present Illness portion of the Data Base. The complete
audit also includes portions II—V, which incorporate audit of the entire Data Base, Problem
List, Initial Plans, Progress Notes, and Discharge Note. Plans for Correction would also be
included for other deficiencies discovered.
    The audit is designed to determine whether or not the record contains evidence that a
given datum was sought for.

INDEX

Little, Brown's Paperback Book Series

Basic Medical Sciences

Boyd & Hoerl	Basic Medical Microbiology
Colton	Statistics in Medicine
Hine & Pfeiffer	Behavioral Science
Kent	General Pathology: A Programmed Text
Levine	Pharmacology
Peery & Miller	Pathology
Richardson	Basic Circulatory Physiology
Roland et al.	Atlas of Cell Biology
Selkurt	Physiology
Sidman & Sidman	Neuroanatomy: A Programmed Text
Siegel, Albers, et al.	Basic Neurochemistry
Snell	Clinical Anatomy for Medical Students
Snell	Clinical Embryology for Medical Students
Streilein & Hughes	Immunology: A Programmed Text
Valtin	Renal Function
Watson	Basic Human Neuroanatomy

Clinical Medical Sciences

Clark & MacMahon	Preventive Medicine
Eckert	Emergency-Room Care
Grabb & Smith	Plastic Surgery
Green	Gynecology
Gregory & Smeltzer	Psychiatry
Judge & Zuidema	Methods of Clinical Examination
Keefer & Wilkins	Medicine
MacAusland & Mayo	Orthopedics
Nardi & Zuidema	Surgery
Niswander	Obstetrics
Thompson	Primer of Clinical Radiology
Ziai	Pediatrics

Manuals and Handbooks

Alpert & Francis	Manual of Coronary Care
Arndt	Manual of Dermatologic Therapeutics
Berk et al.	Handbook of Critical Care
Children's Hospital Medical Center, Boston	Manual of Pediatric Therapeutics
Condon & Nyhus	Manual of Surgical Therapeutics
Friedman & Papper	Problem-Oriented Medical Diagnosis
Gardner & Provine	Manual of Acute Bacterial Infections
Iversen & Clawson	Manual of Acute Orthopaedic Therapeutics
Massachusetts General Hospital	Diet Manual
Massachusetts General Hospital	Manual of Nursing Procedures
Neelon & Ellis	A Syllabus of Problem-Oriented Patient Care
Papper	Manual of Medical Care of the Surgical Patient
Shader	Manual of Psychiatric Therapeutics
Snow	Manual of Anesthesia
Spivak & Barnes	Manual of Clinical Problems in Internal Medicine: Annotated with Key References
Wallach	Interpretation of Diagnostic Tests
Washington University Department of Medicine	Manual of Medical Therapeutics
Zimmerman	Techniques of Patient Care

Little, Brown and Company
34 Beacon Street
Boston, Massachusetts 02106